There is only the fight to recover what has been lost
And found and lost again and again: and now, under conditions
That seem unpropitious. But perhaps neither gain nor loss.
For us, there is only the trying.

T. S. Eliot, "Four Quartets"

Gregers Krabbe

Operational Calculus

With 79 Figures

Springer-Verlag New York · Heidelberg · Berlin 1970

Professor GREGERS KRABBE

Division of Mathematical Sciences

Purdue University, Lafayette, Indiana, U.S.A.

Library of Congress Catalog Card Number 77-79088

Title No. 1577

Printed in the United States of America

To Lucie Murcianni Durand

Preface

Since the publication of an article by G. DOETSCH in 1927 it has been known that the Laplace transform procedure is a reliable substitute for HEAVISIDE's operational calculus*. However, the Laplace transform procedure is unsatisfactory from several viewpoints (some of these will be mentioned in this preface); the most obvious defect: the procedure cannot be applied to functions of rapid growth (such as the function $t \mapsto \exp(t^2)$). In 1949 JAN MIKUSIŃSKI indicated how the unnecessary restrictions required by the Laplace transform can be avoided by a direct approach, thereby gaining in notational as well as conceptual simplicity; this approach is carefully described in MIKUSIŃSKI's textbook "Operational Calculus" [M 1].

The aims of the present book are the same as MIKUSIŃSKI's [M 1]: a direct approach requiring no un-necessary restrictions. The present operational calculus is essentially equivalent to the "calcul symbolique" of distributions having left-bounded support (see 6.52 below and pp. 171 to 180 of the textbook "Théorie des distributions" by LAURENT SCHWARTZ). The present operational calculus is not only applicable to the traditional sort of problems — but also to non-standard problems as well: a problem is called **"non-standard"** if it has no solution unless the initial conditions are at $t = 0-$ (non-standard problems imply degenerate systems of equations). For example, let us solve for i_2 the system of equations

$$\frac{d}{dt} i_1 + i_1 + 2i_2 = 1, \tag{1}$$

$$\frac{d^2}{dt^2} i_1 + 5i_1 + 3\frac{d}{dt} i_2 = 0 \tag{2}$$

subject to the initial conditions $i_2(0-) = 0 = i_1(0-) = i_1'(0-)$: this is a non-standard problem; the answer

$$i_2(t) = \frac{1}{2} - \frac{6}{7} e^{-5t} - \frac{9}{14} e^{2t} \tag{3}$$

* See 5.42.7.

is such that $i_2(0) \neq 0$ (a similar problem is worked out in 2.56). If the usual Laplace transform procedure is applied to the system (1)—(2) subject to the initial conditions

$$i_2(0) = 0 = i_1(0) = i_1'(0),$$

then the answer is again (3), which gives the contradiction $i_2(0) \neq 0$. The same contradiction occurs in any calculus (e.g., MIKUSIŃSKI's) involving initial conditions at $t = 0$ or at $t = 0+$.

Although the distributional Laplace transformation is more general than the usual Laplace transformation, it likewise requires artificial restrictions and is ill-adapted to some initial-value problems; for example, some types of problems amount to finding a function i such that $i(0+) = 0$ and whose distributional derivative equals the Dirac Delta (= the unit impulse): the Laplace transform procedure gives the wrong answer (namely, HEAVISIDE's unit step function U; the correct answer $= U - 1$). This problem lies outside the scope of MIKUSIŃSKI's calculus.

This book is addressed both to engineering students and to mathematics students; in consequence, special emphasis has been given to examples illustrating the practical usefulness of the theory. It could hardly have been otherwise, since it is the choice of practical problems that has mainly determined the theoretical material chosen for this book; for example, electric circuit problems are usually formulated in electrical engineering by means of equations involving the differentiation and anti-differentiation symbols

$$\frac{d}{dt} \text{ and } \int;$$

in order to deal with impulsive currents having non-zero initial parts, these symbols will be given a sufficiently general meaning in § 7 (typically, at least one of the currents is a linear combination of the Dirac Delta with a function: see 7.15 and 7.15.2).

Other applications can be classified as follows. *First*: the traditional sort of problem usually solved by Laplace transformation techniques. *Second*: non-standard problems (e.g., 2.56, 4.13.2, the problem at the beginning of § 4, and 7.35—36). *Third*: problems involving generalized functions (e.g., 6.30, 6.84, 6.55, 7.15, 7.15.2, 7.20, and 14.28). *Fourth*: problems involving functions that are not locally integrable (e.g., 8.4—7, 8.12, and 8.67.5). *Fifth*: problems involving functions of rapid growth (e.g., 3.38—39); these problems are only Laplace-transformable in the sense of the Gel'fand-Shilov-Ehrenpreis theory of ultradistributions (alternatively, the Ditkin-Berg extension of the Laplace transformation could be applied to such problems).

In contrast to other textbooks, the approach adopted here is based on the properties of a commutative sub-algebra of the algebra $\mathscr{L}(\mathscr{D}_+, \mathscr{D}_+)$ of all linear operators in the space $\mathscr{D}_+$ of Schwartz test-functions. Our commutative sub-algebra is isomorphic to SCHWARTZ's convolution algebra $\mathscr{D}'_+$ (see 6.52) and is larger than the algebras discussed in research papers by WESTON (1957) and RJABCEV (1958: see 5.42.0 and 5.42.2). Although our commutative sub-algebra is not a field and is smaller than MIKUSIŃSKI's field of convolution quotients, it is large enough to form the basis of an operational calculus applicable to the kind of diffusion problems that are usually solved by Laplace transform techniques: I discovered this fact to my great surprise; surprise because HEAVISIDE's form $(A \exp(x \sqrt{D}) + B \exp(-x \sqrt{D}))$ of the general solution of the heat equation implies a group extension of the heat semi-group, which in turn seems to require a field (from the 1947 paper of MIKUSIŃSKI it seems that this is the situation that caused him to introduce his field extension). Although this book considers only generalized functions on the real line, the approach can easily be extended to higher dimensions.

The present book is an expanded version of a set of lecture notes that I wrote* for a course which I taught several times; that course (on operational calculus) is offered to advanced seniors and graduate students.

The author is indebted to MICHAEL GOLOMB (who initiated the above course) for many helpful conversations relating to operational calculus.

The text contains a number of improvements suggested by T. K. BOEHME, HARRY POLLARD, MURRAY PROTTER, HARRIS SHULTZ, and JIM KINCAID. This book was also helped along by two non-scientists: YALE ALTMAN and (last but not least) my wife HARRIET.

Lafayette, Indiana,
December 1969

GREGERS KRABBE

* Under National Science Foundation Grant GP-1665.

Organization

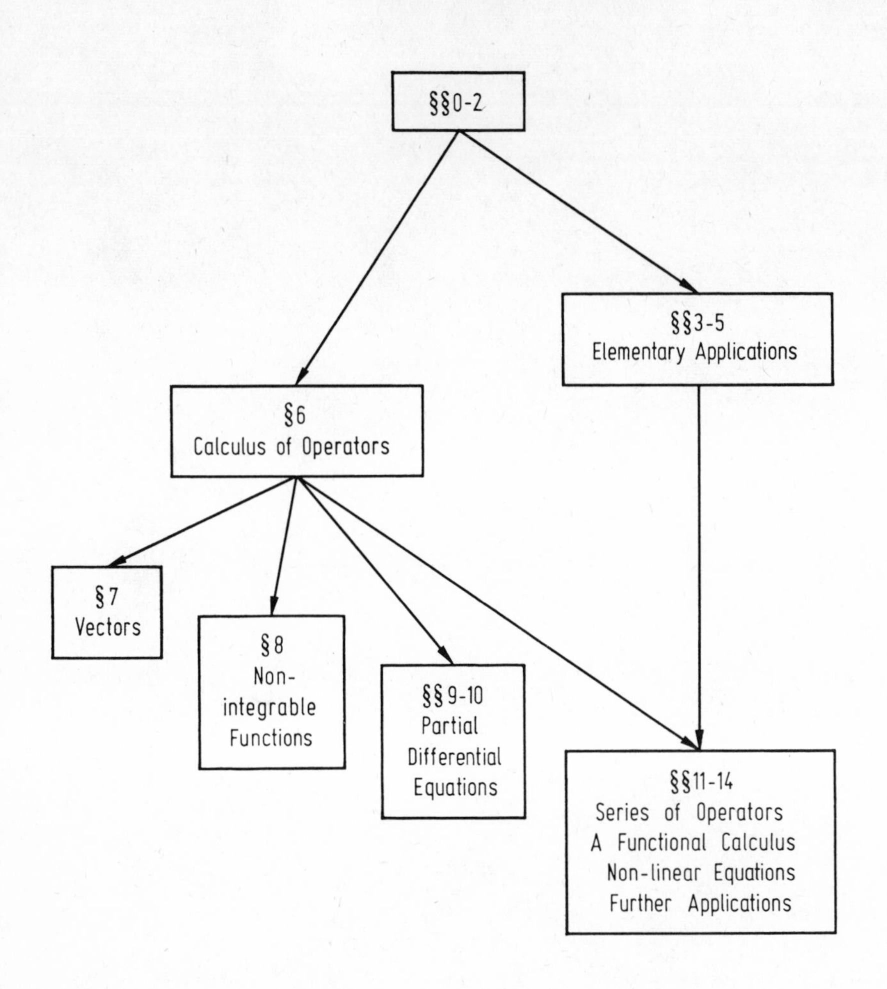

Contents

Chapter 1

Chapter 2

Chapter 3

Chapter 4

Chapter 5

Appendix

Note to the Reader

The basic material is summarized in Chapter 2: some readers might therefore skip lightly over the preliminaries in Chapter 1. The illustrative examples are intended for readers chiefly interested in applications; such readers are encouraged to skip the proofs and read the statements whose practical importance is indicated by the sign **z** placed nearby in the left-hand margin. A *single system of numbering and cross-referencing* is used throughout: each item (formula, definition, theorem, example, remark, lemma, exercise, etc.) is given a number that appears near the left-hand margin. *The numbering is consecutive*; for example, 5.31 is followed by 5.32.0, 5.32.1, 5.32.2, 5.32.3, and 5.33 (some of these numbers refer to formulas, others to paragraphs). Numbers referring to definitions are placed in square brackets; for example, [14.16] refers to Definition 14.16.

Tables of formulas are found on pp. 47–50; a summary of results and a more comprehensive table of formulas is found on pp. 331–344.

Chapter 1

This chapter contains the foundations of operational calculus; the four cornerstones are: BOREL's theorem ("multiplication property", see 1.26), LERCH's theorem (1.24), HEAVISIDE's shifting rule ("translation property", see 1.32), and the "derivation property" (1.42).

§ 0. Operators

This introductory section deals with the basic facts that will be used in the rest of this book; most of the mathematical prerequisites have been gathered here, thereby enabling the other sections to be less theoretical.

Test-functions and Operators

0.0 **Definition.** A *test-function* is an infinitely differentiable function that vanishes to the left of some point.

0.1 *Remarks.* A function $\varphi(\,)$ is infinitely differentiable if each one of the derivatives

$$\varphi'(t),\ \varphi''(t),\ \ldots,\ \varphi^{(k)}(t),\ \ldots \tag{1}$$

exists for all real values of t. An infinitely differentiable function $\varphi(\,)$ is a test-function if (and only if) there exists a real number α such that $\varphi(t)=0$ for $t \leq \alpha$; the number α depends on the function $\varphi(\,)$. Let $\mathsf{T}_0(\,)$ be the function defined by

$$\mathsf{T}_0(t) = \begin{cases} 0 & (t \leq 0) \\ 1 & (t > 0); \end{cases} \tag{0.2}$$

it is often called *Heaviside's unit step function.*

0.3 Example. The equation

$$\varphi(t) = \frac{1}{\sqrt{\pi}}\, t^{-3/2} e^{-1/t}\, \mathsf{T}_0(t) \qquad (-\infty < t < \infty)$$

defines a function $\varphi(\,)$. From [0.2] it follows that $\varphi(t) = 0$ for $t \leq 0$; some routine calculations (15.0 in the Appendix) show that all the derivatives (1) exist at $t = 0$; since $\varphi(\,)$ is infinitely differentiable for all values of $t \neq 0$, the function $\varphi(\,)$ is a test-function. The graph of $\varphi(\,)$ and of each one of its derivatives is a continuous curve whose junction with the real axis has a horizontal tangent:

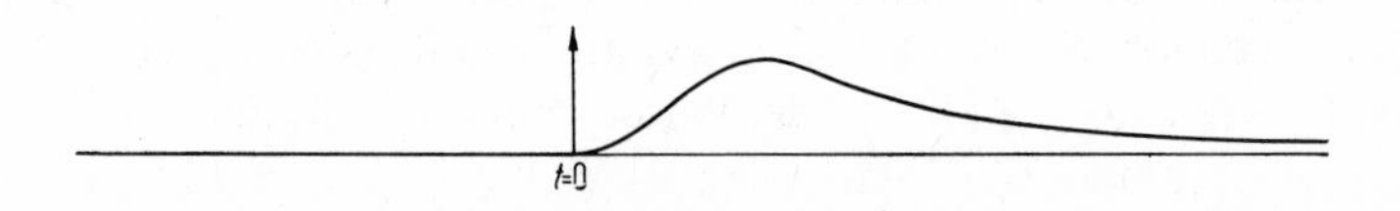

If α is any real number, the function $\varphi_\alpha(\,)$ defined by $\varphi_\alpha(t) = \varphi(t - \alpha)$ $(-\infty < t < \infty)$ is easily seen to be another test-function.

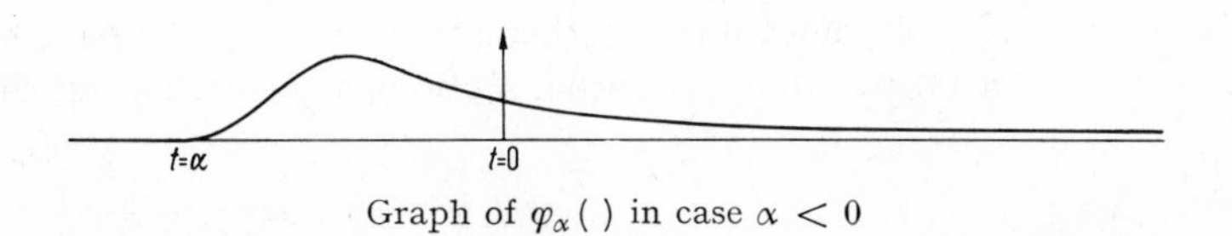

Graph of $\varphi_\alpha(\,)$ in case $\alpha < 0$

0.4 Definitions. Let V be a mapping of test-functions into functions: the function that V assigns to a test-function $\varphi(\,)$ we shall denote by $V \cdot \varphi(\,)$. An *operator* is a mapping of test-functions into test-functions.

0.5 Let V be a mapping of test-functions into functions: if $V \cdot \varphi(\,)$ is a test-function whenever $\varphi(\,)$ is a test-function, then V is an operator.

0.6 For example, let D be the mapping that assigns to each test-function its derivative:

$$D \cdot \varphi(\,) = \varphi'(\,) \qquad \text{for every test-function } \varphi(\,). \tag{0.7}$$

Obviously, $\varphi'(\,)$ is a test-function; therefore, *D is an operator.*

0.8 Another example. If $f(\,)$ is a function, we shall denote by f the mapping that assigns to each test-function $\varphi(\,)$ the function $f \cdot \varphi(\,)$ defined by

$$f \cdot \varphi(\tau) = \int_0^\infty \varphi'(\tau - u)\, f(u)\, du \qquad (-\infty < \tau < \infty); \tag{0.9}$$

it turns out that f is an operator whenever $f(\,)$ is locally integrable on the interval $(0, \infty)$.

0.10 **Equality.** Let V_1 and V_2 be operators: the relation $V_1 = V_2$ means that $V_1 \cdot \varphi() = V_2 \cdot \varphi()$ for every test-function $\varphi()$.

0.11 **The product** of V_1 and V_2 is the mapping $V_1 V_2$ defined by

$$V_1 V_2 \cdot \varphi() = V_1 \cdot (V_2 \cdot \varphi)() \tag{0.12}$$

for every test-function $\varphi()$. The **sum** of V_1 and V_2 is the mapping $V_1 + V_2$ defined by

$$(V_1 + V_2) \cdot \varphi() = V_1 \cdot \varphi() + V_2 \cdot \varphi() \tag{0.13}$$

for every test-function $\varphi()$.

0.14 *Remarks.* Let V_1, V_2, and V_3 be operators: a moment's thought will show that $V_1 V_2$ and $V_1 + V_2$ are operators; further, *multiplication is associative*:

$$(V_1 V_2) V_3 = V_1 (V_2 V_3). \tag{0.15}$$

Convolution

Let $f()$ and $g()$ be functions. When τ is a real number, we shall write

$$f * g(\tau) \stackrel{\text{def}}{=} \int_{-\infty}^{\infty} f(\tau - u)\, g(u)\, du; \tag{0.16}$$

in other words, the left-hand side of the above equation will henceforth be used as an abbreviation of the right-hand side. Let E be the set of all the real numbers τ such that this integral exists.

0.17 **Definition.** The *convolution* of $f()$ and $g()$ is the function $f * g()$ that assigns to each number τ in E the number $f * g(\tau)$. The set E is called the *domain* of $f * g()$.

0.18 *Remarks.* A real number τ is in the domain of $f * g()$ if (and only if) the integral 0.16 exists. Suppose that there exists a number α such that

$$f(t) = 0 \quad (\text{for } t \leq \alpha). \tag{0.19}$$

Let us prove the following two facts. *If τ is in the domain of $f * g()$ then*

$$f * g(\tau) = \int_{-\infty}^{\omega} f(\tau - u)\, g(u)\, du \qquad (\text{when } \omega \geq \tau - \alpha) \tag{0.20}$$

and

(0.21) $$f * g(\tau) = \int_{-\infty}^{\tau-\alpha} f(\tau - u)\, g(u)\, du.$$

We shall also verify the following property. *If α and β are two numbers such that*

(1) $$f(t) = 0 \ (\text{for } t \leq \alpha) \quad \textit{and} \quad g(u) = 0 \ (\text{for } u \leq \beta),$$

then

(0.22) $$f * g(\tau) = 0 \quad (\text{for } \tau \leq \alpha + \beta).$$

Note that 0.21 follows immediately by setting $\omega = \tau - \alpha$ in 0.20. To verify 0.20, note that the equation

(2) $$f * g(\tau) = \int_{-\infty}^{\omega} f(\tau - u)\, g(u)\, du + \int_{\omega}^{\infty} f(\tau - u)\, g(u)\, du$$

comes directly from [0.16]. Suppose that $\omega \geq \tau - \alpha$: since $u \geq \omega$ in the second integral, we have $u \geq \tau - \alpha$, whence $\tau - u \leq \alpha$, whence $f(\tau - u) = 0$ (by 0.19). We have proved that the inequality $\omega \geq \tau - \alpha$ implies that the integrand vanishes for all the values of u involved in the second integral in (2): therefore, this second integral vanishes. This proves 0.20; to verify 0.22, note that $u \leq \tau - \alpha$ in the integral in 0.21: if $\tau \leq \alpha + \beta$ then $u \leq (\alpha + \beta) - \alpha$ (when $u \leq \tau - \alpha$), so that $u \leq \beta$ and $g(u) = 0$ (by (1)); consequently, the integrand vanishes for all the values of u involved in the integral in 0.21: Conclusion 0.22 is at hand.

Note to the reader: you may find that reference numbers hinder your train of thought; if you follow the reasoning, do not stop to look up the statement that a reference number refers to. In other words, *if reference numbers hinder you, ignore them* (the same for superfluous explanations).

Entering Functions

0.23 **Definitions.** A function $g(\,)$ will be called *locally integrable* if it has at most a finite number of discontinuities in every finite interval and if

$$\int_a^b |g(u)|\, du < \infty$$

whenever $-\infty < a < b < \infty$. An "*entering function*" is a locally integrable function that vanishes to the left of some point.

0.24 *Remarks.* A locally integrable function $g(\,)$ is an entering function if (and only if) there exists a real number β such that $g(t) = 0$ for $t \leq \beta$; this number β depends on the function. Clearly, test-functions are entering functions. For example, if α is any real number, the function $g(\,)$ defined by

$$g(t) = \frac{\mathsf{T}_0(t-\alpha)}{\sqrt{t-\alpha}} \qquad (-\infty < t < \infty)$$

is an entering function; the function $\mathsf{T}_0(\,)$ is defined in [0.2].

If $h(\,)$ is an entering function whose derivative $h'(\,)$ is "*continuous*" (that is, continuous on the whole real line), the Fundamental Theorem of Calculus implies that

$$(0.25) \qquad \int_{-\infty}^{\tau} h'(u)\,du = \int_{\beta}^{\tau} h'(u)\,du = h(\tau) \qquad (-\infty < \tau < \infty);$$

here β is one of the numbers such that $h(t) = 0$ for all $t \leq \beta$; the last equation comes from the fact that $h(\beta) = 0$.

Let c be a number; the function $c\,\mathsf{T}_0(\,)$ is defined by

$$c\,\mathsf{T}_0(t) = \begin{cases} 0 & (t \leq 0) \\ c & (t > 0). \end{cases}$$

If $g(\,)$ is an entering function, we can set $\alpha = 0$ in 0.21 to obtain

$$(0.26) \qquad \boxed{c\,\mathsf{T}_0 * g(\tau) = \int_{-\infty}^{\tau} c\,\mathsf{T}_0(\tau-u)\,g(u)\,du = c\int_{-\infty}^{\tau} g(u)\,du}$$

for each τ in the domain of $c\,\mathsf{T}_0 * g(\,)$: since the integral exists for all real values of τ, Equation 0.26 **holds for all real** τ. If $\varphi(\,)$ is a test-function and $-\infty < \tau < \infty$, we have

$$(1) \qquad c\,\mathsf{T}_0 \cdot \varphi(\tau) = c\int_{0}^{\infty} \varphi'(\tau-u)\,du \qquad \text{by [0.9]}$$

$$(2) \qquad = -c\int_{\tau}^{-\infty} \varphi'(x)\,dx$$

$$(3) \qquad = c\int_{-\infty}^{\tau} \varphi'(x)\,dx = c\varphi(\tau) \qquad \text{by 0.25;}$$

Equation (2) is derived from (1) by the change of variable $x = \tau - u$. From (1)—(3) it follows that

$$(0.27) \qquad c\,\mathsf{T}_0 \cdot \varphi(\,) = c\varphi(\,) \qquad \text{for every test-function } \varphi(\,).$$

In the particular case $c = 1$, the function $c\,\mathsf{T}_0()$ is the function $\mathsf{T}_0()$ defined by [0.2]: the equations

$$\boxed{V\,\mathsf{T}_0 = V = \mathsf{T}_0\,V} \qquad \text{for every operator } V \tag{0.28}$$

result immediately from 0.27 and [0.10—11].

0.29 **Theorem.** *Let* $g()$ *and* $\psi()$ *be entering functions. If* $\psi()$ *is continuous, then* $g * \psi()$ *is a continuous entering function.*

Proof. See 15.14 in the Appendix.

0.30 **Theorem.** *Let* $F_1()$, $F_2()$, *and* $H()$ *be entering functions. If either* $H()$ *or* $F_2()$ *is continuous, then*

$$F_1 * [F_2 * H]\,(\tau) = [F_1 * F_2] * H\,(\tau) \qquad (-\infty < \tau < \infty). \tag{0.31}$$

Proof. A repeated application of Definition 0.16 gives

$$F_1 * [F_2 * H]\,(\tau) = \int_{-\infty}^{\infty} \mathrm{d}x \int_{-\infty}^{\infty} v_\tau(x, y)\,\mathrm{d}y\,, \tag{1}$$

where

$$v_\tau(x, y) = F_1(\tau - x)\,F_2(x - y)\,H(y)\,.$$

It will be shown in the Appendix (15.15) that the two integrals

$$\int_{-\infty}^{\infty} \mathrm{d}x \int_{-\infty}^{\infty} v_\tau(x, y)\,\mathrm{d}y \quad \text{and} \quad \int_{-\infty}^{\infty} \mathrm{d}y \int_{-\infty}^{\infty} v_\tau(x, y)\,\mathrm{d}x$$

exist and are equal for every real τ. Consequently, Equation (1) implies that

$$F_1 * [F_2 * H]\,(\tau) = \int_{-\infty}^{\infty} \mathrm{d}y \left[\int_{-\infty}^{\infty} F_1(\tau - x)\,F_2(x - y)\,\mathrm{d}x\right] H(y)\,.$$

The change of variable $u = x - y$ now yields $x = y + u$ and

$$F_1 * [F_2 * H]\,(\tau) = \int_{-\infty}^{\infty} \mathrm{d}y \left[\int_{-\infty}^{\infty} F_1(\tau - y - u)\,F_2(u)\,\mathrm{d}u\right] H(y)$$

$$= \int_{-\infty}^{\infty} [\{F_1 * F_2\}\,(\tau - y)]\,H(y)\,\mathrm{d}y = \{F_1 * F_2\} * H\,(\tau)\,.$$

0.32 *Remarks.* Unless otherwise specified, a "*number*" is a complex number. A point where a function $f(\,)$ is continuous will often be called a "*continuity point*" of $f(\,)$. If τ is a continuity point of a function $f(\,)$, the "*value*" of $f(\,)$ at the point τ is the number $f(\tau)$; the symbol $f(\tau)$ may be meaningless when τ is not a continuity point of the function $f(\,)$.

0.33 **Equality of functions.** If $f(\,)$ and $g(\,)$ are functions, the relation $f(\,) = g(\,)$ will be used to indicate that $f(\tau) = g(\tau)$ whenever τ is a continuity point of both $f(\,)$ and $g(\,)$. If both functions are continuous, then $f(\,) = g(\,)$ if (and only if) $f(\tau) = g(\tau)$ for all real values of τ.

Z *On a first reading:* in order to feel motivated to read the material remaining in this chapter, some readers may find it profitable to go to Chapter 2 (which contains a summary of the basic results). In fact, application-oriented readers may skip the rest of this chapter.

0.34 Let $h_1(\,)$ and $h_2(\,)$ be two functions. If τ is a continuity point of $h_1 * h_2(\,)$, then τ is in the domain of $h_1 * h_2(\,)$: the equations

$$h_1 * h_2(\tau) = \int_{-\infty}^{\infty} h_1(\tau - u)\, h_2(u)\, du = h_2 * h_1(\tau)$$

are from Definition 0.16, the change of variable $y = \tau - u$, and another application of Definition 0.16. Consequently,

$$h_1 * h_2(\,) = h_2 * h_1(\,). \tag{0.35}$$

0.36 **Theorem.** *Suppose that* $g(\,)$ *is an entering function. If* $\psi(\,)$ *is a function whose derivative* $\psi'(\,)$ *is continuous, then the derivative* $[g * \psi]'(\,)$ *of the function* $g * \psi(\,)$ *satisfies the equation*

$$[g * \psi]'(\,) = g * \psi'(\,). \tag{0.37}$$

Proof. Setting $c = 1$ in 0.26 we see that

$$T_0 * \psi'(t) = \int_{-\infty}^{t} \psi'(u)\, du = \psi(t) \qquad (-\infty < t < \infty): \tag{1}$$

the second equation is from 0.25. If $-\infty < \tau < \infty$ then

$$\begin{aligned} \psi * g(\tau) &= [T_0 * \psi'] * g(\tau) && \text{by (1)} \\ &= T_0 * [\psi' * g]\,(\tau) && \text{by 0.31;} \end{aligned}$$

consequently, 0.26 (with $c = 1$) implies

(2) $$\psi * g(\tau) = \int_{-\infty}^{\tau} [\psi' * g](u)\, du \qquad (-\infty < \tau < \infty).$$

Taking derivatives of both sides of (2), we obtain

(3) $$[\psi * g]'(\tau) = \frac{d}{d\tau} \int_{-\infty}^{\tau} [\psi' * g](u)\, du = [\psi' * g](\tau):$$

the last equation is a consequence of the continuity of the function $\psi' * g(\,)$ (whose continuity is in turn a consequence of 0.29 and the continuity of $\psi'(\,)$). The equations

$$[g * \psi]'(\,) = [\psi * g]'(\,) = [\psi' * g](\,) = [g * \psi'](\,)$$

come from 0.35, (3), and 0.35: Conclusion 0.37 is now at hand.

0.38 **Definition.** Given an entering function $g(\,)$, we shall denote by g^* the mapping that assigns to any test-function $\varphi(\,)$ the function $g * \varphi(\,)$:

(0.39) $$g^* \cdot \varphi(\,) = g * \varphi(\,) \quad \text{for any test-function } \varphi(\,).$$

0.40 **Theorem.** *If n is any integer ≥ 0, then*

(0.41) $$[g^* \cdot \varphi]^{(n)}(\,) = [g * \varphi^{(n)}](\,).$$

Proof. Since this obviously holds for $n = 0$, we may proceed by induction: assume that 0.41 holds for $n =$ some integer k, and observe that

$$\begin{aligned} [g^* \cdot \varphi]^{(k+1)}(\,) &= \{[g \cdot \varphi]^{(k)}\}'(\,) \\ &= [g * \varphi^{(k)}]'(\,) && \text{by 0.41 with } n = k \\ &= g * \varphi^{(k+1)}(\,) && \text{by 0.37 with } \psi = \varphi^{(k)}. \end{aligned}$$

Therefore, 0.41 holds for $n = k + 1$ whenever it holds for $n = k$: this concludes the proof of 0.40. The last equation utilizes the fact that $\psi'(\,) = \varphi^{(n+1)}(\,)$ is continuous (because $\varphi^{(n+2)}(\,)$ exists for any integer $n \geq 0$).

0.42 **Theorem.** *The function $g^* \cdot \varphi(\,)$ is infinitely differentiable.*

Proof. Note that $\varphi^{(n)}(\,)$ is a continuous function (since its derivative exists everywhere); from 0.29 it therefore follows that the function $g * \varphi^{(n)}(\,)$ is continuous, and 0.41 now guarantees the existence of the nth derivative of $g^* \cdot \varphi(\,)$.

0.43 **Theorem.** *If $g(\,)$ is an entering function, then g^* is an operator.*

Proof. Let $\varphi(\,)$ be any test-function: in view of 0.5 and [0.39], it will suffice to verify that $g * \varphi(\,)$ is a test-function. Indeed, $g * \varphi(\,)$ is infinitely differentiable (0.42); further, since both $g(\,)$ and $\varphi(\,)$ are entering functions, it follows from 0.22 that the function $g * \varphi(\,)$ vanishes to the left of some point: from [0.0] we now see that $g * \varphi(\,)$ is a test-function.

0.44 **Theorem.** *If $g(\,)$ is an entering function, then the operators g^* and D commute:*

$$Dg^* = g^* D. \tag{0.45}$$

Proof. Let $\varphi(\,)$ be any test-function: we have

$$
\begin{aligned}
[Dg^*] \cdot \varphi(\,) &= D \cdot [g^* \cdot \varphi](\,) && \text{from [0.11]} \\
&= [g * \varphi]'(\,) && \text{from [0.7] and [0.39]} \\
&= [g * (D \cdot \varphi)](\,) && \text{from 0.37 and [0.7]} \\
&= [g^* \cdot (D \cdot \varphi)](\,) && \text{from [0.39];}
\end{aligned}
$$

the conclusion is now an immediate consequence of [0.10—12].

0.46 **Theorem.** *If $f(\,)$ and $g(\,)$ are entering functions, then*

$$(f * g)^* = f^* g^*. \tag{0.47}$$

Proof. Let $\varphi(\,)$ be any test-function: we have

$$
\begin{aligned}
(f * g)^* \cdot \varphi(\,) &= (f * g) * \varphi(\,) && \text{by [0.39]} \\
&= f * (g * \varphi)(\,) && \text{by 0.31} \\
&= f * (g^* \cdot \varphi)(\,) = f^* \cdot (g^* \cdot \varphi)(\,);
\end{aligned}
$$

the last two equations are from [0.39] and the fact that $g^* \cdot \varphi(\,)$ is a test-function (0.42). The conclusion is immediate from [0.10—12].

Exercises

0.48.0 Prove the associative property 0.15.
0.48.1 Prove 0.28.

§ 1. Perfect Operators

Let $f_1()$ and $f_2()$ be entering functions; the relation $f_1() = f_2()$ has been defined in [0.33]. From the definitions [0.39] and [0.16] it follows readily that

$$(1.0) \qquad f_1() = f_2() \quad \text{implies} \quad f_1^* = f_2^*.$$

Conversely,

$$(1.1) \qquad f_1^* = f_2^* \quad \text{implies} \quad f_1() = f_2();$$

see 15.18.2 in the Appendix. The equations

$$(1.2) \qquad f_1^* f_2^* = (f_1 * f_2)^* = (f_2 * f_1)^* = f_2^* f_1^*$$

are from 0.47, 0.35, 1.0, and 0.47.

Numbers referring to definitions are placed in square brackets; these definitions being often obvious from the context, the reader may well ignore such square-bracketed numbers (and as many other numerical references as he can dispense with).

Recall Definition 0.4: if V is an operator and if $\varphi()$ is a test-function, then $V \cdot \varphi()$ is the test-function that the operator V assigns to $\varphi()$.

1.3 **Definitions.** An operator V is said to be *additive* if the equality

$$V \cdot [\varphi_1 + \varphi_2]() = V \cdot \varphi_1() + V \cdot \varphi_2()$$

holds for any test-functions $\varphi_1()$ and $\varphi_2()$. An operator V is called *perfect* if it is an additive operator such that

$$(1.4) \qquad V\psi^* = \psi^* V \qquad \text{for any test-function } \psi().$$

1.5 *Remarks.* The operator $V\psi^*$ is the product (as defined by [0.11]) of the operators V and ψ^* (defined by [0.39]): an operator is perfect if it is additive and commutes with every operator of the form ψ^*.

1.6 **Theorem.** *If V is a perfect operator then*

$$(1.7) \qquad V\varphi^* = (V \cdot \varphi)^* \qquad \text{for every test-function } \varphi().$$

Proof. To begin with, note that

$$(1) \qquad \varphi_1^* \cdot \varphi_2() = \varphi_2^* \cdot \varphi_1() \qquad \text{for any test-functions } \varphi_1() \text{ and } \varphi_2();$$

indeed, the equations

$$\varphi_1^* \cdot \varphi_2(\,) = \varphi_1 * \varphi_2(\,) = \varphi_2 * \varphi_1(\,) = \varphi_2^* \cdot \varphi_1(\,)$$

are from [0.39], [0.35], and [0.39]. Next, let $\psi(\,)$ be any test-function: we have

$$\begin{aligned} [V\varphi^*] \cdot \psi(\,) &= V \cdot [\varphi^* \cdot \psi](\,) && \text{by } [0.12] \\ &= V \cdot [\psi^* \cdot \varphi](\,) && \text{by } (1) \\ &= [V\psi^*] \cdot \varphi(\,) && \text{by } [0.12] \\ &= [\psi^* V] \cdot \varphi(\,) && \text{by } 1.4 \\ &= \psi^* \cdot [V \cdot \varphi](\,) && \text{by } [0.12] \\ &= [V \cdot \varphi]^* \cdot \psi(\,) && \text{by } (1). \end{aligned}$$

In view of [0.10], Conclusion 1.7 is at hand.

1.8 **Theorem.** *Products of perfect operators are perfect.*

Proof. Let V_1 and V_2 be perfect operators: it is easily verified that V_1V_2 is additive (see 1.79.1). Let $\psi(\,)$ be any test-function: the equations

$$[V_1 V_2]\,\psi^* = V_1[V_2\psi^*] = V_1[\psi^* V_2] = [V_1\psi^*]\,V_2 = \psi^*[V_1V_2]$$

are from the associative property (0.15) and two applications of 1.4.

1.9 **Theorem.** *Perfect operators commute.*

Proof. Let F and G be perfect operators: we want to establish that $GF = FG$. Let $\varphi(\,)$ be any test-function; we begin by noting that

$$(G\varphi^*)\,F = (G \cdot \varphi)^*\,F \qquad \text{by } 1.7 \tag{2}$$

$$= F\,(G \cdot \varphi)^* \qquad \text{by } 1.4 \text{ (with } \psi = G \cdot \varphi)$$

$$= F\,(G\varphi^*) \qquad \text{by } 1.7. \tag{3}$$

Next, observe that

$$\begin{aligned} [(GF) \cdot \varphi]^* &= (GF)\,\varphi^* && \text{by 1.7 and 1.8} \\ &= G\,(F\varphi^*) && \text{by } 0.15 \\ &= G\,(\varphi^* F) && \text{by } 1.4 \\ &= (G\varphi^*)\,F && \text{by } 0.15 \\ &= F\,(G\varphi^*) && \text{by } (2)-(3) \\ &= (FG)\,\varphi^* && \text{by } 0.15 \\ &= [(FG) \cdot \varphi]^* && \text{by 1.7 and 1.8}. \end{aligned}$$

Consequently, $[(GF) \cdot \varphi]\,(\,) = [(FG) \cdot \varphi]\,(\,)$ (by 1.1): the conclusion $GF = FG$ is now immediate from [0.10].

Algebra of Perfect Operators

Let V, F, and G be perfect operators; it is easily verified that

$$V(F + G) = VF + VG = FV + GV = (F + G)\,V \tag{1.10}$$

(see 1.79.4). In fact, perfect operators can be added and multiplied according to the usual algebraic laws (more precisely, perfect operators form what is called in algebra a "ring").

1.11 *Remark.* The sum of perfect operators is a perfect operator (see 1.79.5)

1.12 **Theorem.** *D is a perfect operator.*

Proof. The additivity [1.3] of D is obvious; since $\psi(\,)$ is an entering function, the property $D\psi^* = \psi^* D$ is obtained by setting $g = \psi$ in 0.45.

1.13 **Theorem.** *If $h(\,)$ is an entering function, then h^* is a perfect operator.*

Proof. The additivity of h^* is immediate from [0.39], and the equation $h^*\psi^* = \psi^* h^*$ comes directly from 1.2.

The Operator of an Entering Function

Z 1.14 **Definition.** Let $h(\,)$ be an entering function: we set

$$[\![h]\!] = Dh^*, \tag{1}$$

and call it *the operator* of the function $h(\,)$.

1.15 *Remarks.* Thus, the operator of $h(\,)$ is the product of the two perfect operators D and h^*; it follows from 1.12—13 and 1.8 that this operator $[\![h]\!]$ **is a perfect operator.** Note that

$$[\![h]\!] = Dh^* = h^* D \qquad \text{(by (1) and 1.9)}. \tag{1.16}$$

If $\varphi(\,)$ is a test-function then

$$\begin{aligned} [\![h]\!] \cdot \varphi(\,) &= [(h^* D) \cdot \varphi](\,) && \text{by 1.16} \\ &= [h^* \cdot (D \cdot \varphi)](\,) && \text{by [0.12]} \\ &= h * \varphi'(\,) && \text{by [0.39] and [0.7];} \end{aligned}$$

from 0.35 and [0.16] we may now conclude that the equation

$$(1.17) \qquad [\![h]\!] \cdot \varphi(\tau) = \int_{-\infty}^{\infty} \varphi'(\tau - u)\, h(u)\, du \qquad (-\infty < \tau < \infty)$$

holds for any test-function $\varphi(\,)$. Let $\mathsf{T}_0(\,)$ be the function defined by [0.2]; the equations

$$[\![\mathsf{T}_0]\!] \cdot \varphi(\tau) = \int_0^{\infty} \varphi'(\tau - u)\, du = \mathsf{T}_0 \cdot \varphi(\tau)$$

$(-\infty < \tau < \infty)$ are from 1.17 and [0.9]. From [0.10] it therefore follows that

$$(1.17.1) \qquad [\![\mathsf{T}_0]\!] = \mathsf{T}_0 .$$

1.18 **Theorem.** *If V is a perfect operator, then*

$$(1.19) \qquad [\![V \cdot \varphi]\!] = V\, [\![\varphi]\!] \qquad \text{for every test-function } \varphi(\,).$$

Proof. The equations

$$[\![V \cdot \varphi]\!] = (V \cdot \varphi)^* D = (V \varphi^*) D = V (\varphi^* D) = V\, [\![\varphi]\!]$$

are from 1.16, 1.7, 0.15, and 1.16.

1.20 **The operator $[\![\mathsf{T}_0]\!]$ of the function $\mathsf{T}_0(\,)$ is the identity-operator:**

$$(1.21) \qquad \boxed{V [\![\mathsf{T}_0]\!] = V = [\![\mathsf{T}_0]\!]\, V} \qquad \text{for any operator } V:$$

see 0.28 and 1.17.1.

Z 1.22 **About equality.** Let $f(\,)$ and $g(\,)$ be entering functions. From 1.0 and 1.16 it obviously follows that

$$(1.23) \qquad f(\,) = g(\,) \text{ implies } [\![f]\!] = [\![g]\!].$$

Conversely,

$$(1.24) \qquad [\![f]\!] = [\![g]\!] \text{ implies } f(\,) = g(\,).$$

In order to prove 1.24, let us multiply by T_0^* both sides of the hypothesis $([\![f]\!] = [\![g]\!])$; we obtain

$$\mathsf{T}_0^* \, [\![f]\!] = \mathsf{T}_0^* \, [\![g]\!];$$

that is, by [1.14]:

$$\mathsf{T}_0^* \, (Df^*) = \mathsf{T}_0^* (Dg^*),$$

and the associativity property (0.15) now gives

$$(\mathsf{T}_0^* D) \, f^* = (\mathsf{T}_0^* D) \, g^*,$$

whence

$$[\![\mathsf{T}_0]\!] \, f^* = [\![\mathsf{T}_0]\!] \, g^* \qquad \text{(by 1.16)}. \tag{2}$$

From (2) and 1.21 it now follows that $f^* = g^*$: the conclusion $f(\,) = g(\,)$ is therefore immediate from 1.1. This concludes the proof of 1.24. From 1.17 it follows directly that

$$h(\,) = f(\,) + g(\,) \text{ implies } [\![h]\!] = [\![f]\!] + [\![g]\!]. \tag{1.25}$$

1.26 **Multiplication Property.** *The product of the operators of two entering functions* $f()$ *and* $g()$ *equals the product of the operator* D *with the operator of the function* $f * g(\,)$:

$$\boxed{[\![f]\!] \, [\![g]\!] = D [\![f * g]\!]}. \tag{1.27}$$

Proof:

$$\begin{aligned} [\![f]\!] \, [\![g]\!] = (Df^*) \, (g^* D) &= D \, (f^* g^*) \, D && \text{by 1.16} \\ &= D \, [(f * g)^* \, D] && \text{by 1.2} \\ &= D \, [\![f * g]\!] && \text{by 1.16}. \end{aligned}$$

1.28 **Consequence:** the equations

$$\boxed{[\![g]\!] = [\![\mathsf{T}_0]\!] \, [\![g]\!] = D [\![\mathsf{T}_0 * g]\!]} \tag{1.29}$$

are from 1.21 and 1.27.

Translation Property

If $-\infty < \alpha < \infty$ the function $\mathsf{T}_\alpha(\,)$ is defined by

$$\mathsf{T}_\alpha(t) = \begin{cases} 0 & (t \le \alpha) \\ 1 & (t > \alpha). \end{cases} \tag{1.30}$$

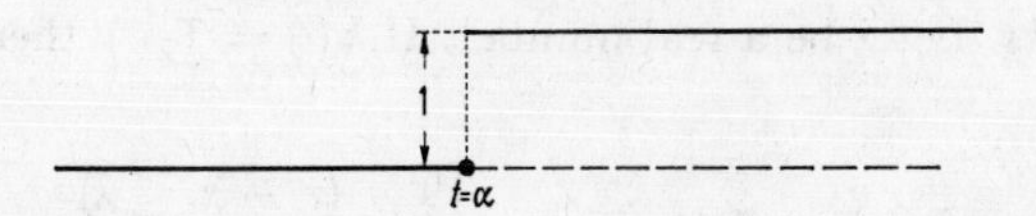

The **translate** of an entering function $h(\,)$ is the function $h_\alpha(\,)$ defined by the equation

$$h_\alpha(\tau) = h(\tau - a) \qquad (-\infty < \tau < \infty). \tag{1.31}$$

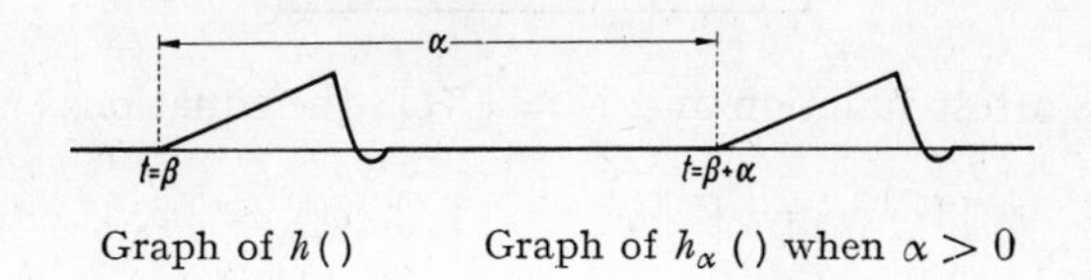

Graph of $h(\,)$ Graph of $h_\alpha(\,)$ when $\alpha > 0$

As we shall see, *its operator* $[\![h_\alpha]\!]$ *is the product of the two operators* $[\![\mathsf{T}_\alpha]\!]$ *and* $[\![h]\!]$:

$$\boxed{[\![h_\alpha]\!] = [\![\mathsf{T}_\alpha]\!]\,[\![h]\!]}\,. \tag{1.32}$$

Proof of 1.32:

(1) $\mathsf{T}_\alpha * h(\tau) = \int_{-\infty}^{\tau-\alpha} \mathsf{T}_\alpha(\tau - u)\, h(u)\, du$ by 0.21

(2) $= \int_{-\infty}^{\tau-\alpha} h(u)\, du$ by [1.30]

(3) $= \int_{-\infty}^{\tau} h(x - \alpha)\, dx$

(4) $= \int_{-\infty}^{\tau} h_\alpha(x)\, dx = \mathsf{T}_0 * h_\alpha(\tau)$ by 1.31 and 0.26;

Equation (3) is obtained from (2) by the change of variable $x = u + \alpha$. From (1)—(4) we have

(5) $\mathsf{T}_\alpha * h(\,) = \mathsf{T}_0 * h_\alpha(\,)$:

the equations

$$[\![\mathsf{T}_\alpha]\!]\,[\![h]\!] = D\,[\![\mathsf{T}_\alpha * h]\!] = D\,[\![\mathsf{T}_0 * h_\alpha]\!] = [\![h_\alpha]\!]$$

are from 1.27, (5), and 1.29. This concludes the proof of 1.32.

1.33 *Remarks.* Let λ be a real number. If $h(\ ) = \mathsf{T}_\lambda(\ )$ then [1.31] gives

$$h_\alpha(\tau) = \mathsf{T}_\lambda(\tau - \alpha) = \begin{cases} 0 & (\tau - \alpha \le \lambda) \\ 1 & (\tau > \alpha + \lambda), \end{cases}$$

so that $h_\alpha(\ ) = \mathsf{T}_{\alpha+\lambda}(\ )$; from 1.32 it now follows that

$$\boxed{[\![\mathsf{T}_\alpha]\!]\,[\![\mathsf{T}_\lambda]\!] = [\![\mathsf{T}_{\alpha+\lambda}]\!]} \qquad (-\infty < \lambda < \infty). \tag{1.34}$$

If $\varphi(\ )$ is a test-function and $V = [\![\mathsf{T}_\alpha]\!]$, the equations

$$[\![V \cdot \varphi]\!] = V\,[\![\varphi]\!] = [\![\mathsf{T}_\alpha]\!]\,[\![\varphi]\!] = [\![\varphi_\alpha]\!]$$

are from 1.19 and 1.32; from 1.24 it therefore follows that $V \cdot \varphi(\ ) = \varphi_\alpha(\ )$, which (in view of [1.31]) implies

$$[\![\mathsf{T}_\alpha]\!] \cdot \varphi(\tau) = \varphi(\tau - \alpha) \qquad (-\infty < \tau < \infty). \tag{1.35}$$

Thus,

$$\boxed{[\![\mathsf{T}_\alpha]\!] \cdot \varphi() = \varphi_\alpha()} \quad \text{for any test-function } \varphi();$$

in other words, the operator $[\![\mathsf{T}_\alpha]\!]$ assigns to each test-function $\varphi()$ its translate $\varphi_\alpha()$. Recall that $-\infty < \alpha < \infty$ in 1.30—35.

$\mathcal{K}$-functions

1.36 **Definition.** A function $h()$ will be called a $\mathcal{K}$-*function* if it has at most a finite number of discontinuities in every finite interval $(0, \lambda)$ and

$$\int_0^\lambda |h(u)|\,du < \infty \quad \text{whenever } 0 < \lambda < \infty.$$

1.37 **The entering function** of a $\mathcal{K}$-function $h()$ is the function $\{h(t)\}(\ )$ defined by

$$\{h(t)\}(\tau) = \begin{cases} 0 & (\tau \le 0) \\ h(\tau) & (\tau > 0). \end{cases} \tag{1.38}$$

For example, the graph of the entering function $\{t^{-1/2}\}()$ has the following shape

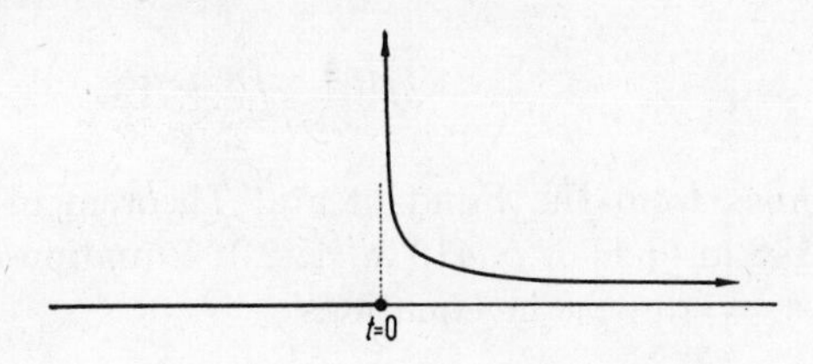

Finally, note the graph of the entering function $\{\cos 3t\}()$:

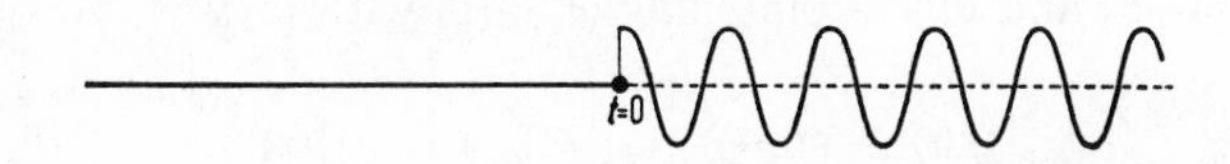

1.39 *Remarks.* A $\mathcal{K}$-function need not be defined on the interval $(-\infty, 0]$, nor need it be defined at its points of discontinuity; note that any locally integrable function [0.23] is a $\mathcal{K}$-function.

Since $\{h(t)\}()$ is obviously an entering function, its operator (as defined in 1.14) is a perfect operator. We shall write

$$\langle h(t)\rangle \overset{\text{def}}{=} [\![\{h(t)\}]\!] = D\{h(t)\}^* = \{h(t)\}^* D: \tag{1.40}$$

the last two equations are from 1.16; thus, $\langle h(t)\rangle$ is the operator of the entering function $\{h(t)\}()$. From 1.17 and [1.38] we see that the equation

$$\langle h(t)\rangle \cdot \varphi(\tau) = \int_0^\infty \varphi'(\tau - u)\, h(u)\, du \qquad (-\infty < \tau < \infty) \tag{1.41}$$

holds for any test-function $\varphi()$.

1.42 **Derivation Property.** *Consider a function $f()$ that is continuous in the interval $(0, \infty)$ and satisfies the equation $f(0-) = f(0+)$; if its derivative $f'()$ is a $\mathcal{K}$-function, then*

$$D\langle f(t) - f(0-)\rangle = \langle f'(t)\rangle. \tag{1.43}$$

Proof. From [1.36] it follows that the function $f'()$ is continuous on any interval $[0, \tau]$ with the possible exception of finitely-many points;

consequently,

$$(1)\qquad \int_0^\tau f'(u)\,du = f(\tau) - f(0+)$$

$$(2)\qquad = \{f(t) - f(0-)\}(\tau) \qquad (\text{all } \tau > 0):$$

Equation (1) comes from the Fundamental Theorem of Calculus (see Theorem 6.6 p. 143 in [J 1] or p. 418 in [K 2]); Equation (2) is from our hypothesis $f(0-) = f(0+)$. The equations

$$(3)\qquad \int_0^\tau f'(u)\,du = \int_{-\infty}^\tau \{f'(t)\}(u)\,du = \mathsf{T}_0 * \{f'(t)\}(\tau)$$

are from [1.38] and 0.26. Combining (1)—(2) with (3):

$$(4)\qquad \{f(t) - f(0-)\}(\tau) = \mathsf{T}_0 * \{f'(t)\}(\tau) \qquad (\text{all } \tau > 0).$$

On the other hand,

$$(5)\qquad \mathsf{T}_0 * \{f'(t)\}(\tau) = 0 \qquad (\text{for } \tau \leq 0):$$

this comes from 0.22. From (4) and (5) we have

$$\{f(t) - f(0-)\}(\,) = \mathsf{T}_0 * \{f'(t)\}(\,),$$

whence

$$[\![\{f(t) - f(0-)\}]\!] = [\![\mathsf{T}_0 * \{f'(t)\}]\!]; \qquad (\text{by } [1.23]);$$

left-multiplying by D both sides of this last equation, we obtain

$$D[\![\{f(t) - f(0-)\}]\!] = D\,[\![\mathsf{T}_0 * \{f'(t)\}]\!] = [\![\{f'(t)\}]\!] = \langle f'(t)\rangle:$$

the two last equations are from 1.29 and [1.40]. Conclusion 1.43 is now immediate from [1.40].

1.44 *Remarks.* If α is a real number, the right-hand limit $f(\alpha+)$ is defined by

$$f(\alpha+) \overset{\text{def}}{=} \lim_{\tau\to\alpha} f(\tau) \qquad \text{with } \tau > \alpha;$$

the left-hand limit $f(\alpha-)$ is defined analogously. A function $f(\,)$ is said to be continuous *in* the interval $(0, \infty)$ if

$$f(\alpha-) = f(\alpha) = f(\alpha+) \qquad (\text{all } \alpha > 0).$$

1.45 Definition 1.38 gives

$$\{t\}(\tau) = \begin{cases} 0 & (\tau \le 0) \\ \tau & (\tau > 0) \end{cases} \qquad \text{and} \qquad \{1\}(\tau) = \begin{cases} 0 & (\tau \le 0) \\ 1 & (\tau > 0). \end{cases}$$

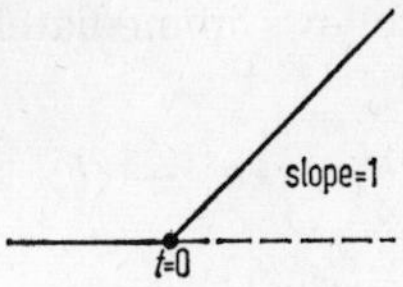

slope=1

$t=0$

Graph of $\{t\}(\ )$

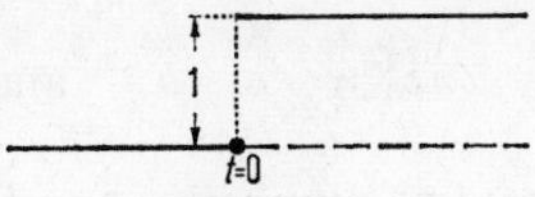

1

$t=0$

Graph of $\{1\}(\ )$

Note that $\{t\}'(\ ) = \{1\}(\ )$. The operator-equation

$$D\langle t\rangle = \langle 1\rangle \tag{1.46}$$

is obtained by setting $f(t) = t$ in 1.43.

1.47 **About equality.** Let $F(\ )$ and $G(\ )$ be $\mathcal{K}$-functions. In view of [1.38] and [0.33], the relation

$$\{F(t)\}(\ ) = \{G(t)\}(\ ) \tag{1}$$

means that $F(\tau) = G(\tau)$ at all the points $t = \tau > 0$ where both $F(\ )$ and $G(\ )$ are continuous: let us verify that (1) is equivalent to the operator-equation

$$\langle F(t)\rangle = \langle G(t)\rangle. \tag{2}$$

Indeed, (1) implies

$$[\![\{F(t)\}]\!] = [\![\{G(t)\}]\!] \qquad \text{by 1.23,} \tag{3}$$

whence (2) follows from [1.40]; conversely, (2) implies (3) (by [1.40]), which implies (1) (by 1.24).

Numbers and Operators

Let c be a number. The equation $h(t) = c$ defines a $\mathcal{K}$-function, and [1.38] gives

$$\{c\}(\tau) = \begin{cases} 0 & (\tau \le 0) \\ c & (\tau > 0). \end{cases} \tag{4}$$

The operator $\langle c\rangle$ of the function $\{c\}(\,)$ is determined by the equation $\langle c\rangle = \llbracket\{c\}\rrbracket$ (see [1.40]). Since $\{c\}(\,) = c\,\mathsf{T}_0(\,)$, we have $\langle c\rangle = \llbracket c\mathsf{T}_0\rrbracket$: this equation enables us to conclude from 1.17.1 and 0.27 that

$$\langle c\rangle \cdot \varphi(\,) = \llbracket c\,\mathsf{T}_0\rrbracket \cdot \varphi(\,) = c\varphi(\,) \tag{1.48}$$

for every test-function $\varphi(\,)$. From 1.48 it follows immediately that the equations

$$\langle ab\rangle = \langle a\rangle\,\langle b\rangle \quad \text{and} \quad \langle a+b\rangle = \langle a\rangle + \langle b\rangle \tag{1.49}$$

hold for any two numbers a and b. In the particular case $c = 1$, Definitions (4) and [0.2] show that $\{1\}(\,) = \mathsf{T}_0(\,)$, whence

$$\langle 1\rangle = \llbracket \mathsf{T}_0\rrbracket\,. \tag{1.50}$$

Z 1.51 **Agreement.** *When c is a number, we shall often remove the angular brackets* that occur in the expression $\langle c\rangle$.

1.52 **Indeed,** the equations 1.49 imply that there is no algebraic reason to distinguish between the number c and the operator $\langle c\rangle$; further,

$$a = b \;\; \textit{if (and only if)} \;\; \langle a\rangle = \langle b\rangle\,. \tag{1.53}$$

It could be said that the correspondence $c \mapsto \langle c\rangle$ (which assigns to each number c the perfect operator $\langle c\rangle$) is an *imbedding* of the complex number field into the ring of perfect operators.

1.54 **In consequence** of 1.50 and [1.51], we have

$$\boxed{1 = \langle 1\rangle = \llbracket \mathsf{T}_0\rrbracket} \tag{1.55}$$

Let V be a perfect operator; from [1.51] and 1.9 we see that

$$cV = \langle c\rangle\, V = V\langle c\rangle = Vc \qquad \text{for any number } c. \tag{1.56}$$

In particular, 1.55 and 1.21 give

$$1V = V = V1. \tag{1.57}$$

Subtraction of operators is defined as follows:

$$V - V_1 \overset{\text{def}}{=} V + \langle -1\rangle\, V_1. \tag{1.58}$$

Let a be a number; setting $V_1 = \langle a\rangle$ in [1.58], we can use 1.49 to obtain

$$V - a = V + \langle -a\rangle. \tag{1.59}$$

1.60 **Theorem.** *Let $g()$ be a $\mathcal{K}$-function, and let a be a number. The operator $\langle ag(t)\rangle$ is the product of the operator $\langle a\rangle$ with the operator $\langle g(t)\rangle$:*

$$\langle ag(t)\rangle = a\langle g(t)\rangle; \tag{1.61}$$

further,

$$\langle g(t) - a\rangle = \langle g(t)\rangle - a. \tag{1.62}$$

Proof. Let $\varphi()$ be a test-function and $-\infty < \tau < \infty$:

$$\begin{aligned}
[\langle ag(t)\rangle \cdot \varphi](\tau) &= a \int_0^\infty \varphi'(\tau - u)\, du && \text{by [1.41]}\\
&= a\,[\langle g(t)\rangle \cdot \varphi]\,(\tau) && \text{by [1.41]}\\
&= \langle a\rangle\,[\langle g(t)\rangle \cdot \varphi]\,(\tau) && \text{by 1.48}\\
&= [\langle a\rangle\,\langle g(t)\rangle] \cdot \varphi(\tau) &&\\
&= [a\langle g(t)\rangle] \cdot \varphi(\tau) && \text{by 1.56;}
\end{aligned}$$

therefore, $\langle ag(t)\rangle \cdot \varphi() = a\langle g(t)\rangle \cdot \varphi()$: Conclusion 1.61 is now at hand. To prove 1.62, note that [1.38] gives

$$\{g(t) - a\}(\tau) = \{g(t)\}(\tau) - \{a\}(\tau) \qquad (-\infty < \tau < \infty),$$

so that

$$[\![\{g(t) - a\}]\!] = [\![\{g(t)\}]\!] + [\![\{-a\}]\!] \qquad \text{(by 1.25)},$$

whence

$$\langle g(t) - a\rangle = \langle g(t)\rangle + \langle -a\rangle \qquad \text{(by [1.40])}:$$

Conclusion 1.62 is now immediate from 1.59.

1.63 *Remark.* Let $f()$ be as in 1.43: from 1.62 we see that

$$\langle f(t) - f(0-)\rangle = \langle f(t)\rangle - f(0-);$$

Equation 1.43 may therefore be written

$$\langle f'(t)\rangle = D\langle f(t)\rangle - f(0-)\, D; \tag{1.64}$$

this is because of 1.10 and 1.56.

Ratios of Perfect Operators

1.65 An operator V is called *invertible* when it is a perfect operator such that the equality $VF = 1$ holds for some perfect operator F; *such an* *operator F is unique*: indeed, if F_0 is a perfect operator such that $VF_0 = 1$, the equations

$$F_0 = F_0 1 = F_0(VF) = (F_0 V)\, F = (VF_0)\, F = 1F = F$$

are from 1.57, the hypothesis $VF = 1$, associativity (0.15), commutativity (1.9), and our hypothesis $VF_0 = 1$.

1.66 **Definition.** When V is an invertible operator, we denote by V^{-1} the unique perfect operator F such that $VF = 1$. When H is a perfect operator, we write

$$\frac{H}{V} \overset{\text{def}}{=} V^{-1}H = HV^{-1}: \tag{1.67}$$

in other words, the left-hand side of the first equation is an abbreviation of its right-hand side; the second equation comes from 1.9. In particular, when $H = 1$, then

$$\frac{1}{V} = V^{-1} \quad \text{and} \quad V^{-1}V = VV^{-1} = 1. \tag{1.68}$$

1.69 **For example,** it follows from 1.46, 1.55, and 1.12 that

$$\boxed{D^{-1} = \langle t \rangle}. \tag{1.70}$$

Let V, V_1, and V_2 be perfect operators; the following equalities are easily verified:

$$\frac{1}{V_1 V_2} = \frac{1}{V_1}\,\frac{1}{V_2}, \tag{1.71}$$

$$\frac{1}{V^m} = \left(\frac{1}{V}\right)^m \qquad \text{(any integer } m \geq 1\text{)}. \tag{1.72}$$

Further, if H and F are perfect operators, then

$$\frac{RH}{RV} = \frac{H}{V} \qquad \text{(if } R \text{ is invertible)}, \tag{1.73}$$

and

$$\frac{H}{V_1} + \frac{F}{V_2} = \frac{HV_2 + FV_1}{V_1 V_2}. \tag{1.74}$$

1.75 *Remark. If V is invertible, then $V \neq 0$.* Indeed, if $V = 0$ then we have the contradiction: $1 = V^{-1}V = V^{-1}0 = 0$.

1.76 **Theorem.** *If y and V are perfect operators such that the equation $Vy = R$ holds for some invertible operator R, then V is invertible and $y = R/V$.*

Proof. Right-multiplying by R^{-1} both sides of our hypothesis $Vy = R$, we obtain

$$VyR^{-1} = RR^{-1} = 1,$$

whence $V[yR^{-1}] = 1$: therefore, V is invertible: the conclusion now comes from 1.77 below.

1.77 **Theorem.** *Let y and H be operators. If V is invertible, then*

$$Vy = H \text{ implies } y = \frac{H}{V}. \tag{1.78}$$

Proof. Left-multiply both sides of the hypothesis ($Vy = H$) by V^{-1}; this gives $(V^{-1}V)\,y = V^{-1}H$; the conclusion is immediate from 1.68 and [1.67].

Exercises

1.79.0 Prove the following two equalities:

$$\langle t^{-1/2}\rangle = 2D\langle t^{1/2}\rangle \text{ and } D^{-1}\langle t^{-1/2}\rangle = 2\langle t^{1/2}\rangle.$$

1.79.1 Suppose that V_1 and V_2 are additive operators: use the following two definitions

$$(V_1V_2)\cdot\varphi(\,) = V_1\cdot(V_2\cdot\varphi)(\,) \qquad \text{(see [0.12])}, \tag{1}$$

$$(V_1+V_2)\cdot\varphi(\,) = V_1\cdot\varphi(\,) + V_2\cdot\varphi(\,) \qquad \text{(see [0.13])} \tag{2}$$

to prove that V_1V_2 and $V_1 + V_2$ are additive. *Hints*:

$$\begin{aligned}
(V_1V_2)\cdot[\varphi_1+\varphi_2] &= V_1\cdot(V_2\cdot[\varphi_1+\varphi_2]) && \text{by (1)}\\
&= V_1\cdot(V_2\cdot\varphi_1 + V_2\cdot\varphi_2) && \text{by [1.3]}\\
&= V_1\cdot(V_2\cdot\varphi_1) + V_1\cdot(V_2\cdot\varphi_2) && \text{by [1.3]}\\
&= (V_1V_2)\cdot\varphi_1 + (V_1V_2)\cdot\varphi_2 && \text{by (1)},
\end{aligned}$$

and

$$\begin{aligned}
(V_1+V_2)\cdot[\varphi_1+\varphi_2] &= V_1\cdot[\varphi_1+\varphi_2] + V_2\cdot[\varphi_1+\varphi_2] && \text{by (2)}\\
&= V_1\cdot\varphi_1 + V_1\cdot\varphi_2 + V_2\cdot\varphi_1 + V_2\cdot\varphi_2 && \text{by [1.3]}\\
&= [V_1\cdot\varphi_1 + V_2\cdot\varphi_1] + [V_1\cdot\varphi_2 + V_2\cdot\varphi_2]\\
&= (V_1+V_2)\cdot\varphi_1 + (V_1+V_2)\cdot\varphi_2 && \text{by (2)}.
\end{aligned}$$

1.79.2 Suppose that F, G, and V are additive operators; prove that $V(F+G) = VF + VG$. *Hints:*

$$\begin{aligned}[V(F+G)]\cdot\varphi &= V\cdot[(F+G)\cdot\varphi] && \text{by (1)}\\ &= V\cdot[F\cdot\varphi + G\cdot\varphi] && \text{by (2)}\\ &= V\cdot[F\cdot\varphi] + V\cdot[G\cdot\varphi] && \text{by [1.3]}\\ &= (VF)\cdot\varphi + (VG)\cdot\varphi && \text{by (1)}\\ &= (VF+VG)\cdot\varphi && \text{by (2).}\end{aligned}$$

1.79.3 With the same hypotheses as in 1.79.2, prove that $FV + GV = (F+G)\,V$. *Hints:*

$$\begin{aligned}(FV+GV)\cdot\varphi &= (FV)\cdot\varphi + (GV)\cdot\varphi && \text{by (2)}\\ &= F\cdot(V\cdot\varphi) + G\cdot(V\cdot\varphi) && \text{by (1)}\\ &= (F+G)\cdot(V\cdot\varphi) && \text{by (2)}\\ &= [(F+G)V]\cdot\varphi && \text{by (1).}\end{aligned}$$

1.79.4 Prove 1.10. *Hints:* use 1.79.2—3 and 1.9.

1.79.5 Prove that the sum of perfect operators is a perfect operator. *Hints:* use 1.79.1 and

$$\begin{aligned}[V_1+V_2]\,\psi^* &= V_1\psi^* + V_2\psi^* && \text{by 1.79.3}\\ &= \psi^* V_1 + \psi^* V_2 = \psi^*[V_1+V_2] && \text{by 1.79.2.}\end{aligned}$$

1.79.6 Prove 1.23. *Hint:* use 1.17.

1.79.7 Prove 1.25. *Hint:* use 1.17.

1.79.8 Prove 1.49. *Hints:* the second equation is immediate from 1.48 and 1.25; further,

$$\begin{aligned}(\langle a\rangle\langle b\rangle)\cdot\varphi(\,) &= \langle a\rangle\cdot(\langle b\rangle\cdot\varphi)(\,) && \\ &= \langle a\rangle\cdot(b\varphi)(\,) && \text{by 1.48}\\ &= a(b\varphi)(\,) && \text{by 1.48}\\ &= ab\varphi(\,) = \langle ab\rangle\cdot\varphi(\,) && \text{by 1.48.}\end{aligned}$$

1.79.9 Prove 1.53. *Hints:* from 1.24 it follows that the equation $\langle a\rangle = \langle b\rangle$ implies $\{a\}(\,) = \{b\}(\,)$, which implies $\{a\}(1) = \{b\}(1)$; the conclusion $a = b$ is obtained by noting that $\{c\}(1) = c$.

1.79.10 Prove 1.71. *Hints:* show that $[V_1^{-1}V_2^{-1}]\,V_2V_1 = 1$: this implies $[V_2V_1]^{-1} = V_1^{-1}V_2^{-1}$: Conclusion 1.71 now comes from 1.9.

1.79.11 Prove 1.72. *Hints:* from 1.71 it follows that

$$\frac{1}{V^2} = \frac{1}{VV} = \frac{1}{V}\,\frac{1}{V} = \left(\frac{1}{V}\right)^2.$$

1.79.12 Prove 1.73. *Hints:*

$$\begin{aligned} \frac{RH}{RV} = RH(RV)^{-1} &= HR(RV)^{-1} && \text{by [1.67] and 1.9} \\ &= HR(R^{-1}V^{-1}) && \text{by 1.71} \\ &= H(RR^{-1})\,V^{-1} && \text{by 0.15.} \end{aligned}$$

1.79.13 Prove 1.74. *Hints:*

$$\begin{aligned} \frac{HV_2 + FV_1}{V_1V_2} &= [HV_2 + FV_1]\,(V_1V_2)^{-1} && \text{by 1.67} \\ &= HV_2(V_1^{-1}V_2^{-1}) + FV_1(V_1^{-1}V_2^{-1}) && \text{by 1.71} \\ &= H(V_2V_2^{-1})\,V_1^{-1} + FV_2^{-1} && \text{by 0.15, 1.9, and 1.68.} \end{aligned}$$

Chapter 2

The present chapter involves differentiation and integration on the interval $[0, \infty)$; it begins with a review of the central facts. The table of formulas (3.7—33) can be used in most of the applications; in order to cope with more complicated problems, § 4 is devoted to techniques of partial fraction decomposition. The extremely simple worked-out examples in this chapter are intended to reveal the procedure that will be applied later on to more complicated situations — and, later still, generalized.

Some of our problems admit only non-classical solutions having discontinuities on the interval $[0, \infty)$: see 2.50 and 2.56. The non-classical example 2.50 has been included in order to make §§ 2.47—51 better disclose the approach and viewpoint, the "philosophy" of this book: this type of problem really belongs to a later chapter, since the physical meaning of the operator D is only validated in Chapter 3. Example 2.56 deals with a non-standard problem; although of some practical importance, such problems are not readily solved by classical methods. The worked out problems 3.38 and 3.39 are classical, but not Laplace-transformable.

§ 2. The Basic Facts

A $\mathcal{K}$-*function* is a function that is locally integrable on the interval $(0, \infty)$ (see [1.36]). *The entering function of a* $\mathcal{K}$-function $h()$ is the function $\{h(t)\}()$ defined by

$$\{h(t)\}(\tau) = \begin{cases} 0 & (\tau \leq 0) \\ h(\tau) & (\tau > 0); \end{cases} \tag{2.0}$$

the operator (see [1.14]) of the function $\{h(t)\}()$ is denoted $\langle h(t) \rangle$; it will be convenient to write

$$\boxed{h \overset{\text{def}}{=} \langle h(t) \rangle}:$$

in other words, we shall use h as an abbreviation of $\langle h(t)\rangle$. Consequently,

$$(2.1) \qquad h = \langle h(t)\rangle = [\![\{h(t)\}]\!] .$$

We shall call h the **canonical operator** of the function $h()$. From 1.41 we see that h is the mapping that assigns to each test-function $\varphi()$ the function $h \cdot \varphi()$ defined by

$$(2.2) \qquad h \cdot \varphi(\tau) = \int_0^\infty \varphi'(\tau - u)\, h(u)\, \mathrm{d}u \qquad (-\infty < \tau < \infty):$$

in fact, the range of integration is finite (see 15.28—30). Thus, to any $\mathcal{K}$-function $h()$ there corresponds the perfect operator h: this correspondence has various properties to be described presently. If $F()$ and $G()$ are $\mathcal{K}$-functions, then

$$\{F(t)\}() = \{G(t)\}()$$

if (and only if) $F(\tau) = G(\tau)$ at each point $t = \tau > 0$ where both $F()$ and $G()$ are continuous. From 1.23—24 it follows that

$$(2.3) \qquad \boxed{F = G \quad \text{implies} \quad \{F(t)\}() = \{G(t)\}()}$$

and

$$(2.4) \qquad \boxed{\{F(t)\}() = \{G(t)\}() \quad \text{implies} \quad F = G}:$$

see 1.47. If a, b, and c are numbers, we can combine 1.61—62 and 1.25 with [2.1] to obtain

$$(2.5) \qquad \boxed{\langle aF(t) + bG(t) + c\rangle = aF + bG + c}.$$

Let us discuss convolution. We can set $\alpha = \beta = 0$ in 0.22 to obtain

$$(2.6) \qquad \{F(t)\} * \{G(t)\}(\tau) = 0 \qquad (\text{for } \tau \leq 0).$$

In case τ is in the domain of $\{F(t)\} * \{G(t)\}()$, we can set $\alpha = 0$ in 0.21 to obtain

$$(2.7) \qquad \{F(t)\} * \{G(t)\}(\tau) = \int_0^\tau F(\tau - u)\, G(u)\, \mathrm{d}u \qquad (\text{for } \tau > 0).$$

Equations 2.6—7 imply that

$$\text{(2.7.1)} \qquad \left\{\int_0^t F(t-u)\,G(u)\,du\right\}(\,) = \{F(t)\} * \{G(t)\}(\,):$$

it can be shown* that the above function is locally integrable; we may therefore apply 1.23 and [2.1] to obtain

$$\text{(1)} \qquad \left\langle \int_0^t F(t-u)\,G(u)\,du \right\rangle = \llbracket\{F(t)\} * \{G(t)\}\rrbracket .$$

On the other hand, it follows from 1.27 and [2.1] that

$$\text{(2)} \qquad D\llbracket\{F(t)\} * \{G(t)\}\rrbracket = \langle F(t)\rangle\,\langle G(t)\rangle = FG;$$

since D is invertible (1.70), we may multiply by D^{-1} both sides of (2), which gives

$$\text{(3)} \qquad \llbracket\{F(t)\} * \{G(t)\}\rrbracket = FD^{-1}G.$$

Combining (1) and (3):

$$\text{(2.8)} \qquad \boxed{\left\langle \int_0^t F(t-u)\,G(u)\,du \right\rangle = FD^{-1}G = \langle F(t)\rangle\, D^{-1}\langle G(t)\rangle};$$

note that

$$FD^{-1}G = D^{-1}FG = \frac{FG}{D}.$$

2.8.1 *Remark.* If $h(\tau) = 0$ for $\tau \leq 0$, it follows immediately from [2.0] that

$$\text{(2.8.2)} \qquad h(\,) = \{h(t)\}(\,);$$

from 1.23 it therefore follows that

$$\llbracket h\rrbracket = \llbracket\{h(t)\}\rrbracket = h:$$

the last equation is from [2.1]. In conclusion:

$$\text{(2.8.3)} \qquad h = \llbracket h\rrbracket \qquad \textbf{(whenever } h(\tau) = 0 \textbf{ for } \tau \leq 0\textbf{)}.$$

2.8.4 *Remark.* From 2.6 and 2.8.3 we see that

$$[\{F(t)\} * \{G(t)\} = \llbracket\{F(t)\} * \{G(t)\}\rrbracket;$$

---

* We shall not prove this; for all practical purposes, Theorem 2.61 is sufficient

consequently, (3) gives

$$\boxed{\{F(t)\} * \{G(t)\} = FD^{-1}G = \left\langle \int_0^t F(t-u)\, G(u)\, du \right\rangle} : \tag{2.9}$$

the second equation is from 2.8.

Translation Property

If $-\infty < \alpha < \infty$, the function $\mathsf{T}_\alpha(\,)$ is defined by

$$\mathsf{T}_\alpha(\tau) = \begin{cases} 0 & (\tau \leq \alpha) \\ 1 & (\tau > \alpha). \end{cases} \tag{2.10}$$

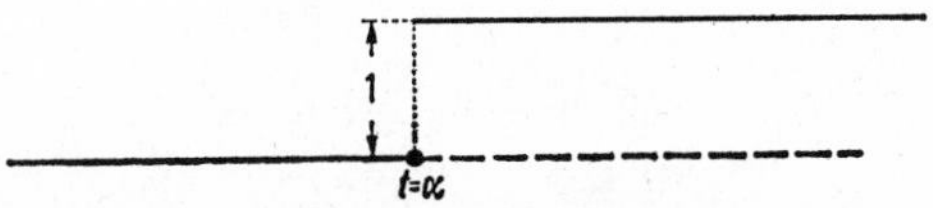

The operator $[\![\mathsf{T}_\alpha]\!]$ of the function $\mathsf{T}_\alpha(\,)$ satisfies the equation

$$[\![\mathsf{T}_\alpha]\!] \cdot \varphi(\tau) = \varphi(\tau - \alpha) \qquad (-\infty < \tau < \infty) \tag{4}$$

for any test-function $\varphi(\,)$ (see 1.35). If $F(\,)$ is a $\mathcal{K}$-function, 1.32 gives

$$[\![\mathsf{T}_\alpha]\!] \langle F(t) \rangle = [\![\{F(t)\}_\alpha]\!], \tag{5}$$

where $\{F(t)\}_\alpha(\,)$ is the function defined as in [1.31]:

$$\{F(t)\}_\alpha(\tau) = \{F(t)\}(\tau - \alpha) \qquad (-\infty < \tau < \infty). \tag{6}$$

If $\tau \leq \alpha$, then $\tau - \alpha \leq 0$ and

$$\{F(t)\}(\tau - \alpha) = 0 = \mathsf{T}_\alpha(\tau)\, F(\tau - \alpha) \qquad (\text{for } \tau \leq \alpha), \tag{7}$$

and

$$\{F(t)\}(\tau - \alpha) = F(\tau - \alpha) = \mathsf{T}_\alpha(\tau)\, F(\tau - \alpha) \qquad (\text{for } \tau > \alpha). \tag{8}$$

Equations (7)—(8) imply that

$$\{F(t)\}_\alpha(\tau) = \begin{cases} 0 & (\tau \le \alpha) \\ F(\tau - \alpha) & (\tau > \alpha) \end{cases} \tag{2.10.1}$$

and

$$\{F(t)\}_\alpha(\tau) = \mathsf{T}_\alpha(\tau)\, F(\tau - \alpha) \qquad (-\infty < \tau < \infty). \tag{9}$$

From (9) it follows that

$$\{F(t)\}_\alpha() = \{\mathsf{T}_\alpha(t)\, F(t - \alpha)\}(), \tag{2.10.2}$$

whence

$$\begin{aligned} [\![\{F(t)\}_\alpha]\!] &= [\![\{\mathsf{T}_\alpha(t)\, F(t-\alpha)\}]\!] && \text{by 1.23} \\ &= \langle \mathsf{T}_\alpha(t)\, F(t-\alpha)\rangle && \text{by [2.1]}. \end{aligned}$$

Thus, Equation (5) gives

$$[\![\mathsf{T}_\alpha]\!]\, \langle F(t)\rangle = \langle \mathsf{T}_\alpha(t)\, F(t-\alpha)\rangle. \tag{10}$$

If $\alpha \ge 0$ then $\{\mathsf{T}_\alpha(t)\}() = \mathsf{T}_\alpha()$: consequently, [2.1] and 1.23 give

$$\mathsf{T}_\alpha = \langle \mathsf{T}_\alpha(t)\rangle = [\![\mathsf{T}_\alpha]\!] \qquad (\text{when } \alpha \ge 0). \tag{2.11}$$

Z Combining 2.11 with (5) and (10):

$$\boxed{\mathsf{T}_\alpha \langle F(t)\rangle = \langle \mathsf{T}_\alpha(t)\, F(t-\alpha)\rangle} \qquad (\text{when } \alpha \ge 0); \tag{2.12}$$

note that $\mathsf{T}_\alpha \langle F(t)\rangle = \mathsf{T}_\alpha F$ is the product (as defined in [0.11]) of the operators T_α and F.

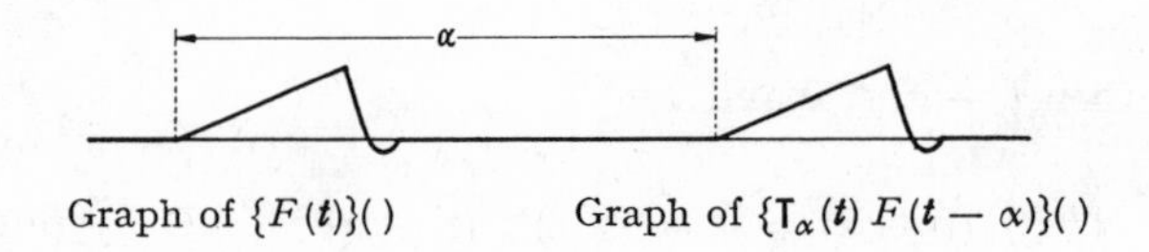

Graph of $\{F(t)\}()$ Graph of $\{\mathsf{T}_\alpha(t)\, F(t-\alpha)\}()$

From 2.11 and (4) we see that the equation

$$\mathsf{T}_\alpha \cdot \varphi(\tau) = \varphi(\tau - \alpha) \qquad (-\infty < \tau < \infty) \tag{2.13}$$

holds for any test-function $\varphi(\,)$ *in case* $\alpha \geq 0$. From 2.13 it follows immediately that

$$\mathsf{T}_\alpha \mathsf{T}_\lambda = \mathsf{T}_{\alpha+\lambda} \quad (\text{when } \alpha \geq 0 \text{ and } \lambda \geq 0). \tag{2.14}$$

From 1.34 and 2.11 it results directly that

$$\mathsf{T}_\alpha [\![\mathsf{T}_{-\alpha}]\!] = \mathsf{T}_0 \quad (\text{when } \alpha \geq 0). \tag{2.15}$$

Numbers and Operators

Recall our agreement 1.51: if c is a number, we often write

$$c \text{ **instead of** } \langle c \rangle; \tag{2.16}$$

in particular,

$$\boxed{1 = \langle 1 \rangle = \langle \mathsf{T}_0(t) \rangle = \mathsf{T}_0} \quad (\text{see } 2.11) \tag{2.17}$$

and

$$\boxed{V1 = V = 1V \quad \text{for any operator } V}. \tag{2.18}$$

In view of 2.17 and 1.68, Equation 2.15 can be written

$$\boxed{[\![\mathsf{T}_{-\alpha}]\!] = \frac{1}{\mathsf{T}_\alpha}} \quad (\text{when } \alpha \geq 0). \tag{2.19}$$

Until further notice, let V be a perfect operator. If $\varphi(\,)$ is a test-function, then

$$[\![V \cdot \varphi]\!] = V [\![\varphi]\!] \quad (\text{see } 1.19). \tag{2.20}$$

Further, if c is a number, then 1.56 gives

$$cV \cdot \varphi(\,) = c(V \cdot \varphi)(\,) \quad (\text{by } 1.48), \tag{2.21}$$

and

$$cV = \langle c \rangle V = V \langle c \rangle = Vc \quad (\text{by } [2.16]). \tag{2.22}$$

The equations

$$VD^n F = D^n FV = D^n VF = VFD^n \tag{2.23}$$

hold for any integer $n \geq 1$ and any $\mathcal{K}$-function $F(\,)$: as we saw in 1.10, perfect operators can be manipulated according to the usual algebraic laws.

Functions Having no Jumps on $[0, \infty)$

Z 2.24 **Derivation Property.** *If $F(\,)$ has no jumps on $[0, \infty)$, then*

$$DF - F(0-)\,D = \langle F'(t)\rangle \qquad \text{(when } F'(\,) \text{ is a } \mathcal{K}\text{-function)}. \tag{2.25}$$

2.26 *Remarks.* A function $F(\,)$ *has no jumps on* $[0, \infty)$ if (and only if) $F(\alpha-) = F(\alpha+)$ for any $\alpha \geq 0$. Recall that DF is the product (as defined in [0.11]) of the operator D with the operator F; Formula 2.25 states that the operator

$$DF - F(0-)\,D$$

is the canonical operator of the derivative of the function $F()$. In case $F()$ is continuous on the interval $[0, \infty)$, Property 2.25 is an immediate consequence of 1.42 and 1.64; otherwise, a small reasoning is needed (see 15.18.3 in the Appendix).

If the derivative $F'(\,)$ of $F(\,)$ is continuous on the interval $(0, \infty)$ and if $F(0-) = F(0+)$, then $F'(\,)$ is a $\mathcal{K}$-function and $F(\,)$ has no jumps on $[0, \infty)$: all continously differentiable functions on $(0, \infty)$ have the Derivation Property (provided that their right- and left-hand limits coincide at $t = 0$). The graph of the function

$$F(\,) = \{\mathsf{T}_\alpha(t) \sin(t - \alpha)\}(\,)$$

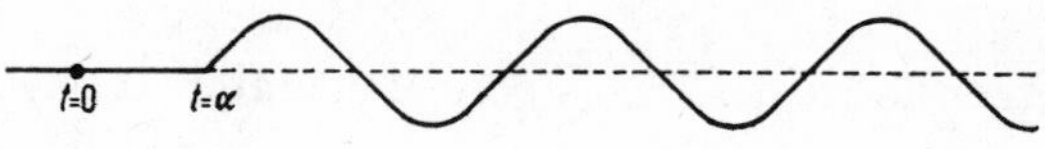

clearly indicates that $F(\,)$ has no jumps; note that $F'(\,)$ is a $\mathcal{K}$-function, although $F'(\alpha)$ has no meaning. Formula 2.25 does not apply to functions $F(\,)$ having jumps on $[0, \infty)$; in fact, if $F(\,)$ has jumps at $t = \alpha_k > 0$ $(k = 1, 2, 3, \ldots, n)$, then

$$\langle F'(t)\rangle = DF - F(0-)\,D - \sum_{k=1}^{n} [F(\alpha_k+) - F(\alpha_k-)]\, D\mathsf{T}_\alpha\text{:}$$

see 5.12.0.

2.27 **Integration Property.** *If $h()$ is an entering function, then*

(2.28) $$\left\langle \int_{-\infty}^{t} h(u)\, du \right\rangle = D^{-1}h + \int_{-\infty}^{0} h(u)\, du\,.$$

Proof. Obviously,

$$\int_{-\infty}^{\tau} h(u)\, du = \int_{0}^{\tau} h(u)\, du + \int_{-\infty}^{0} h(u)\, du \qquad (\text{all } \tau > 0);$$

from 2.4—5 it therefore follows that

$$\left\langle \int_{-\infty}^{t} h(u)\, du \right\rangle = \left\langle \int_{0}^{t} h(u)\, du \right\rangle + \left\langle \int_{-\infty}^{0} h(u)\, du \right\rangle:$$

Conclusion 2.28 is now immediate from 2.32; the angular brackets on the right-hand side of 2.28 have been omitted because of our convention [2.16].

Various Consequences

If $F()$ has no jumps on $[0, \infty)$, we can solve 2.25 for F:

(2.29) $$\boxed{F = D^{-1}\langle F'(t)\rangle + F(0-)} \qquad (\text{when } F'() \text{ is a } \mathcal{K}\text{-function}).$$

Setting $F(t) = t^{1/2}$ in 2.29, we see that

(2.30) $$\langle t^{1/2}\rangle = 2^{-1}D^{-1}\langle t^{-1/2}\rangle.$$

If $F(t) = t^{\alpha}$ for some number $\alpha \geq 0$, then 2.25 gives

(2.31) $$D\langle t^{\alpha}\rangle = \alpha\langle t^{\alpha-1}\rangle.$$

If $F(t) = 1$ then $\{F(t)\}() = \mathsf{T}_0()$ and

$$F = \langle F(t)\rangle = \langle 1\rangle = 1 \qquad (\text{by [2.1] and 2.17});$$

substituting into 2.9, we can use 2.18 to write

(2.32) $$\boxed{D^{-1}G = \left\langle \int_{0}^{t} G(u)\, du \right\rangle}:$$

obviously, $D^{-1}G$ is the product of the operator D^{-1} with the operator G.

Substituting $F(t) = G(t) = t^{-1/2}$ into 2.9, we obtain

$$\langle t^{-1/2}\rangle\, D^{-1}\langle t^{-1/2}\rangle = \int_0^t \frac{du}{\sqrt{(t-u)\,u}} = 2\int_0^{\pi/2} d\lambda = \pi:$$

the middle equation is from the change of variable $u = t\sin^2\lambda$ (we have omitted the last three angular brackets); consequently,

$$\boxed{\langle t^{-1/2}\rangle^2 = \pi D}\,. \tag{2.33}$$

If a is a number and $F(t) = e^{at}$, then $F(0-) = 1$ and $F = \langle e^{at}\rangle$: the Derivation Property (2.24) yields

$$D\langle e^{at}\rangle - D = \langle a\, e^{at}\rangle = a\langle e^{at}\rangle:$$

the last equation is from 2.5. Thus,

$$(D-a)\,\langle e^{at}\rangle = D;$$

since D is invertible, we may use 1.76 to obtain

$$\boxed{\langle e^{at}\rangle = \frac{D}{D-a}}\,. \tag{2.34}$$

Let α and β be numbers; since

$$\{e^{\alpha t}\cos\beta t\}(\,) = 2^{-1}\{e^{(\alpha+i\beta)t} + e^{(\alpha-i\beta)t}\}(\,), \tag{2.35}$$

equations 2.5 and 2.34 give

$$\langle e^{\alpha t}\cos\beta t\rangle = \frac{2^{-1}D}{D-(\alpha+i\beta)} + \frac{2^{-1}D}{D-(\alpha-i\beta)} = \frac{D\,(D-\alpha)}{(D-\alpha)^2+\beta^2}: \tag{2.36}$$

the second equation is from 1.74. A similar reasoning gives

$$\boxed{\left\langle \frac{e^{\alpha t}}{\beta}\sin\beta t\right\rangle = \frac{D}{(D-\alpha)^2+\beta^2}}\,. \tag{2.37}$$

Note the following particular case of 2.36:

$$\boxed{\frac{D^2}{D^2+\beta^2} = \langle\cos\beta t\rangle}\,.$$

2.38 **Theorem.** *If n is any integer ≥ 0, then*

$$(2.39)\qquad \frac{D}{(D-a)^{n+1}} = \left\langle \frac{t^n}{n!}\, e^{at} \right\rangle .$$

Proof. Since 2.39 holds for $n = 0$ (by 2.34), we proceed by induction: assume that 2.39 holds for $n = m$, and observe that the equations

$$(1)\qquad \frac{D}{(D-a)^{m+2}} = \left[\frac{1}{D-a}\right] \frac{D}{(D-a)^{m+1}} = [\langle e^{at}\rangle D^{-1}] \left\langle \frac{t^m}{m!}\, e^{at} \right\rangle$$

come from 2.34 and our induction assumption (2.39 with $n = m$). On the other hand, 2.8 gives

$$(2)\qquad \langle e^{at}\rangle D^{-1} \left\langle \frac{t^m}{m!}\, e^{at}\right\rangle = \int_0^t e^{a(t-u)} \frac{u^m}{m!}\, e^{au}\, du = \frac{t^{m+1}}{(m+1)!}\, e^{at}:$$

we have omitted the angular brackets in the last two equations. From (1)—(2) it now follows that 2.39 holds for $n = m + 1$ whenever it holds for $n = m$. This proves that 2.39 holds for any integer $n \geq 0$.

2.40 **Particular case.** If $a = 0$ then 2.39 becomes

$$(2.41)\qquad \frac{1}{D^n} = \left\langle \frac{t^n}{n!} \right\rangle \qquad \text{(for any integer } n \geq 0\text{)},$$

which implies

$$(2.42)\qquad (n!)\, D^{-n} = \langle t^n \rangle \quad \text{and} \quad D^n \langle t^n \rangle = n!;$$

the last operator-equation corresponds to the function equation $\{n!\}(\,) = \{t\}^{(n)}(\,)$.

Three Integral Equations

2.43 **Warning:** *angular brackets will often be omitted from the right-hand side of operator-equations,* as in Equations (1)—(3) below.

2.44 **First problem.** To find a $\mathcal{K}$-function $y(\,)$ such that

$$(1)\qquad \left\langle \int_0^t \sin(t-u)\, y(u)\, du \right\rangle = t^2.$$

If $y(\,)$ is such a function, it follows from 2.8 and (1) that $\langle \sin t\rangle D^{-1} y = t^2$, whence

$$\frac{y}{D^2+1} = t^2 \qquad \text{(by 2.37 with } \alpha = 0\text{)}.$$

Solving for y:

$$y = D^2 t^2 + t^2 = 2 + t^2 \qquad \text{(by 2.42)}. \tag{2}$$

Reversing our steps, we see that Equation (1) is satisfied when $y(\,)$ is the function $\{2 + t^2\}(\,)$.

2.45 **Second problem** Given a number, b, let us find a function $y(\,)$ having no jumps on $[0, \infty)$ such that $y'(\,)$ is a $\mathcal{K}$-function,

$$y(0-) = 0, \quad \text{and} \quad 3y = b - \int_0^t \frac{y'(u)\,du}{\sqrt{t-u}}.$$

If $y(\,)$ is such a function, 2.8 gives

$$3y = b - \langle t^{-1/2}\rangle D^{-1}\langle y'(t)\rangle = b - \langle t^{-1/2}\rangle y:$$

the last equation is from 2.29 and our condition $y(0-) = 0$. Thus,

$$b = (3 + \langle t^{-1/2}\rangle)\, y;$$

multiplying by $(3 - \langle t^{-1/2}\rangle)$ both sides, we obtain

$$3b - b\langle t^{-1/2}\rangle = (3^2 - \langle t^{-1/2}\rangle^2)\, y = (9 - \pi D)\, y:$$

the last equation is from 2.33. Solving for y:

$$y = -\, D^{-1}\frac{3bD}{\pi D - 9} + \frac{bD}{\pi D - 9}\, D^{-1}\langle t^{-1/2}\rangle.$$

Consequently, 2.5 and 2.34 give

$$y = \frac{-3b}{\pi} D^{-1}\langle e^{9t/\pi}\rangle + \frac{b}{\pi}\langle e^{9t/\pi}\rangle D^{-1}\langle t^{-1/2}\rangle.$$

From 2.32 and 2.8 it now follows that

$$y = -\frac{3b}{\pi}\int_0^t e^{9u/\pi}\,du + \frac{b}{\pi}\int_0^t e^{9(t-u)/\pi}\frac{du}{\sqrt{u}}.$$

2.46 **Abel's integral equation.** Suppose that $F(\,)$ is a function having no jumps on $[0, \infty)$ and whose derivative $F'(\,)$ is a $\mathcal{K}$-function: let us find a $\mathcal{K}$-function $y(\,)$ satisfying the operator-equation

$$F = \int_0^t \frac{y(u)\,du}{\sqrt{t-u}}. \tag{3}$$

If $y(\,)$ is such a function, it follows from 2.8 that

$$F = \langle t^{-1/2}\rangle D^{-1} y;$$

multiplying by $\langle t^{-1/2}\rangle$ both sides, we may use 2.33 to obtain

$$\pi y = \langle t^{-1/2}\rangle F = \langle t^{-1/2}\rangle \left(D^{-1}\langle F'(t)\rangle + F(0-)\right):$$

the second equation is from 2.29. Thus,

$$y = \pi^{-1}\langle t^{-1/2}\rangle D^{-1}\langle F'(t)\rangle + \pi^{-1}F(0-)\langle t^{-1/2}\rangle,$$

so that 2.8 gives

$$y = \pi^{-1}\int_0^t \frac{F'(u)\,du}{\sqrt{t-u}} + \pi^{-1}F(0-)\,t^{-1/2}.$$

Differential Equations: Procedure

If $y(\,)$ is a $\mathcal{K}$-function, its canonical operator $y = \langle y(t)\rangle$ is a perfect operator; the product of the perfect operators D and y is therefore a perfect operator Dy; if the left-hand limit $y(0-)$ exists, then the equation

(2.47)
$$\boxed{\frac{\partial}{\partial t}\,y \overset{\text{def}}{=} Dy - y(0-)\,D}$$

defines a perfect operator $\partial y/\partial t$ called the $[0,\infty)$-*derivative* of the $\mathcal{K}$-function $y(\,)$. *If* $y'(\,)$ *is a* $\mathcal{K}$*-function and if* $y(\,)$ *has no jumps on* $[0,\infty)$, it then follows from 2.25 (Derivation Property) that $\partial y/\partial t$ is the canonical operator of the function $y'(\,)$:

(2.48)
$$\frac{\partial}{\partial t}\,y = y' = \langle y'(t)\rangle \qquad \text{if} \quad y(\,) \text{ has no jumps on } [0,\infty).$$

Recall 2.26: a function $y()$ is said to **have no jumps on** $[0,\infty)$ if (and only if)

$$y(\alpha-) = y(\alpha+) \qquad (\text{for any } \alpha \geq 0).$$

2.49 *Remarks.* The $\mathcal{K}$-function $\{1\}(\,) = \mathsf{T}_0(\,)$ has a jump at $t = 0$: since $\mathsf{T}_0(0-) = 0$ and

$$\mathsf{T}_0 = \langle 1\rangle = 1,$$

Definition [2.47] gives

$$\frac{\partial}{\partial t}\,\mathsf{T}_0 = D\mathsf{T}_0 = D1 = D \qquad (\text{by } 2.17\text{—}18);$$

note that

$$\mathsf{T}_0' = \left\langle \frac{\mathrm{d}}{\mathrm{d}t}\, \mathsf{T}_0(t) \right\rangle = 0.$$

As we shall see (in § 6), the operator D corresponds to an impulse of magnitude $= 1$ applied at $t = 0$. Let us now illustrate how Definition [2.47] will be applied.

2.50 **A non-classical problem.** Let a be any number. The initial-value problem

(1) $$y(0-) = a \quad \text{with} \quad m \frac{\partial}{\partial t} y = f$$

governs the velocity y of a particle of mass m subjected to a force f. In view of [2.47], the system of equations (1) implies the equation $m(Dy - aD) = f$; that is,

(2) $$y = a + m^{-1} D^{-1} f.$$

Let us consider the case where f is the impulse D of magnitude $= 1$ applied at the time $t = 0$ (see 6.41): Equation (2) becomes

(3) $$y = a + m^{-1},$$

which, by 2.1 and 2.3, implies

(4) $$\{y(t)\}(\,) = \{a + m^{-1}\}(\,).$$

If $y(\,)$ is continuous on $(0, \infty)$ and satisfies (1), it follows from (4) that $y(\tau) = a + m^{-1}$ for all $\tau > 0$: all such solutions of (1) have the same value on the interval $(0, \infty)$. Reversing our steps, we see that the equation

(5) $$y(\tau) = \begin{cases} a & (\tau \leq 0) \\ a + m^{-1} & (\tau > 0) \end{cases}$$

implies (4); since (4) $\Rightarrow$ (3) $\Rightarrow$ (2) and (2) $\Rightarrow$ (1) when $y(0-) = a$ (which is clearly the case), Equation (5) defines a solution of (1).

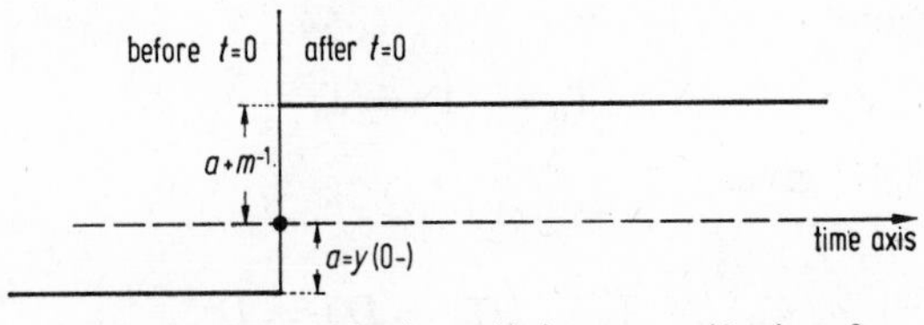

Graph of the velocity $y(\,)$ in case $y(0-) < 0$

Some authors call $y(0+)$ the **"starting"** value (to contrast with the *initial* value $y(0-)$). If the initial velocity $y(0-) = -m^{-1}$ then $y(\tau) = 0$ for all $\tau > 0$ (see (5)): *the particle is brought to a stop at* $t = 0+$ by means of the impulse of magnitude $= 1$ applied at $t = 0$.

In case f is the operator of a $\mathcal{K}$-function $f(\,)$, we can combine (2) and 2.32 to obtain

$$y = a + m^{-1} \int_0^t f(u)\, du.$$

2.51 **A classical problem.** Let $y(\,)$ be a $\mathcal{K}$-function such that

$$(1) \qquad y(0-) = 3 \quad \text{with} \quad \frac{\partial}{\partial t} y + 2^{-1} y = t^{1/2} + t^{-1/2};$$

angular brackets are again going to be omitted on the right-hand side (as in 2.43). In view of [2.47], the system of equations (1) implies the equation

$$(2) \qquad Dy + 2^{-1} y = 3D + t^{1/2} + t^{-1/2}.$$

Solving (2) for y, we obtain

$$y = \frac{3D}{D + 2^{-1}} + \frac{1}{D + 2^{-1}} [t^{1/2} + t^{-1/2}];$$

since $t^{-1/2} = 2D\langle t^{1/2}\rangle$ (by 2.30), we have

$$(3) \quad y = \frac{3D}{D + 2^{-1}} + \frac{1}{D + 2^{-1}} [1 + 2D] \langle t^{1/2}\rangle = 3\langle e^{-t/2}\rangle + 2\langle t^{1/2}\rangle:$$

the last equation is from 2.34 and 1.73. From 2.5 and 2.3 it now follows that any solution $y(\,)$ of (1) which is continuous on $(0, \infty)$ satisfies the equation

$$(4) \qquad y(\tau) = 3e^{-\tau/2} + 2\tau^{1/2} \qquad (\text{all } \tau > 0).$$

Reversing our steps, we see that the equation

$$y(t) = 3e^{-t/2} + 2t^{1/2} \qquad (-\infty < t < \infty)$$

defines a continuous $\mathcal{K}$-function satisfying (4) and $y(0-) = 3$; since (4) $\Rightarrow$ (3) $\Rightarrow$ (2) and (2) $\Rightarrow$ (1) when $y(0-) = 3$, it satisfies our initial problem (1).

Higher Derivatives

2.52 **Definition.** If $y(\,)$ is a $\mathcal{K}$-function, we set

$$\text{(2.53)} \qquad \frac{\partial^m}{\partial t^m} y = D^m y - y(0-)\,D^m - y'(0-)\,D^{m-1} - \cdots - y^{(m-1)}(0-)\,D$$

for any integer $m \geq 1$. In the particular case $m = 2$:

$$\text{(2.54)} \qquad \frac{\partial^2}{\partial t^2} y \overset{\text{def}}{=} D^2 y - y(0-)\,D^2 - y'(0-)\,D.$$

If $y(\,)$ is the function $\mathsf{T}_\alpha(\,)$ then $y(0-) = y'(0-) = 0$, so that

$$\frac{\partial^2}{\partial t^2} \mathsf{T}_\alpha = D^2 \mathsf{T}_\alpha \qquad (\text{if } \alpha \geq 0)$$

as we shall see (6.82), the operator $D^2\mathsf{T}_\alpha$ corresponds to a concentrated couple of moment $= -1$.

Let $y(\,)$ and $y'(\,)$ be functions having no jumps at $t = 0$: if $y''(\,)$ is continuous in the interval $(0, \infty)$, then both $y(\,)$ and $y'(\,)$ have no jumps on $[0, \infty)$, whence

$$\begin{aligned} \langle y''(t)\rangle &= D\langle y'(t)\rangle - y'(0-)\,D && \text{by } 2.24 \\ &= D\,[Dy - y(0-)\,D] - y'(0-)\,D && \text{by } 2.24 \\ &= \frac{\partial^2}{\partial t^2} y && \text{by } [2.54]: \end{aligned}$$

thus, **the operator defined by [2.54] is the canonical operator of the second derivative of the function $y(\,)$.**

2.55 **Important property.** *Let m be any integer ≥ 1. Suppose that $y(\,)$, $y'(\,), \ldots, y^{(m-1)}(\,)$ have no jumps at $t = 0$; if $y^{(m)}(\,)$ is continuous in $(0, \infty)$ then*

$$\frac{\partial^m}{\partial t^m} y = \left\langle \frac{d^m}{dt^m} y(t) \right\rangle = \langle y^{(m)}(t)\rangle = y^{(m)};$$

this is proved by an induction reasoning similar to the reasoning used in the case $m = 2$ discussed above.

2.56 **A non-standard problem.** Given a $\mathcal{K}$-function $F(\,)$, consider the problem

$$\text{(1)} \qquad x(0-) = x'(0-) = y(0-) = 0$$

with

$$(2) \qquad \frac{\partial x}{\partial t} + x + 2y = F, \quad \text{and} \quad \frac{\partial^2 x}{\partial t^2} + 5x + 3\frac{\partial y}{\partial t} = 0.$$

A problem will be called "non-standard" if it has no solutions when the initial values of the problem are replaced by starting values. If $F(0+) \neq 0$, the problem (1)—(2) is non-standard.

In view of [2.53], the system (1)—(2) implies

$$(3) \qquad (D+1)\,x + 2y = F,$$

$$(4) \qquad (D^2+5)\,x + 3Dy = 0,$$

and this system can be solved by ordinary algebra:

$$y = -F\frac{D^2+5}{(D+5)\,(D-2)} = -F - F\frac{15-3D}{(D+5)\,(D-2)},$$

which can be put in the form

$$(5) \qquad y = -F + \frac{30}{7}\{F(t)\} * \{e^{-5t}\} - \frac{9}{7}\{F(t)\} * \{e^{2t}\} \quad \text{(see 4.13.2)}.$$

Thus, if $y(\,)$ is continuous on $(0, \infty)$ and satisfies (1)—(2) for some function $x(\,)$, it follows from (5) and 2.3 that $y(\tau) = H(\tau)$ for all $\tau > 0$, where

$$H(\tau) = -F(\tau) + \int_0^\tau F(\tau-u)\left[\frac{30}{7}e^{-5u} - \frac{9}{7}e^{2u}\right]du. \qquad \text{(see 2.9)}.$$

Reversing our steps, we see that the function $y(\,) = \{H(t)\}(\,)$ satisfies the initial condition $y(0-) = 0$ and (2); however, since

$$y(0+) - y(0-) = -F(0+),$$

it has a jump of magnitude $-F(0+)$ at $t = 0$.

2.57 **Comments.** The above type of problem arises in practical situations in engineering. Given any three numbers a, b, and c, the system obtained by adjoining the starting conditions

$$x(0+) = a, \qquad x'(0+) = b, \qquad y(0+) = c$$

to (2) has *no solution* for arbitrary $\mathcal{K}$-functions $F(\,)$. Nevertheless, our procedure yields solutions when (2) is adjoined arbitrary initial conditions: see 4.13.2. G. Doetsch has discussed this problem [S 2]: to him is due the first satisfactory discussion of this situation.

Exercises

2.58.0 Find a $\mathcal{K}$-function $y(\,)$ such that

$$\int_0^t \frac{y(u)\,du}{\sqrt{t-u}} = 1 + t + t^2.$$

Answer: $\dfrac{1}{\pi\sqrt{t}}\left(1 + 2t + \dfrac{8}{3}t^2\right).$

2.58.1 Find a $\mathcal{K}$-function $y(\,)$ such that

$$\int_0^t (t-u)^2\, y(u)\, du = -t + \sin t.$$

Answer: $-2^{-1}\{\cos t\}(\,).$

2.58.2 Find a $\mathcal{K}$-function $y(\,)$ such that

$$\int_0^t \cos(t-u)\, y(u)\, du = \sin t.$$

Answer: $y(t) = 1$ for $-\infty < t < \infty.$

2.58.3 Let c_0, c_1 and β be given numbers. Find a $\mathcal{K}$-function $y(\,)$ such that

$$y(0-) = c_0, \quad y'(0-) = c_1, \quad \text{and} \quad \frac{\partial^2}{\partial t^2}y + \beta^2 y = 1.$$

Answer: $y = (c_0 - \beta^{-2})\cos\beta t + (c_1/\beta)\sin\beta t + \beta^{-2}.$

2.58.4 Find a $\mathcal{K}$-function $y(\,)$ such that

$$y(0-) = 1, \quad y'(0-) = 0, \quad \text{and} \quad \frac{\partial^2}{\partial t^2}y + 3\frac{\partial}{\partial t}y + 2y = 2.$$

Answer: $y(t) = 1$ for $-\infty < t < \infty.$

2.58.5 Prove the following operator-equations:

$$\langle t^{-1/2}\rangle = \frac{4}{3}D^2\langle t^{3/2}\rangle,$$

and

$$D^{-2}\langle t^{-1/2}\rangle = \frac{4}{3}\langle t^{3/2}\rangle.$$

Use 2.9 to prove that

$$F\langle t\rangle G = \{F(t)\} * \{G(t)\}.$$

2.58.6 Given two numbers a and b, find a function $y(\,)$ such that $y(0-) = 0$, having no jumps on $[0, \infty)$, such that $y'(\,)$ is a $\mathcal{K}$-function, and such that

$$\int_0^t \frac{y'(u)\,du}{\sqrt{t-u}} = \pi a b^2.$$

Answer: $y(t) = 2ab^2 t^{1/2} \quad (\infty < t < \infty)$.

Continuity of the Convolution Integral

The proofs of the following theorems 2.61—66 may be omitted on a first reading.

2.59 **Definition.** Let (a, b) be an interval; a function will be called *regulated on* (a, b) if it is defined on the interval (a, b) and has finite limits on both sides of each point in (a, b).

2.60 Thus, a function $f(\,)$ is regulated on (a, b) if (and only if)

$$|f(t-)| + |f(t)| + |f(t+)| < \infty \qquad (a < t < b).$$

Clearly, all continuous functions are regulated.

2.61 **Theorem.** *Let $f(\,)$ and $g(\,)$ be entering functions; if at least one of these functions is regulated on $(-\infty, \infty)$, then $f * g(\,)$ is a continuous function.*

Proof: see 15.14 in the Appendix.

2.62 Let $F(\,)$ and $G(\,)$ be $\mathcal{K}$-functions; if at least one of these functions is regulated on $(0, \infty)$, we can apply 2.61 with $f(\,) = \{F(t)\}(\,)$ and $g(\,) = \{G(t)\}(\,)$ to conclude that the function $\{F(t)\} * \{G(t)\}(\,)$ is continuous; in particular,

$$\{F(t)\} * \{G(t)\}(0+) = \{F(t)\} * \{G(t)\}(0-) = 0: \tag{2.63}$$

the last equation is from 2.6.

2.64 Since most of our worked-out problems are stated and solved by means of operator-equations, it is appropriate to note the following two theorems.

2.65 **Theorem.** *If* $y_k(\,)$ $(k = 0, 1, 2, \ldots, n)$ *are* $\mathcal{K}$*-functions, then*

$$(1)\qquad y_0 + y_1 + \cdots + y_{n-1} = y_n$$

if (and only if) the equation

$$(2)\qquad y_0(\tau) + y_1(\tau) + \cdots + y_{n-1}(\tau) = y_n(\tau)$$

holds at each point $\tau > 0$ *where all the functions* $y_k(\,)$ $(k = 0, 1, 2, \ldots, n)$ *are continuous.*

Proof. Let $y(\,)$ be the function defined by

$$(3)\qquad y(t) = y_0(t) + y_1(t) + \cdots + y_{n-1}(t) \qquad (-\infty < t < \infty);$$

from 2.4 we see that

$$(4)\qquad y = \langle y_0(t) + y_1(t) + \cdots + y_{n-1}(t)\rangle$$

$$(5)\qquad = y_0 + y_1 + \cdots + y_{n-1} \qquad \text{(by 2.5 and [2.1])}.$$

From (4)—(5) it follows that our hypothesis (1) implies $y = y_n$, whence $\{y(t)\}(\,) = \{y_n(t)\}(\,)$ (by 2.3): Conclusion (2) now comes from (3) and 1.47. Conversely, let us take (2) as a hypothesis: from (2) and (3) it follows that $\{y(t)\}(\,) = \{y_n(t)\}(\,)$, whence $y = y_n$ (by 2.4): Conclusion (1) is now immediate from (4)—(5).

2.66 **Theorem.** *Let* $y(\,), F(\,), G(\,)$, *and* $H(\,)$ *be* $\mathcal{K}$*-functions; suppose that at least one of the two functions* $F(\,)$ *and* $G(\,)$ *is regulated on* $(0, \infty)$. *If* c *is a number, then the operator-equation*

$$(2.67)\qquad y = \left\langle \int_0^t F(t-u)\,G(u)\,du \right\rangle + H + c$$

implies that the equation

$$(2.68)\qquad y(\tau) = \int_0^\tau F(\tau-u)\,G(u)\,du + H(\tau) + c$$

holds at each point $\tau > 0$ *where both* $y(\,)$ *and* $H(\,)$ *are continuous. Conversely, if* 2.68 *holds for all* $\tau > 0$, *then* 2.67 *holds.*

Proof. From 2.62 we see that the function

$$(6)\qquad h(\,) \overset{\text{def}}{=} \{F(t)\} * \{G(t)\}(\,)$$

is continuous: it is therefore a $\mathcal{K}$-function. On the other hand, (6), [2.1], and 2.9 imply that

$$h = \left\langle \int_0^t F(t-u)\,G(u)\,du \right\rangle;$$

therefore, 2.67 can be written in the form

$$y = h + H + \langle c \rangle, \tag{7}$$

which (by 2.65) implies that the equation

$$y(\tau) = h(\tau) + H(\tau) + \{c\}(\tau) \tag{8}$$

holds at every point $\tau > 0$ where all three functions $y(\ )$, $h(\ )$, and $H(\ )$ are continuous. Since $h(\ )$ is continuous everywhere, Conclusion 2.68 is immediate from (8), and since

$$h(\tau) = \int_0^\tau F(\tau - u)\, G(u)\, du:$$

see (6) and 2.7.1. Conversely, 2.68 implies

$$\langle y(t)\rangle = \left\langle \int_0^t F(t-u)\, G(u)\, du + H(t) + c \right\rangle \qquad \text{by } 2.4$$

$$= \left\langle \int_0^t F(t-u)\, G(u)\, du \right\rangle + H + c \qquad \text{by } 2.5,$$

whence 2.67 is an immediate consequence of 2.1.

§ 3. Elementary Applications

We shall make a list of several important properties; the letters a, b, c will denote numbers. Recall that

$$\mathsf{T}_\alpha(\tau) = \begin{cases} 0 & (\tau \le \alpha) \\ 1 & (\tau > \alpha); \end{cases} \tag{3.1}$$

obviously,

$$\mathsf{T}_\alpha(t) = \mathsf{T}_0(t - \alpha) \qquad (-\infty < t < \infty). \tag{3.2}$$

Note that the identity-operator is the operator of the function $\mathsf{T}_0(\ )$:

$$\boxed{\mathsf{T}_0 = 1}. \qquad \text{(see 2.17).} \tag{3.3}$$

Let β be a real number; from [3.1] it follows that the function $c(\mathsf{T}_\alpha - \mathsf{T}_\beta)(\ )$ satisfies the equation

$$c(\mathsf{T}_\alpha - \mathsf{T}_\beta)(\tau) = \begin{cases} c & (\alpha < \tau \le \beta) \\ 0 & (\text{otherwise}). \end{cases} \tag{3.4}$$

The following example is an illustration of the Translation Property (2.12):

$$\mathsf{T}_\alpha\langle F(t)\rangle = \langle \mathsf{T}_\alpha(t)\, F(t-\alpha)\rangle \qquad \text{(when } \alpha \geq 0). \tag{3.5}$$

3.6 **Example.** Let us find a $\mathcal{K}$-function $y(\ )$ such that

$$y(0-) = 0 \quad \text{and} \quad \frac{\partial}{\partial t}\, y + y = \begin{cases} 1 & (0 < t < \alpha) \\ 0 & \text{(otherwise)}: \end{cases} \tag{1}$$

in accordance with 2.43, the angular brackets have been omitted from the right-hand side of the last equation. The equations

$$(D+1)\, y = Dy + y = \frac{\partial}{\partial t}\, y + y = \mathsf{T}_0 - \mathsf{T}_\alpha = 1 - \mathsf{T}_\alpha \tag{2}$$

are from 1.10, [2.47], (1), 3.4, and 3.3. Solving for y:

$$y = (1 - \mathsf{T}_\alpha)\, \frac{1}{D+1} = (1 - \mathsf{T}_\alpha)\left(1 - \frac{D}{D+1}\right)$$

$$= (1 - \mathsf{T}_\alpha)\,(1 - \langle e^{-t}\rangle) \qquad \text{by 2.34.}$$

Consequently, the problem (1) implies the operator-equation

$$\begin{aligned} y &= 1 - \langle e^{-t}\rangle - \mathsf{T}_\alpha + \mathsf{T}_\alpha\langle e^{-t}\rangle \\ &= \langle 1 - e^{-t} - \mathsf{T}_\alpha(t) + \mathsf{T}_\alpha(t)\, e^{-(t-\alpha)}\rangle \qquad \text{by 2.5 and 3.5.} \end{aligned} \tag{3}$$

Reversing our steps, we see that the equation

$$y(t) = 1 - e^{-t} - \mathsf{T}_\alpha(t) + \mathsf{T}_\alpha(t)\, e^{-t+\alpha} \qquad (-\infty < t < \infty) \tag{4}$$

implies (3) (by 2.65); since (3) $\Rightarrow$ (2) and $y(0-) = 0$, we have (4) $\Rightarrow$ (1). In view of [3.1], Equation (4) can also be written

$$y(\tau) = \begin{cases} 1 - \exp(-\tau) & (\tau \leq \alpha) \\ 1 - \exp(-\tau) - 1 + \exp(-\tau + \alpha) & (\tau > \alpha). \end{cases}$$

Table of Formulas

(3.7) $$\frac{\partial y}{\partial t} = Dy - y(0-)\,D \qquad \text{(see [2.47])}.$$

The following five equations (3.8.0)—(3.8.4) hold when $y(\,)$ and $y'(\,)$ are $\mathcal{K}$-functions such that $y(\,)$ has no jumps on $[0, \infty)$:

(3.8.0) $$\frac{\partial y}{\partial t} = y' = \langle y'(t)\rangle = \left\langle \frac{\mathrm{d}}{\mathrm{d}t}\, y(t)\right\rangle \qquad \text{(see 2.48)},$$

(3.8.1) $$y' = Dy - y(0-)\,D \qquad \text{(see 2.48, 3.7)},$$

(3.8.2) $$y = D^{-1}y' + y(0-) \qquad \text{(see 2.48, 3.7)},$$

(3.8.3) $$D^{-1}y' = y - y(0-) \qquad \text{(see 2.48, 3.7)},$$

(3.8.4) $$Dy = y' + y(0-)\,D \qquad \text{(see 2.48, 3.7)}.$$

Next, the two definitions:

(3.9) $$\frac{\partial^2}{\partial t^2}\, y = D^2 y - y(0-)\,D^2 - y'(0-)\,D \qquad \text{(see [2.54])},$$

(3.10) $$\frac{\partial^3}{\partial t^3}\, y = D^3 y - y(0-)\,D^3 - y'(0-)\,D^2 - y''(0-)\,D.$$

Both preceding formulas are particular cases of the general definition:

(3.11) $$\frac{\partial^m}{\partial t^m}\, y = D^m y - \sum_{k=0}^{m-1} y^{(k)}(0-)\,D^{m-k}$$

— see [2.53].

The angular brackets have been omitted from the right-hand side of each one of the following equations:

(3.12) $$\mathsf{T}_\alpha \langle F(t)\rangle = \mathsf{T}_\alpha(t)\, F(t-\alpha) \qquad \text{(if } \alpha \geq 0);$$

see 2.12; that is,

$$\mathsf{T}_\alpha F = \mathsf{T}_\alpha(t)\, F(t-\alpha) \qquad \text{(by [2.1])},$$

or, equivalently,

$$\mathsf{T}_\alpha F = \mathsf{T}_0(t-\alpha)\, F(t-\alpha) \qquad \text{(if } \alpha \geq 0).$$

In view of 2.10.1 and 2.10.2, Formula 3.12 can also be written

$$\mathsf{T}_\alpha F = \begin{cases} 0 & (t \leq \alpha) \\ F(t-\alpha) & (t > \alpha) \end{cases}$$

— provided, of course, that $\alpha \geq 0$. Next, the formula

$$D^{-1}h + \int_{-\infty}^{0} h(u)\,du = \int_{-\infty}^{t} h(u)\,du: \tag{3.13}$$

comes from 2.28.

$$D^{-1}G = \int_0^t G(u)\,du \qquad \text{(see 2.32)}, \tag{3.14}$$

$$FD^{-1}G = \langle F(t)\rangle\, D^{-1}\langle G(t)\rangle = \int_0^t F(t-u)\,G(u)\,du: \tag{3.15}$$

see 2.8. Further, we have

$$FD^{-1}G = \{F(t)\} * \{G(t)\} = \int_0^t F(t-u)\,G(u)\,du: \tag{3.15.1}$$

see 2.9.

$$\langle t^{-1/2}\rangle^2 = \pi D \qquad \text{(see 2.33)}, \tag{3.16}$$

$$D^{-2}H = \int_0^t (t-u)\,H(u)\,du \qquad \text{(see 3.40.0)}, \tag{3.17}$$

$$D^{-1} = \frac{1}{D} = t \qquad \text{(see 2.41)}, \tag{3.18}$$

(3.19) $$n!\,D^{-n} = t^n \qquad \text{(see 2.41)},$$

(3.20) $$D^{-n} = \frac{1}{D^n} = \frac{t^n}{n!} \qquad \text{(see 2.41)},$$

(3.21) $$\frac{D}{(D-a)} = e^{at} \qquad \text{(see 2.34)},$$

(3.22) $$\frac{n!\,D}{(D-a)^{n+1}} = t^n e^{at} \qquad \text{(see 2.39)},$$

(3.23) $$\frac{D}{(D-a)^m} = e^{at}\,\frac{t^{m-1}}{(m-1)!} \qquad \text{(see 2.39)},$$

(3.24) $$\frac{1}{D-a} = D^{-1} e^{at} = \frac{e^{at}-1}{a} \qquad \text{(see 3.40.1)},$$

(3.25) $$\frac{D}{(D-c)^2+\beta^2} = \left(\frac{e^{ct}}{\beta}\right)\sin\beta t \qquad \text{(see 3.40.2)},$$

(3.25) $$\frac{D}{(D-c)^2-\beta^2} = \left(\frac{e^{ct}}{\beta}\right)\sinh\beta t \qquad \text{(see 3.40.2)},$$

(3.26) $$\frac{D^2}{(D^2+\beta^2)^2} = \left(\frac{t}{2\beta}\right)\sin\beta t \qquad \text{(see 3.40.3)},$$

(3.27) $$\frac{D}{D^2+\beta^2} = \beta^{-1}\sin\beta t \qquad \text{(see 3.40.4)},$$

(3.28) $$\frac{D^2}{D^2+\beta^2} = \cos\beta t \qquad \text{(see 3.40.5)},$$

(3.28) $$\frac{D^2}{D^2 - \beta^2} = \cosh \beta t$$ (see 3.40.5),

(3.29) $$\frac{D^3}{(D^2 + \beta^2)^2} = \frac{1}{2\beta} \sin \beta t + \frac{t}{2} \cos \beta t$$ (see 3.40.6),

(3.30) $$\frac{1}{D^2 + \beta^2} = \frac{1 - \cos \beta t}{\beta^2}$$ (see 3.40.7),

(3.30.1) $$\left(\frac{1}{D + \alpha}\right)^m = \frac{1}{\alpha^m} - \frac{e^{-\alpha t}}{\alpha^m} \left[1 + \alpha t + \cdots + \frac{(\alpha t)^{m-1}}{(m-1)!}\right]$$

(see 4.32.1),

(3.30.2) $$\frac{(D^2 - \beta^2) D}{(D + \beta^2)^2} = t \cos \beta t$$ (see 4.52.10),

(3.31) $$\frac{(D - \alpha) D}{(D - \alpha)^2 + \beta^2} = e^{\alpha t} \cos \beta t$$ (see 2.36),

(3.32) $$\frac{D}{[(D - \alpha)^2 + \beta^2]^2} = \frac{e^{\alpha t}}{2\beta^2} \left[-t \cos \beta t + \frac{1}{\beta} \sin \beta t\right]$$

(see 4.46),

(3.33) $$\frac{D}{(D^2 + a^2)(D^2 + b^2)} = \frac{a \sin bt - b \sin at}{ab(a^2 - b^2)}$$ (see 4.16).

Additional formulas can be obtained from Laplace transform tables; for example, by replacing by D the complex variable p in the tables on pp. 383—404 of the textbook [V 1]. For more information on obtaining operator-equations from Laplace transform tables, see 5.39. A more complete table is found on pp. 339—343 of the present book.

Illustrative Examples

3.34 **A classical situation.** Rather than solving the initial-value problem

(1) $y(0-) = 2, \qquad y'(0-) = -4 \quad \text{with} \quad y'' + 2y' + 5y = 0,$

consider instead the problem

(2) $y(0-) = 2, \qquad y'(0-) = -4 \quad \text{with} \quad \dfrac{\partial^2 y}{\partial t^2} + 2\dfrac{\partial y}{\partial t} + 5y = 0.$

In view of 2.55, all classical solutions* of (1) are solutions of (2): by solving (2) rather than (1) we have nothing to lose (in fact, (1) $\Leftrightarrow$ (2) in all classical situations: see 8.67).

From (2) and [3.11] it follows that the equations (2) imply the equation

(3) $D^2 y + 2Dy + 5y = 2D^2 + (-4)D + 2(2)D;$

solving for y:

$$y = \frac{2D^2}{D^2 + 2D + 5} = \frac{2(D+1)D - 2D}{(D+1)^2 + 2^2};$$

the second equality is for the purpose of using 3.31 and 3.25:

(4) $$y = 2e^{-t}\cos 2t - 2\left(\frac{e^{-t}}{2}\right)\sin 2t.$$

Reversing our steps, we see that (4) $\Rightarrow$ (3) and (3) $\Rightarrow$ (2) (provided that $y(0-) = 2$ and $y'(0-) = -4$); since the equation

(5) $$y(t) = 2e^{-t}\cos 2t - 2\left(\frac{e^{-t}}{2}\right)\sin 2t \quad (-\infty < t < \infty)$$

implies $y(0-) = 2$, $y'(0-) = -4$ and (4), it follows that (5) $\Rightarrow$ (2). Finally, since $y''(\ )$ is continuous, (5) defines a solution of (1) (since (5) $\Rightarrow$ (2) and by 2.55).

3.35 **A simple electric circuit.** Let E be the voltage applied to a simple circuit containing an inductance L, a resistance R, and a capacitance C; the equation

(3.36) $$L\frac{\partial i}{\partial t} + Ri - E = \frac{-1}{C}\int_{-\infty}^{t} i(u)\,du$$

---

* The term "classical solution" is defined in 8.65.

governs the current i. Suppose that $L = 0$; from 3.13 we see that the equation 3.36 is equivalent to

$$(6) \qquad CRi - CE = -D^{-1}i - \int_{-\infty}^{0} i(u)\, du:$$

the last term on the left-hand side is the *initial charge* of the capacitor. We shall suppose that the capacitor has no initial charge:

$$(7) \qquad CRi + D^{-1}i = CE,$$

and try to find i in case the circuit is subjected to a voltage of magnitude $7R$ applied between the times $t = \alpha$ and $t = \lambda$: this means that $E = 7R(\mathsf{T}_\alpha - \mathsf{T}_\lambda)$ in the above equation (see 3.4).

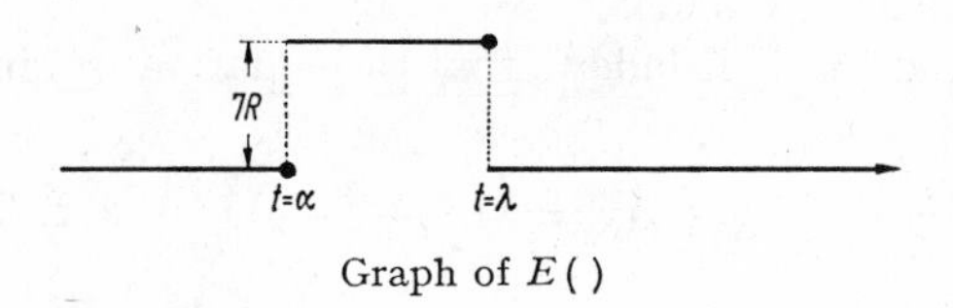

Graph of $E(\ )$

Solving Equation (7) for i:

$$i = \frac{7RC(\mathsf{T}_\alpha - \mathsf{T}_\lambda)}{RC + D^{-1}} = 7(\mathsf{T}_\alpha - \mathsf{T}_\lambda)\frac{D}{D + 1/RC} = 7(\mathsf{T}_\alpha - \mathsf{T}_\lambda)\langle e^{-t/RC}\rangle:$$

the last equation is from 3.21. Consequently,

$$i = 7[\mathsf{T}_\alpha\langle e^{-t/RC}\rangle - \mathsf{T}_\lambda\langle e^{-t/RC}\rangle]$$

$$(8) \qquad = 7[\mathsf{T}_\alpha(t)\, e^{-(t-\alpha)/RC} - \mathsf{T}_\lambda(t)\, e^{-(t-\lambda)/RC}] \qquad \text{(by 3.12)}.$$

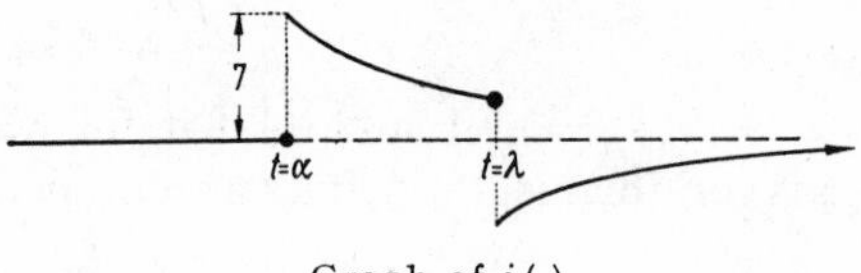

Graph of $i(\ )$

3.37 A system of differential equations. In the electrical circuit

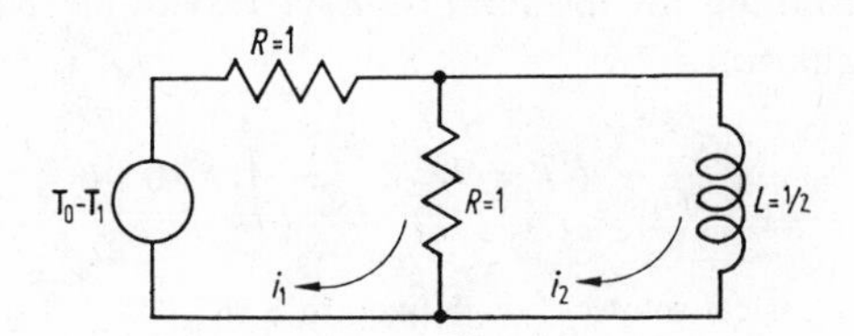

the applied voltage is the unit-area pulse $\mathsf{T}_0 - \mathsf{T}_1$: the problem is to find the current i_2 assuming $i_2(0-) = 0$. The voltage law gives

$$(1) \qquad 2i_1 - i_2 = \mathsf{T}_0 - \mathsf{T}_1 \quad \text{and} \quad -i_1 + i_2 + 2^{-1}\frac{\partial i_2}{\partial t} = 0.$$

Let $i_1(\,)$ and $i_2(\,)$ be two $\mathcal{K}$-functions satisfying the system of equations (1); from [3.7] and our initial condition $i_2(0-) = 0$ we obtain

$$\frac{\partial i_2}{\partial t} = Di_2;$$

it now follows from (1) and 3.3 that

$$(2) \qquad 2i_1 - i_2 = 1 - \mathsf{T}_1 \quad \text{and} \quad -i_1 + (1 + 2^{-1}D)\, i_2 = 0.$$

Solving the system (2) for i_2:

$$(3) \quad i_2 = (1 - \mathsf{T}_1)\,\frac{1}{D+1} = (1 - \mathsf{T}_1)\,\frac{\mathrm{e}^{-t} - 1}{-1} = (1 - \mathsf{T}_1)\,\langle 1 - \mathrm{e}^{-t}\rangle:$$

the middle equation is from 3.24. From (3) and 3.12 it now follows

$$(4) \qquad i_2(t) = 1 - \mathrm{e}^{-t} - \mathsf{T}_1(t)\,(1 - \mathrm{e}^{-(t-1)}) \quad \text{(see (3)–(4) in 3.6).}$$

3.38 **A non Laplace-transformable problem:**

$$(5) \qquad t\exp(t^2) = \int_0^t \cos(t-u)\, y(u)\, du.$$

The equations

$$\frac{D}{D^2+1}\, y = \frac{D^2}{D^2+1}\, D^{-1}y = \langle \cos t\rangle\, D^{-1}y = \langle t\mathrm{e}^{t^2}\rangle$$

are from 3.28, 3.15, and (5). Solving for y:

$$(6) \qquad y = (D + D^{-1})\langle t\mathrm{e}^{t^2}\rangle = D\langle t\mathrm{e}^{t^2}\rangle + \int_0^t u\mathrm{e}^{u^2}\, du:$$

the second equation is from 3.14. If $F(t) = t\exp(t^2)$ then $F(0-) = 0$ and

$$DF = F' + F(0-)\, D = F' \qquad \text{(by 3.8.4);}$$

consequently, $D\langle t e^{t^2}\rangle = \exp(t^2) + 2t^2 \exp(t^2)$; substituting into (6):

$$y = \exp(t^2) + 2t^2 \exp(t^2) + 2^{-1} \int_0^t e^{u^2} \, d(u^2),$$

from which we see that the equation

$$y(t) = \exp(t^2) + 2t^2 \exp(t^2) + 2^{-1} [\exp(t^2) - 1] \qquad (-\infty < t < \infty)$$

defines the unique solution of the problem (5).

3.39 Another non Laplace-transformable problem. If $G(\,)$ is any $\mathcal{K}$-function, the equation

$$y = \int_0^t e^{t-u} y(u) \, du + G$$

implies

$$y = \frac{D}{D-1} D^{-1} y + G \qquad \text{(by 3.15 and 3.21)}.$$

Solving for y:

$$y = \frac{D-1}{D-2} G = \left(1 + \frac{1}{D-2}\right) G = G + \frac{D}{D-2} D^{-1} G \tag{7}$$

$$= G + \int_0^t e^{2(t-u)} G(u) \, du: \tag{8}$$

the last equation is from 3.21 and 3.15. The function $G(t) = 2(t-1) \exp(t^2)$ is not Laplace-transformable; from (7)—(8) we see that

$$y = 2(t-1) \exp(t^2) + e^{2t} \int_0^t e^{-2u} \, 2(u-1) \, e^{u^2} \, du;$$

the change of variable $\lambda = -2u + u^2$ now gives

$$y(t) = 2(t-1) \exp(t^2) + e^{2t} [\exp(-2t + t^2) - 1] \qquad (-\infty < t < \infty).$$

Exercises

3.40.0 Prove 3.17. *Hints:* use 3.15, 3.18, and

$$D^{-2} H = t D^{-1} H = \int_0^t (t-u) \, H(u) \, du.$$

3.40.1 Prove 3.24. *Hints:* use 3.21 and

$$\frac{1}{D-a} = a^{-1}\left[\frac{D}{D-a} - 1\right].$$

3.40.2 Prove 3.25. *Hints:* see 2.37 and use the formula analogous to 2.35; also, repeat the procedure with

$$e^{ct}\sinh bt = 2^{-1}\left[e^{(c+b)t} - e^{(c-b)t}\right].$$

3.40.3 Prove 3.26. *Hints:* note that

$$2it\sin bt = te^{ibt} - te^{-ibt},$$

and proceed as in 3.40.2, using 3.22.

3.40.4 Prove 3.27.

3.40.5 Prove 3.28. *Hints:* note that

$$\cosh bt = D\langle b^{-1}\sinh bt\rangle \qquad \text{(by 2.25)}$$

and use 3.25; proceed analogously for cos.

3.40.6 Prove 3.29. *Hints:* multiply by D both sides of 3.26 and use the Derivation Property (2.24).

3.40.7 Prove 3.30. *Hints:* multiply by D^{-1} both sides of 3.27 and use Formula 3.14.

3.40.8 Proceed as in 3.34 to solve the following classical problems:

(.9) $\qquad y(0-) = 1, \quad y'(0-) = 0, \quad y'' + 3y' + 2y = 2.$

Answer: $y(t) = 1 \quad (-\infty < t < \infty)$.

(.10) $\qquad y(0-) = y'(0-) = 0$ with $y'' + 4y = \cos 2t$.

Answer: $(t/4)\sin 2t$.

3.40.11 Find a $\mathcal{K}$-function $y(\,)$ such that

$$y = \sin t + \int_0^t 2\cos(t-u)\,y(u)\,du.$$

Answer: $y(t) = te^t \quad (-\infty < t < \infty)$.

3.40.12 Find a $\mathcal{K}$-function $y(\,)$ such that

$$y = 6t + \int_0^t \sin(t-u)\,y(u)\,du.$$

Answer: $y(t) = t^3 + 6t$.

3.40.13 Find a $\mathcal{K}$-function $y(\,)$ such that

$$\int_0^t \cos(t-u)\, y(u)\, du = 5 \sin t + 6t.$$

Answer: $y(t) = 11 + 3t^2$.

3.40.14 Proceed as in 3.34 to solve the following classical problems:

(.15) $\quad y(0-) = y'(0-) = 0 \quad \text{with} \quad y'' + 4y' + 4y = e^{-2t}.$

Answer: $(t^2/2) \exp(-2t)$.

(.16) $\quad y(0-) = y'(0-) = 0 \quad \text{with} \quad y'' + b^2 y = \begin{cases} b^2 & (0 < t < \alpha) \\ 0 & (\text{otherwise}). \end{cases}$

Hint: use 3.30.

Answer: $1 - \cos bt - \mathsf{T}_\alpha(t)\,(1 - \cos[bt - b\alpha])$.

(.17) $\quad y(0-) = y'(0-) = 0 \quad \text{with} \quad y'' - 2ay' + a^2 y = F.$

Answer: $\displaystyle\int_0^t \frac{(t-u)^2}{2}\, e^{a(t-u)}\, F(u)\, du.$

(.18) $\quad y(0-) = y'(0-) = \cdots = y^{(m-1)}(0-) = 0 \quad \text{with} \quad y^{(m)} = F.$

Hints: use 3.11, 3.19, and 3.15.

Answer: $\displaystyle\int_0^t \frac{(t-u)^{m-1}}{(m-1)!}\, F(u)\, du.$

§ 4. Partial Fraction Decomposition

Before starting on the subject-matter of this section, let us examine an example which displays the remarkable power of operational calculus. The system of equations

$$R_1 i_1 + L_1 \frac{\partial}{\partial t} i_1 + M \frac{\partial}{\partial t} i_2 = E_1, \tag{1}$$

$$M \frac{\partial}{\partial t} i_1 + R_2 i_2 + L_2 \frac{\partial}{\partial t} i_2 = 0 \tag{2}$$

governs the currents i_1 and i_2 in the inductively coupled network sketched below

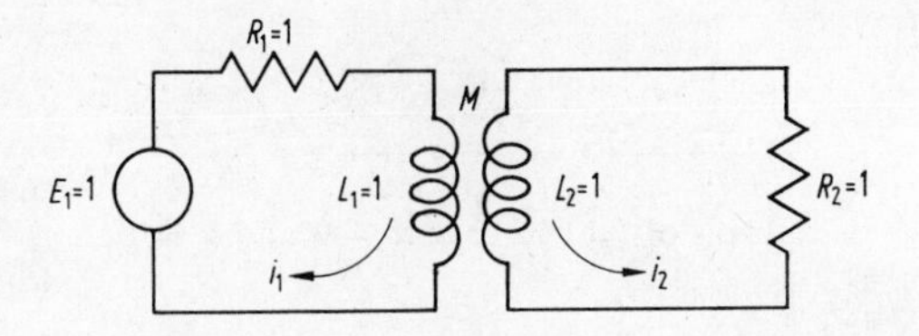

Suppose that $E_1 = L_1 = L_2 = M = R_1 = R_2 = 1$ and let $a = i_1(0-)$ and $b = i_2(0-)$ be the initial values of the currents. From [3.7] it follows that (1)—(2) imply the system

$$i_1 + Di_1 + Di_2 = 1 + cD \qquad (\text{with } c = a + b), \tag{3}$$

$$Di_1 + i_2 + Di_2 = cD. \tag{4}$$

Subtracting (4) from (3), we obtain $i_1 - i_2 = 1$, which implies that $i_1(0+) \neq i_2(0+)$; consequently, if we adjoin to (3)—(4) the starting values $i_1(0+) = i_2(0+) = 0$, *then the resulting problem has no solution*: this is a non-standard problem (2.56). Note that the equation $i_1 - i_2 = 1$ does **not** imply the equation $i_1(0-) - i_2(0-) = 1$: see 2.3—4.

This is one of the reasons why *initial values should be understood as values taken on at the time* $t = 0-$.

Let us solve the system (3)—(4) by Cramer's Rule:

$$i_2 = \frac{\begin{vmatrix} 1 + D & 1 + cD \\ D & cD \end{vmatrix}}{\begin{vmatrix} 1 + D & D \\ D & 1 + D \end{vmatrix}} = \frac{D(c-1)}{1 + 2D};$$

therefore, $i_2 = 2^{-1}(c - 1)\langle e^{-t/2}\rangle$ (recall that $c = a + b$). Reversing our steps, we see that the equation

$$i_2(\tau) = \begin{cases} b & (\tau \le 0) \\ 2^{-1}(a + b - 1)\exp(-\tau/2) & (\tau > 0) \end{cases}$$

determines a solution of (1)—(2) such that $i_2(0-) = b$: the function $i_2(\,)$ has a jump on $[0, \infty)$.

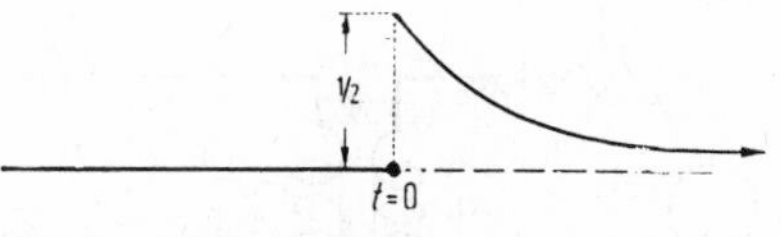

Graph of $i_2(\,)$ in case $a = 0$ and $b = 1$

The road ahead. Our main problem-solving tools are operations (on perfect operators) that correspond to ordinary differentiation and integration; for example, the operation $\partial/\partial t$ which acts on operators but corresponds to d/dt. It is often required to find a *function* whose canonical operator is the solution of a given operator-problem: we shall presently discuss several formulas and techniques to find such functions.

4.0 To begin with the simplest case, let $z_0 = 0, z_1, z_2, \ldots, z_n$ be distinct numbers (that is, $z_k \neq z_m$ when $k \neq m$), and let G be the product of $n + 1$ operators of the form $(D - z)$:

$$G = D^r (D - z_1)(D - z_2) \cdots (D - z_n) \qquad (r = 0, 1).$$

If $a_0, a_1, \ldots, a_\nu$ are given numbers, and

$$H = a_0 + a_1 D + \cdots + a_\nu D^\nu \qquad (\text{with } \nu \leq n),$$

then

$$\frac{H}{G} = rs + t^r \left[\frac{HD^r}{G}\right]_{D=0} + \sum_{k=1}^{n} \frac{D}{D - z_k} \left[\frac{H}{G}\left(\frac{D - z_k}{D}\right)\right]_{D=z_k} \tag{4.1}$$

where

$$s = \frac{a_0 (z_1^{-1} + z_2^{-1} + \cdots + z_n^{-1}) + a_1}{(0 - z_1)(0 - z_2) \cdots (0 - z_n)}. \tag{4.2}$$

Consequently,

$$\frac{H}{G} = rs + b_0 t^r + b_1 \frac{D}{D - z_1} + \cdots + b_n \frac{D}{D - z_n}, \tag{4.3}$$

where

$$b_0 = \left[\frac{HD^r}{G}\right]_{D=0} = \frac{a_0}{(0 - z_1)(0 - z_2) \cdots (0 - z_n)}$$

and

$$b_k = \left[\frac{H}{G}\left(\frac{D - z_k}{D}\right)\right]_{D=z_k} = \frac{a_0 + a_1 z_k + \cdots + a_\nu z_k^\nu}{z_k^r (z_k - z_1)(z_k - z_2) \cdots (z_k - z_n)}, \tag{4.4}$$

the factor $(z_k - z_k)$ being absent from the denominator. Note that

$$b_k = \left[\frac{H}{G_k}\right]_{D=z_k} \qquad (\text{for } k \neq 0),$$

where G_k is the result of replacing $(D - z_k)$ by D in G.

4.5 *Remarks.* The above formulas 4.1—4 are only valid when $r = (1 \pm 1)/2$; they are particular cases of Decomposition Theorem 4.23. To avoid errors, we should always verify beforehand that $\nu \leq n$. From 4.3 and 3.21 it follows that

$$\frac{H}{G} = rs + b_0 t^r + b_1 \exp(z_1 t) + \cdots + b_n \exp(z_n t).$$

4.6 **First example.** Let us solve the initial-value problem

$$y(0-) = 1, \quad y'(0-) = 0, \qquad y'' - y' - 6y = 2.$$

Proceeding as in 3.34, we find that

$$D^2 y - D^2 - Dy + D - 6y = 2;$$

solving for y, we get

$$y = \frac{D^2 - D + 2}{(D-3)(D+2)}.$$

We may now apply 4.1—4 (with $r = 0$) to obtain

$$y = b_0 + b_1 \frac{D}{D-3} + b_2 \frac{D}{D+2}, \tag{1}$$

with

$$b_0 = \left[\frac{H}{G}\right]_{D=0} = \frac{2}{(0-3)(0+2)} = -\frac{1}{3}, \tag{2}$$

$$b_1 = \left[\frac{H}{G}\left(\frac{D+3}{D}\right)\right]_{D=3} = \left[\frac{D^2 - D + 2}{(D+2)\,D}\right]_{D=3} = \frac{8}{15}, \tag{3}$$

and

$$b_2 = \left[\frac{D^2 - D + 2}{(D-3)\,D}\right]_{D=-2} = \frac{4}{5}. \tag{4}$$

From (1)—(4) and 3.21 it now follows that

$$y = -\frac{1}{3} + \frac{8}{15} e^{3t} + \frac{4}{5} e^{-2t}.$$

4.7 A system of equations. To solve the initial-value problem

$$X(0-) = 1, \qquad Y(0-) = 0 = Z(0-),$$

with

$$X' + 2X + Y = 1,$$

$$Y' - Y + Z' + 2Z = 0,$$

$$X' - X + Z' = -1.$$

Proceeding as in 3.34, we find that this problem implies the the following system of equations

$$(D + 2)\, X + Y = 1 + D, \tag{5}$$

$$(D - 1)\, Y + (D + 2)\, Z = 0, \tag{6}$$

$$(D - 1)\, X + DZ = -1 + D. \tag{7}$$

The system (5)—(7) can be solved by any one of the techniques of everyday algebra; for example, by using Cramer's Rule (which in this case involves third order determinants), we obtain

$$X = \frac{(D-1)\,(D^2 + 2D + 2)}{(D-1)\,(D+2)\,(D+1)} = \frac{D^2 + 2D + 2}{(D+2)\,(D+1)}: \tag{8}$$

the second equation is from 1.73. From 4.1—4 we obtain*

$$X = b_0 + b_1 \frac{D}{D+2} + b_2 \frac{D}{D+1},$$

where

$$b_0 = \left[\frac{D^2 + 2D + 2}{(D+2)\,(D+1)}\right]_{D=0} = 1,$$

$$b_1 = \left[\frac{D^2 + 2D + 2}{(D+1)\,D}\right]_{D=-2} = 1,$$

and

$$b_2 = \left[\frac{D^2 + 2D + 2}{(D+2)\,D}\right]_{D=-1} = -1.$$

Consequently,

$$X = 1 + \exp(-2t) - \exp(-t).$$

---

* For a comment on the calculations, see 5.42.3.

Exercises

Solve the following problems:

4.8.0 $$y(0-) = 0, \qquad y' + y = e^{2t}.$$

Answer: $-3^{-1}e^{-t} + 3^{-1}e^{2t}$.

4.8.1 $$y(0-) = 1, \quad y'(0-) = 0, \quad y'' - y' - 6y = 2.$$

Answer: $(8/15)\,e^{3t} + (4/5)\,e^{-2t} - 1/3$.

4.8.2 $$y(0-) = 1, \; y'(0-) = -1, \quad y'' + 3y' + 2y = 12e^{2t}.$$

Answer: $3e^{-2t} - 3e^{-t} + e^{2t}$.

4.8.3 Solve for y the system

$$x' + y' - 4y = 1,$$
$$x + y' - 3y = t^2$$

subject to the conditions

$$y(0-) = 0 = x(0-).$$

Hint: use 4.1.

Answer: $y = t/2 + 1/4 - (1/4)\,e^{2t}$.

Decompositions Obtained by Division

Let us agree that a "*real polynomial*" is a finite sequence of real numbers. If f is a polynomial $(a_0, a_1, \ldots, a_k, \ldots)$, the *degree* of f is the largest value of k such that $a_k \neq 0$; we set

$$f(D) \overset{\text{def}}{=} a_0 + a_1 D + a_2 D^2 + \cdots + a_k D^k + \cdots.$$

Our agreement 2.16 implies that

$$f(z) = a_0 + a_1 z + a_2 z^2 + \cdots + a_k z^k + \cdots$$

whenever z is a number.

4.9 **Division algorithm.** Let h_0 and g_0 be real polynomials; the usual division procedure (sometimes called "algorithm") gives three real polynomials q, h_1, and g_1 such that the degree of the polynomials h_1 is *less* than the degree of g_1 and

$$\frac{h_0(D)}{g_0(D)} = q(D) + \frac{h_1(D)}{g_1(D)}. \tag{4.10}$$

The division procedure is the same as in elementary algebra: see the following example.

4.11 **First application.** To find a $\mathcal{K}$-function $y(\,)$ such that

$$y + \int_0^t \cos(t-u)\, y(u)\, du = 1.$$

If $y(\,)$ is such a function, the equations

$$1 = y + \langle \cos t \rangle D^{-1} y = y + \frac{D^2}{D^2+1} D^{-1} y$$

are from 3.15 and 3.28. Solving for y:

$$(1) \quad y = \frac{D^2+1}{D^2+D+1} = 1 - \frac{D}{D^2+D+1} = 1 - \frac{D}{(D+1/2)^2+3/4};$$

the second equation is obtained by the division algorithm

$$\begin{array}{rl} D^2 + 1 & \underline{|D^2 + D + 1} \\ D^2 + D + 1 & 1 \\ \hline -D + 0 & \end{array}$$

From 3.25 (with $\beta = \sqrt{3}/2$) it now follows that

$$y = 1 - \frac{2}{\sqrt{3}} e^{-t/2} \sin t \frac{\sqrt{3}}{2}.$$

4.12 **Remark.** The division algorithm is often effective when dealing with operators of the form $h_0(D)/V^m$. For example,

$$\frac{h_0(D)}{f(D)^2} = \frac{1}{f(D)}\left(q(D) + \frac{h_1(D)}{f_1(D)}\right) = \frac{q(D)}{f(D)} + \frac{h_1(D)}{f(D)\, f_1(D)}.$$

4.13.0 **An integro-differential equation.** To find a $\mathcal{K}$-function $y(\ )$ having no jumps on $[0, \infty)$, such that $y'(\ )$ is a $\mathcal{K}$-function, and satisfying the two equations

$$y(0-) = 1 \quad \text{and} \quad y = \cos t + 2 \int_0^t \sin(t-u)\, y'(u)\, du.$$

If $y(\ )$ is such a function, we can apply 3.15 and 3.28 to obtain

$$(2)\quad y = \frac{D^2}{D^2+1} + 2\langle \sin t\rangle D^{-1}y' = \frac{D^2}{D^2+1} + 2\frac{D}{D^2+1}(D^{-1}y'):$$

the second equation is from 3.27. On the other hand, 3.8.3 gives $D^{-1}y' = y - 1$ (since $y(0-) = 1$); substituting into (2):

$$y = \frac{D^2 + 2Dy - 2D}{D^2+1};$$

solving for y:

$$y = \frac{D^2 - 2D}{(D-1)^2} = \frac{1}{D-1}\left[\frac{D^2 - 2D}{D-1}\right]$$

$$= \frac{1}{D-1}\left[D - \frac{D}{D-1}\right] \qquad \text{(as in 4.12)}$$

$$= \frac{D}{D-1} - \frac{D}{(D-1)^2} = e^t - te^t.$$

Heaviside's Expansion

4.13.1 Given a $\mathcal{K}$-function $F(\ )$ and two real polynomials h_0 and g_0, the division algorithm yields three real polynomials q, h_1, and g_1 such that

$$(3)\qquad F\frac{h_0(D)}{g_0(D)} = q(D)\,F + FD^{-1}\frac{h_1(D)\,D}{g_1(D)} \qquad \text{(see 4.10).}$$

Suppose that $g_1(D)$ is the product of n operators of the form $D - z$:

$$g_1(D) = (D - z_1)(D - z_2)\cdots(D - z_n);$$

if $z_0 = 0, z_1, z_2, \ldots, z_n$ are distinct numbers, we can apply 4.1 with $r = 0$ and $H = h_1(D)\,D$:

$$\frac{h_1(D)\,D}{g_1(D)} = \sum_{k=1}^{n} \frac{D}{D - z_k}\left[\frac{h_1(D)\,D}{g_1(D)}\left(\frac{D - z_k}{D}\right)\right]_{D=z_k};$$

that is,

$$\frac{h_1(D)\,D}{g_1(D)} = \sum_{k=1}^{n} \frac{D}{D - z_k} \left[\frac{h_1(D)}{E_k}\right]_{D=z_k}, \tag{4}$$

where E_k is the result of omitting the factor $(D - z_k)$ from $g_1(D)$. Combining (3) and (4):

$$F\frac{h_0(D)}{g_0(D)} = q(D)\,F + FD^{-1} \sum_{k=1}^{n} \frac{D}{D - z_k} \left[\frac{h_1(D)}{E_k}\right]_{D=z_k} \tag{5}$$

$$= q(D)\,F + \sum_{k=1}^{n} FD^{-1}\langle \exp(z_k t)\rangle \left[\frac{h_1(D)}{E_k}\right]_{D=z_k};$$

$$= q(D)\,F + \sum_{k=1}^{n} \{F(t)\} * \{\exp(z_k t)\} \left[\frac{h_1(D)}{E_k}\right]_{D=z_k}: \tag{6}$$

the last two equations are from 3.21 and 3.15.1. It is historically appropriate to call the resulting formula

$$F\frac{h_0(D)}{g_0(D)} = q(D)\,F + \sum_{k=1}^{n} \left\langle \int_0^t F(t-u)\exp(z_k u)\,du \right\rangle \left[\frac{h_1(D)}{E_k}\right]_{D=z_k}$$

"Heaviside's expansion."

4.13.2 **A non-standard problem.** Given any three numbers a, b, c, and an arbitrary $\mathcal{K}$-function $F()$, let us solve for y the system

$$x(0-) = a, \quad x'(0-) = b, \quad y(0-) = c, \tag{7}$$

$$\frac{\partial x}{\partial t} + x + 2y = F, \tag{8}$$

$$\frac{\partial^2 x}{\partial t^2} + 5x + 3\frac{\partial y}{\partial t} = 0: \tag{9}$$

this is an extension of the situation considered in 2.56 (the solution $y()$ has a jump at $t = 0$ which depends on the starting value $F(0+)$). From [3.11] it follows that the system (7)—(9) implies

$$(D+1)\,x + 2y = F + aD,$$

$$(D^2+5)\,x + 3Dy = \lambda D + aD^2 \qquad (\text{with } \lambda = 3c + b).$$

Solving for y:

$$y = -F \frac{D^2 + 5}{D^2 + 3D - 10} + \frac{(\lambda + a)\, D^2 + (\lambda - 5a)\, D}{(D + 5)\,(D - 2)}. \tag{10}$$

From 4.10 we see that

$$\frac{D^2 + 5}{D^2 + 3D - 10} = 1 + \frac{15 - 3D}{(D + 5)\,(D - 2)},$$

so that (5)—(6) give

$$F \frac{D^2 + 5}{(D + 5)\,(D - 2)} = F + \{F(t)\} * \{e^{-5t}\} \left[\frac{15 - 3D}{D - 2}\right]_{D=-5} + \{F(t)\} * \{e^{2t}\} \left[\frac{15 - 3D}{D + 5}\right]_{D=2}. \tag{11}$$

That is,

$$F \frac{D^2 + 5}{(D + 5)\,(D - 2)} = F + \{F(t)\} * \left\{\frac{30}{-7} e^{-5t} + \frac{9}{7} e^{2t}\right\}. \tag{12}$$

Further, 4.1 gives

$$\frac{(\lambda + a)\, D^2 + (\lambda - 5a)\, D}{(D + 5)\,(D - 2)} = \frac{2}{7}(5a + 2\lambda) \frac{D}{D + 5} + \frac{3}{7}(\lambda - a) \frac{D}{D - 2}. \tag{13}$$

Combining (10) with (12) and (13), we obtain

$$y = -F + F_1,$$

where

$$F_1 = \int_0^t F(t - u) \left(\frac{30}{7} e^{-5u} - \frac{9}{7} e^{2u}\right) du + \frac{2}{7}(5a + 2\lambda)\, e^{-5t} + \frac{3}{7}(\lambda - a)\, e^{2t}.$$

Reversing our steps, we see that the equation

$$y(\tau) = \begin{cases} c & (\tau \leq 0) \\ -F(\tau) + F_1(\tau) & (\tau > 0) \end{cases}$$

defines a solution of the system (7)—(9).

4.13.3 **An integro-differential equation.** Given a number c, let us find a $\mathcal{K}$-function $y(\ )$ such that

$$y(0-) = c \quad \text{with} \quad \frac{\partial y}{\partial t} + 3y = -2\int_0^t y(u)\,du + \begin{cases} 1 & (\alpha < t < \lambda) \\ 0 & (\text{otherwise}): \end{cases}$$

as usual, angular brackets have been omitted from the right-hand side. If $y(\ )$ is such a function, we may use 3.7, 3.14, and 3.4 to obtain

$$Dy + 3y = -2D^{-1}y + cD + \mathsf{T}_\alpha - \mathsf{T}_\lambda.$$

Solving for y:

$$y = \frac{cD + \mathsf{T}_\alpha - \mathsf{T}_\lambda}{D + 3 + 2D^{-1}} = \frac{cD^2}{(D+1)(D+2)} + (\mathsf{T}_\alpha - \mathsf{T}_\lambda)\frac{D}{(D+1)(D+2)},$$

and a double application of 4.1 now gives

$$y = c\left(\frac{2D}{D+2} - \frac{D}{D+1}\right) + (\mathsf{T}_\alpha - \mathsf{T}_\lambda)\left(\frac{D}{D+1} - \frac{D}{D+2}\right)$$

$$= 2ce^{-2t} - ce^{-t} + \mathsf{T}_\alpha \langle e^{-t} - e^{-2t}\rangle + \mathsf{T}_\lambda \langle e^{-2t} - e^{-t}\rangle \quad (\text{by } 3.21),$$

$$= 2ce^{-2t} - ce^{-t} + \mathsf{T}_\alpha(t)\,[e^{-t+\alpha} - e^{-2t+2\alpha}]$$

$$+ \mathsf{T}_\lambda(t)\,[e^{-2t+2\lambda} - e^{-t+\lambda}]:$$

the last equation is from 3.12.

Decomposition into Two Parts

Let V and G be invertible (1.65) operators; if H is a perfect operator, it follows from 1.73 that

$$(4.14) \qquad \boxed{\frac{H}{VG} = \frac{A}{G} + \frac{B}{V} \quad \text{whenever} \quad H = AV + BG}.$$

4.15 In case A/G and B/V are operators listed in 3.18—33, the first equation in 4.14 gives the function whose operator is H/VG.

4.16 **For example,**

$$\frac{D}{(D^2 + a^2)\ (D^2 + b^2)} = \frac{A}{D^2 + b^2} + \frac{B}{D^2 + a^2} \tag{1}$$

whenever A and B are operators such that

$$D = A\,(D^2 + a^2) + B\,(D^2 + b^2);$$

it is easily seen that this last equation holds when

$$A = -B = [a^2 - b^2]^{-1} D.$$

Consequently, we may use (1) and 3.27 to obtain 3.33.

4.17 **A second example:**

$$\frac{D}{D^2\,(D^2 + 1)} = \frac{A}{D^2 + 1} + \frac{B}{D^2} \quad \text{whenever} \quad D = AD^2 + B\,(D^2 + 1);$$

obviously, this last equation is satisfied by $A = -D$ and $B = D$; therefore,

$$\frac{D}{D^2\,(D^2 + 1)} = -\sin t + t \qquad \text{(by 3.27)}.$$

4.18 **An integral equation.** To find a $\mathcal{K}$-function $y(\,)$ such that

$$y(t) = 4t - 3 \int_0^t \sin(t - u)\, y(u)\, du \qquad (\text{all } t > 0).$$

If $y(\,)$ is such a function, it follows from 2.66 and 3.15 that

$$y = 4t - 3\langle \sin t\rangle\, D^{-1} y = 4D^{-1} - \frac{3D}{D^2 + 1}\, D^{-1} y:$$

the second equation is from 3.18 and 3.27. Solving for y, we obtain

$$y = \frac{4\,(D^2 + 1)}{D\,(D^2 + 4)} = \frac{A}{D^2 + 4} + \frac{B}{D}$$

whenever A and B are such that

$$4\,(D^2 + 1) = AD + B\,(D^2 + 4);$$

this last equation is easily solvable for A and B (say, by comparing coefficients, of like powers of D): we find $A = 3D$ and $B = 1$. Consequently, 3.27 and 3.18 give

$$y = \left\langle \frac{3}{2} \sin 2t \right\rangle + \langle t \rangle.$$

Reversing our steps (as in 3.34), we see that the equation

$$y(t) = \frac{3}{2}\sin 2t + t \qquad (-\infty < t < \infty)$$

determines a solution of our integral equation.

4.19 **Last example.** Observe that the equation

$$\frac{D^2}{(D+1)^2(D^2+D+1)} = \frac{A}{(D+1)^2} + \frac{B}{D^2+D+1} \tag{1}$$

holds whenever

$$D^2 = A(D^2+D+1) + B(D+1)^2 \qquad \text{(see 4.14)}. \tag{2}$$

We find that (2) is satisfied by $A = -D$ and $B = D$; consequently, if $y(\,)$ is a function whose canonical operator is (1), then

$$y = \frac{-D}{(D+1)^2} + \frac{D}{(D+1/2)^2 + 3/4},$$

whence

$$y = -t\mathrm{e}^{-t} + \left(\frac{2\,\mathrm{e}^{-t/2}}{\sqrt{3}}\right)\sin t\,\frac{\sqrt{3}}{2} \qquad \text{(by 3.22 and 3.25)}.$$

Exercises

4.20.0 Using the procedure followed in 4.16—19, verify the equation

$$\frac{1}{D(D^2+1)} = -\sin t + 1.$$

4.20.1 Solve for i_2 the following system of two equations:

$$R_1 i_1 + \frac{1}{C}\int_0^t [i_1(u) - i_2(u)]\,du = \lambda\mathrm{e}^{-at},$$

$$\frac{1}{C}\int_0^t [i_2(u) - i_1(u)]\,du + R_2 i_2 = 0.$$

Answer: $i_2 = \dfrac{\lambda}{R_1 R_2 C(b-a)}\,[\mathrm{e}^{-at} - \mathrm{e}^{-bt}]$, where $b = \dfrac{R_1+R_2}{R_1 R_2 C}$.

4.20.2 Proceed as in 3.34 to solve the following classical problems:

$$y(0-) = 0, \quad y'(0-) = 1, \quad y''(t) + y(t) = \begin{cases} 1 & (0 < t < \alpha) \\ 0 & \text{(otherwise)}. \end{cases} \tag{.3}$$

Answer: $y(t) = \begin{cases} 1 - \cos t + \sin t & (0 < t < \alpha) \\ \sin t - \cos t + \cos(t - \alpha) & (t > \alpha). \end{cases}$

(.4) $$y(0-) = 1, \quad y'(0-) = 5, \quad y'' + 2y' + 5y = 0.$$

Answer: $e^{-t}(\cos 2t + 3 \sin 2t)$.

(.5) $$y(0-) = y'(0-) = 0, \quad y'' + 4y = \sin t.$$

Answer: $(1/3) \sin t - (1/6) \sin 2t$.

(.6) $$y(0-) = y'(0-) = 1, \quad y'' + 4y = \sin t.$$

Answer: $(1/3)(\sin t + \sin 2t) + \cos 2t$.

4.21.0 Solve for $x(\,)$ the system (7)—(9) in 4.13.2.

Answer: $x(\tau) = \begin{cases} b\tau + a & (\tau \leq 0) \\ F_0(\tau) & (\tau > 0), \end{cases}$

where

$$F_0 = \int_0^t F(t-u)\left(\frac{15}{7} e^{-5u} + \frac{6}{7} e^{2u}\right) du + \frac{(5a - 2\lambda)}{7} e^{-5t} + \frac{2(a-\lambda)}{7} e^{2t}.$$

4.21.1 Find the jump at $t = 0$ of the solution $y()$ of the system (7)—(9) in 4.13.2.

Answer: $y(0+) - y(0-) = -F(0+) + a + b + 2c$.

4.21.2 The simple electric circuit

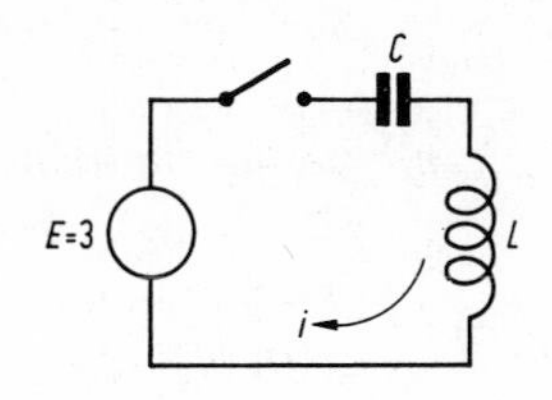

is supposed to be initially at rest (no charge, no current, and no voltage at $t = 0-$); under those circumstances, the equation

$$L \frac{\partial i}{\partial t} - E = \frac{-1}{C} \int_0^t i(u)\, du$$

governs the current i for times $t > 0$. By closing the switch at $t = 0$ a constant voltage $E = 3$ is impressed on the circuit: find the voltage V across the inductor. *Hint*: $V = L\, \partial i/\partial t$; solve the circuit equation for i and find $\partial i/\partial t$.
Answer: $V = 3\langle \cos t/\sqrt{LC}\rangle$.

4.21.3 Find the jump of V (in the preceding problem) at $t = 0$.
Answer: $V(0+) - V(0-) = 3$.

Repeated Linear Factors

Let a be a number; an operator V is a *power* of $D - a$ if the equation $V = (D - a)^m$ holds for some integer $m \geq 0$; powers of $D - a$ are invertible (1.65), since

$$(D - a)^{-m} = D^{-1}\langle e^{at} t^{m-1}/(m-1)!\rangle \qquad \text{(by 3.23)}.$$

Given a function $\varphi(\,)$, we set

$$\text{(4.22)} \qquad \{\varphi \mid (D - a)^m\} \overset{\text{def}}{=} \sum_{s=0}^{m} \frac{1}{(s-1)!}\, \varphi^{(s-1)}(a)\, (D - a)^{s-1}.$$

4.23 **Decomposition Theorem.** *Let* $z_0 = 0, z_1, z_2, \ldots, z_n$ *be distinct numbers, and let* g *be a real polynomial such that*

$$\text{(4.24)} \qquad g(D) = V_0 V_1 \cdots V_n,$$

where V_k *is a power of* $D - z_k$ *(for* $k = 0, 1, \ldots, n$*). If* h *is a real polynomial of degree not exceeding the degree of* g*, then*

$$\text{(4.25)} \qquad \frac{h(D)}{g(D)} = \frac{\{\varphi_0 \mid V_0 D\}}{V_0} + \sum_{k=1}^{n} \frac{\{\varphi_k \mid V_k\}\, D}{V_k},$$

where $\varphi_k(\,)$ $(k = 0, 1, 2, \ldots, n)$ *are the functions defined for almost all complex values of* z *by*

$$\text{(4.26)} \qquad \varphi_k(z) = \left[\frac{h(D)}{g(D)}\left(\frac{V_k}{D}\right)\right]_{D=z}.$$

4.27 *Remarks.* Since $z_0 = 0$, we have $V_0 = D^p$ for some integer $p \geq 0$; thus

$$(4.28) \qquad \varphi_0(z) = \left[\frac{h(D)\, D^p}{g(D)}\right]_{D=z} = \left[\frac{h(D)}{V_1 V_2 \cdots V_n}\right]_{D=z}:$$

the second equation is from 4.24. Combining 4.26 with 4.24, we see that

$$(4.29) \qquad \varphi_k(z) = \left[\frac{h(D)}{G_k}\right]_{D=z} \qquad (k = 1, 2, 3, \ldots, n),$$

where G_k *is the result of replacing* V_k *by* D *in* $g(D)$. Since $1/(s-1)! = 0$ when $s = 0$, we see from [4.22] that

$$\{\varphi \mid (D-a)^m\} = 0 \text{ when } m = 0$$

and

$$(4.30) \qquad \{\varphi \mid (D-a)^m\} = \varphi(a) + \varphi'(a)(D-a) + \cdots + \frac{1}{(m-1)!}\varphi^{(m-1)}(a)(D-a)^{m-1}$$

whenever $m \geq 1$. In particular

$$(4.31) \qquad \boxed{\{\varphi \mid D-a\} = \varphi(a)}.$$

If $g(0) \neq 0$ then $g(D) = V_0 V_1 \cdots V_n$ with $V_0 = D^0 = 1$; Equation 4.25 becomes

$$\frac{h(D)}{g(D)} = \{\varphi_0 \mid D\} + \sum_{k=1}^{n} \frac{\{\varphi_k \mid V_k\}\, D}{V_k};$$

from 4.31 (with $a = 0$) and 4.28 (with $z = z_0 = 0$) we therefore obtain

$$(4.32) \qquad \boxed{\frac{h(D)}{g(D)} = \frac{h(0)}{g(0)} + \sum_{k=1}^{n} \frac{\{\varphi_k \mid V_k\}\, D}{V_k}} \qquad (\text{in case } g(0) \neq 0).$$

The reader interested in proving the Decomposition Theorem might look at 13.19—21. As a first example, let us consider the case where $g(D) = (D+\alpha)^m$ and $h(D) = 1$; Formula 4.32 gives

$$(7) \qquad \frac{1}{(D+\alpha)^m} = \frac{h(D)}{g(D)} = \frac{1}{\alpha^m} + \frac{\{\varphi_1 \mid V_1\}\, D}{V_1}$$

with

(8) $$V_1 = (D + \alpha)^m,$$

so that

(9) $$\varphi_1(z) = \left[\frac{h(D)}{D}\right]_{D=z} = \frac{1}{z} \qquad \text{(by 4.29)}.$$

On the other hand, the equations

(10) $$\{\varphi_1 | V_1\} = \{\varphi_1 | (D + \alpha)^m\} = \sum_{k=0}^{m-1} \frac{1}{k!} \varphi_1^{(k)}(-\alpha)\,(D + \alpha)^k$$

are from (8) and 4.30. Since (9) gives

$$\varphi_1^{(k)}(z) = (-1)^k\, k!\, z^{-k-1},$$

we have

(11) $$\frac{\{\varphi_1 | V_1\} D}{V_1} = -\sum_{k=0}^{m-1} \frac{1}{\alpha^{k+1}} \frac{D}{(D + \alpha)^{m-k}} \qquad \text{(by [10])}$$

$$= -\sum_{k=0}^{m-1} \frac{1}{\alpha^{k+1}} \frac{t^{m-k-1}}{(m - k - 1)!} \mathrm{e}^{-\alpha t} \qquad \text{(by 3.23)}.$$

We can now combine (7) and (11)—(12) to obtain

$$\frac{1}{(D + \alpha)^m} = \frac{1}{\alpha^m} - \frac{\mathrm{e}^{-\alpha t}}{\alpha^m} \sum_{k=0}^{m-1} \frac{(\alpha t)^{m-k-1}}{(m - k - 1)!}.$$

In other words:

(4.32.1) $$\boxed{(D + \alpha)^{-m} = \frac{1}{\alpha^m} - \frac{\mathrm{e}^{-\alpha t}}{\alpha^m} \left[1 + \alpha t + \cdots + \frac{(\alpha t)^{m-1}}{(m - 1)!}\right]}.$$

4.33 **Examples.** Formula 4.25 gives

(1) $$\frac{D + 2}{D^2(D + 1)} = \frac{\{\varphi_0 | D^2 D\}}{D^2} + \frac{\{\varphi_1 | D + 1\} D}{D + 1},$$

where

(2) $$\varphi_0(z) = \left[\frac{D+2}{D+1}\right]_{D=z} = \frac{z+2}{z+1} \qquad \text{(by 4.28)},$$

and

(3) $$\varphi_1(z) = \left[\frac{D+2}{D^2(D)}\right]_{D=z} = \frac{z+2}{z^3} \qquad \text{(by 4.29)}.$$

From 4.30 we see that

(4) $$\{\varphi_0 \mid D^3\} = \varphi_0(0) + \varphi_0'(0)\,D + \varphi_0''(0)\,D/2;$$

in view of (2), an easy calculation now yields

(5) $$\{\varphi_0 \mid D^3\} = 2 - D + D^2 \qquad \text{(by (4))}.$$

Finally

(6) $$\{\varphi_1 \mid D+1\} = \varphi_1(-1) = -1 \qquad \text{(by 4.31 and (3))}.$$

The conclusion

$$\frac{D+2}{D^2(D+1)} = \frac{2}{D^2} - \frac{1}{D} + 1 - \frac{D}{D+1} = t^2 - t + 1 - e^{-t}$$

is immediate from (1), (5), (6), 3.19, and 3.21.

Heaviside Procedure

4.34 Let h_k and f be real polynomials, and let a be a number such that $f(a) \neq 0$. To find the series expansion of $h_k(z)/f(z)$ in powers of $z - a$ we set $x = z - a$:

$$\frac{h_k(z)}{f(z)} = \frac{h_k(x+a)}{f(x+a)} = \frac{\alpha_0 + \alpha_1 x + \alpha_2 x^2 + \cdots}{\lambda_0 + \lambda_1 x + \lambda_2 x^2 + \cdots} = \sum_{r=0}^{\infty} c_r x^r;$$

the last equation is obtained by performing the long division. Since $x = z - a$, we have obtained

$$\frac{h_k(z)}{f(z)} = \sum_{r=0}^{\infty} c_r (z-a)^r.$$

4.35 It is immediately apparent from 4.26 that $\varphi_k(z)$ is such a ratio $h_k(z)/f(z)$ with $f(z_k) \neq 0$; the procedure described in 4.34 therefore gives the series expansion of $\varphi_k(D)$ in powers of $D - z_k$:

(1) $$\varphi_k(D) = \sum_{r=0}^{\infty} c_r (D - z_k)^r.$$

Since $\varphi_k(\,)$ is analytic at the point $z = z_k$, the series expansion (1) is a Taylor series; consequently,

$$c_r = \frac{1}{r!}\,\varphi_k^{(r)}(z_k) \qquad (\text{all } r \geq 0). \tag{2}$$

From 4.30 and (2) it follows that the equation

$$\{\varphi_k \mid (D - z_k)^m\} = c_0 + c_1 (D - z_k) + c_2 (D - z_k)^2 + \cdots + c_{m-1} (D - z_k)^{m-1}$$

holds in case $m \geq 1$.

4.36 **Conclusion:** *the operator* $\{\varphi_k \mid (D - z_k)^m\}$ *can be obtained by discarding from the series expansion* (1) *all but the first* m *terms.*

Illustrative example. From 4.32 we see that

$$\frac{(D^3 + 3D)\,D}{(D+1)\,(D+2)\,(D-1)^3} = 0 + \frac{\{\varphi_1 \mid D+1\}\,D}{D+1} + \frac{\{\varphi_2 \mid D+2\}\,D}{D+2} + \frac{\{\varphi_3 \mid (D-1)^3\}\,D}{(D-1)^3},$$

where $\varphi_1(\,)$, $\varphi_2(\,)$, and $\varphi_3(\,)$ are the functions defined by 4.29:

$$\varphi_1(z) = \left[\frac{(D^3 + 3D)\,D}{(D)\,(D+2)\,(D-1)^3}\right]_{D=z} = \frac{(z^3 + 3z)\,z}{(z)\,(z+2)\,(z-1)^3},$$

$$\varphi_2(z) = \left[\frac{(D^3 + 3D)\,D}{(D+1)\,(D)\,(D-1)^3}\right]_{D=z} = \frac{(z^3 + 3z)\,z}{(z+1)\,(z)\,(z-1)^3},$$

$$\varphi_3(z) = \left[\frac{(D^3 + 3D)\,D}{(D+1)\,(D+2)\,(D)}\right]_{D=z} = \frac{(z^3 + 3z)\,z}{(z+1)\,(z+2)\,(z)}.$$

We can use 4.31 to obtain

$$\{\varphi_1 \mid D+1\} = \varphi_1(-1) \quad \text{and} \quad \{\varphi_2 \mid D+2\} = \varphi_2(-2).$$

In view of 4.35, the operator $\{\varphi_3 \mid (D-1)^3\}$ can be obtained by discarding all but the first 3 terms of the series expansion

$$\varphi_3(D) = \frac{(D^3 + 3D)D}{(D+1)\,(D+2)\,(D)} = \sum_{r=0}^{\infty} c_r (D-1)^r.$$

We proceed as in 4.34. To begin, we set $x = D - 1$ in $\varphi_3(D)$, which gives

$$\varphi_3(D) = \frac{4 + 6x + 3x^2 + x^3}{6 + 5x + x^2}.$$

Next, we perform the long division — the work can be shortened by omitting x:

$$\begin{array}{lllll}
4 + 6 + 3 + 1 & & & & \big| \; 6 + 5 + 1 \\
4 + \frac{10}{3} + \frac{2}{3} & & & & \;\; \frac{2}{3} + \frac{4}{9} + \frac{1}{54} \\
\hline
\quad \frac{8}{3} + \frac{7}{3} + 1 & & & & \\
\quad \frac{8}{3} + \frac{20}{9} + \frac{4}{9} & & & & \\
\hline
\qquad \frac{1}{9} + \frac{5}{9} & & & &
\end{array}$$

Since $x = D - 1$, the result is

$$\varphi_3(D) = \frac{2}{3} + \frac{4}{9}(D-1) + \frac{1}{54}(D-1)^2 + \cdots;$$

consequently, we may use 4.36 to conclude that

$$\{\varphi_3 \mid (D-1)^3\} = \frac{2}{3} + \frac{4}{9}(D-1) + \frac{1}{54}(D-1)^2.$$

Repeated Quadratic and Linear Factors

Let h and g be polynomials such that the degree of h does not exceed the degree of g: there always exist distinct numbers z_k such that $g(D)$ satisfies the hypotheses of Decomposition Theorem 4.23. Consequently, 4.23 gives a completely general and systematic procedure for finding the function whose operator is $h(D)/g(D)$. In cases where some of the numbers z_k are not real, it is often advantageous to utilize the forthcoming Decomposition Theorem 4.39.

The technique indicated in 4.15 gives

$$\frac{4(D^2+1)}{D(D^2+b^2)} = \frac{3D}{D^2+b^2} + \frac{1}{D} = \frac{3}{b}\sin bt + t \quad (\text{when} \;\; b = 2),$$

but this technique is not applicable when $b \neq 2$: the answer can be obtained by using 4.23 with $z_1 = ib$ and $z_2 = -ib$ (when b is a real number), but 4.39 is intended for precisely such cases. In general, algebraic techniques such as the ones in 4.9—19 give almost immediate answers — if they are applicable; in contrast, Decomposition Theorems 4.39 and 4.23 are always applicable but require more computations.

Therefore, *it is best to try out algebraic techniques* such as the ones in 4.9—19 *before having recourse to Decomposition Theorems* 4.23 and 4.39.

4.36.1 **Notation.** If a is a number, there exist two real numbers $\dot{a}$ and $\ddot{a}$ such that $a = \dot{a} + \ddot{a}i$ (the real number $\dot{a}$ is the "*real part*", the real number $\ddot{a}$ is the "*imaginary part*" of a). We set

$$\bar{a} \overset{\text{def}}{=} \dot{a} - \ddot{a}i \quad \text{and} \quad Q(a) \overset{\text{def}}{=} (D - \dot{a})^2 + \ddot{a}^2. \tag{4.37}$$

If V is a power of $Q(a)$, there exists an integer $m \geq 1$ such that $V = Q(a)^m$; for any function $\varphi(\,)$ we set

$$[\varphi \mid V] \overset{\text{def}}{=} \sum_{r=0}^{m-1} \frac{2D}{(D - a)^{m-r}} \left[\frac{1}{r!} \left(\frac{\mathrm{d}}{\mathrm{d}z}\right)^r \frac{\varphi(z)}{(z - \bar{a})^m}\right]_{z=a}. \tag{4.38}$$

4.39 **Decomposition Theorem.** *Let $z_0 = 0, z_1, z_2, \ldots, z_n$ be distinct real numbers, and let a_k $(k = n+1, \ldots, n+s)$ be distinct numbers with non-zero imaginary parts. Suppose that g is a real polynomial such that*

$$g(D) = V_0 V_1 \cdots V_n V_{n+1} \cdots V_{n+s}, \tag{4.40}$$

where

$$V_k \text{ is a power of } \begin{cases} D - z_k & \text{if } k = 0, 1, 2, \ldots, n \\ Q(a_k) & \text{if } k \geq n+1. \end{cases}$$

If h is a real polynomial of degree not exceeding the degree of g, then

$$\frac{h(D)}{g(D)} = \frac{\{\varphi_0 \mid DV_0\}}{V_0} + \sum_{k=1}^{n} \frac{\{\varphi_k \mid V_k\}\, D}{V_k} + \sum_{k=n+1}^{n+s} \mathcal{R}[\varphi_k \mid V_k]. \tag{4.41}$$

4.42 *Remarks.* We shall not prove this theorem. The symbol $\mathcal{R}[\varphi_k | V_k]$ denotes the real part of the function whose canonical operator is the operator $[\varphi_k | V_k]$ defined in 4.38. The functions $\varphi_k(\,)$ are defined as in the previous section:

$$\varphi_0(z) = \left[\frac{h(D)}{V_1 V_2 \cdots V_{n+s}}\right]_{D=z} \quad \text{and} \quad \varphi_k(z) = \left[\frac{h(D)}{G_k}\right]_{D=z}, \tag{4.43}$$

where G_k *is the result of replacing* V_k *by* D *in* $g(D)$. Note that the first decomposition formula (4.25) is included in 4.41.

Consider the case where D is not a factor of the denominator (in other words, suppose that $V_k \neq D$ for all values of k):

$$g(D) = V_0 V_1 \cdots V_{n+s} = V_1 \cdots V_{n+s},$$

and $V_0 = (D - z)^0 = 1$; from 4.31 and 4.43 we see that

$$\frac{\{\varphi_0 \mid D V_0\}}{V_0} = \{\varphi_0 \mid D\} = \varphi_0(0) = \left[\frac{h(D)}{g(D)}\right]_{D=0}.$$

Consequently, the equality

$$(4.44) \qquad \boxed{\frac{h(D)}{g(D)} = \frac{h(0)}{g(0)} + \sum_{k=1}^{n} \frac{\{\varphi_k \mid V_k\} D}{V_k} + \sum_{k=n+1}^{n+s} \mathscr{R}[\varphi_k \mid V_k]}$$

obtains whenever $g(0) \neq 0$.

4.45 **Application.** In case $V = Q(a)^2$, Definition [4.38] becomes

$$[\varphi \mid V] = \left[\frac{2D}{(D-a)^2} \frac{\varphi(z)}{(z-\bar{a})^2} + \frac{2D}{D-a}\left(\frac{\mathrm{d}}{\mathrm{d}z}\right)\frac{\varphi(z)}{(z-\bar{a})^2}\right]_{z=a}.$$

Consequently, if $\varphi(z) = 1$, then

$$(1) \qquad [\varphi \mid Q(a)^2] = 2\exp(at)\left[\frac{t}{(a-\bar{a})^2} - \frac{2}{(a-\bar{a})^3}\right] \qquad \text{(by 3.23)}.$$

Note that

$$(2) \qquad \frac{D}{[(D-\alpha)^2 + \beta^2]^2} = \frac{h(D)}{Q(a)^2},$$

where $h(D) = D$ and $Q(a) = (D-\alpha)^2 + \beta^2$. From 4.44 we have

$$(3) \qquad \frac{h(D)}{Q(a)^2} = \left[\frac{D}{Q(a)^2}\right]_{D=0} + \mathscr{R}[\varphi \mid Q(a)^2];$$

but 4.43 gives

$$(4) \qquad \varphi(z) = \left[\frac{D}{D}\right]_{D=z} = 1.$$

From (3), (4), and (1) we see that

$$(5) \qquad \frac{h(D)}{Q(a)^2} = 0 + \mathscr{R}\left\{\exp(\dot{a}t + \ddot{a}ti)\left[-\frac{t}{2\ddot{a}^2} + \frac{1}{2\ddot{a}^3 i}\right]\right\}.$$

Equations (2) and (5) now give

$$\frac{D}{[(D-\dot{a})^2+\ddot{a}^2]^2}=\frac{\exp(\dot{a}\,t)}{2\,\ddot{a}^2}\Big(-t\cos(\ddot{a}t)+\frac{1}{\ddot{a}}\sin(\ddot{a}t)\Big). \tag{4.46}$$

4.47 **Particular case.** If $V=Q(a)$, Definition [4.38] becomes

$$[\varphi\mid Q(a)]=\frac{2D}{D-a}\Big[\frac{\varphi(z)}{z-\bar{a}}\Big]_{z=a}=2\mathrm{e}^{at}\frac{\varphi(a)}{a-\bar{a}};$$

the last equation is from 3.21. Consequently, if $a=\alpha+\beta i$ (for two real numbers α and β), the above equation gives

$$\boxed{[\varphi\mid(D-\alpha)^2+\beta^2]=\mathrm{e}^{\alpha t}\varphi(\alpha+\beta i)\Big(\frac{\exp\beta it}{\beta i}\Big)}\,. \tag{4.48}$$

Illustrative Examples

4.49 Consider the system

$$X''-X+5Y'=t, \tag{1}$$

$$-2X'+Y''-4Y=-2 \tag{2}$$

subject to the conditions

$$X(0-)=X'(0-)=Y(0-)=Y'(0-)=0. \tag{3}$$

We can proceed as in 4.7 to obtain

$$X=\frac{11D^2-4}{D(D^2+1)(D^2+4)}.$$

From 4.41 we see that

$$X=\frac{\{\varphi_0\mid DD\}}{D}+\mathscr{R}[\varphi_1\mid D^2+1]+\mathscr{R}[\varphi_2\mid D^2+4], \tag{4}$$

where

$$\varphi_0(z)=\frac{11z^2-4}{(z^2+1)(z^2+4)}\qquad\text{(by 4.43)}, \tag{5}$$

$$\varphi_1(z)=\frac{11z^2-4}{z(z)(z^2+4)}\qquad\text{(again by 4.43)}, \tag{6}$$

and

$$(7) \qquad \varphi_2(z) = \frac{11z^2 - 4}{z(z^2+1)\,(z)}.$$

Therefore,

$$(8) \qquad \{\varphi_0 \mid DD\} = \varphi_0(0) + \varphi_0'(0)\,D = -1 + 0 \quad \text{(by 4.30 and (5))},$$

and

$$(9) \qquad [\varphi_1 \mid D^2 + 1] = [\varphi_1 \mid (D-0)^2 + 1^2]$$

$$= e^0 \varphi_1(0+i)\left(\frac{\exp it}{i}\right) \qquad \text{(by 4.48)}$$

$$= 5\left(\frac{e^{it}}{i}\right) \qquad \text{(by (6))}.$$

Further, the equations

$$[\varphi_2 \mid D^2 + 4] = [\varphi_2 \mid (D-0)^2 + 2^2] = e^0 \varphi_2(0+2i)\left(\frac{\exp 2it}{2i}\right)$$

again come from 4.48, and therefore

$$(10) \qquad [\varphi_2 \mid D^2 + 4] = -4\left(\frac{e^{2it}}{2i}\right) \qquad \text{(by (7))}.$$

Combining (8)—(10) with (4), we obtain

$$X = -t + 5 \sin t - 2 \sin 2t.$$

4.50 **A second example.** Let us find the function $y(\,)$ such that

$$y = \frac{3D}{(D^2 - 2D + 5)\,(D^2 + 9)}.$$

From 4.44 we see that

$$(1) \quad y = \frac{0}{(0+5)\,(0+9)} + \mathscr{R}[\varphi_1 \mid D^2 - 2D + 5] + \mathscr{R}[\varphi_2 \mid D^2 + 9],$$

where

$$(2) \qquad \varphi_1(z) = \frac{3z}{(z)\,(z^2+9)} \quad \text{and} \quad \varphi_2(z) = \frac{3z}{(z^2 - 2z + 5)\,(z)} \qquad \text{(see 4.43)}.$$

Therefore, we can use 4.48 to obtain

$$(3) \quad [\varphi_1 \mid D^2 - 2D + 5] = [\varphi_1 \mid (D-1)^2 + 2^2] = e^t \varphi_1(1+2i)\left(\frac{\exp 2it}{2i}\right)$$

and

(4) $$[\varphi_2 \mid D^2+9] = [\varphi_2 \mid (D-0)^2+3^2] = e^0 \varphi_2(0+3i)\left(\frac{\exp 3it}{3i}\right).$$

Substituting (2)—(4) into (1), we employ a short calculation to obtain

$$y = \frac{3e^t}{52}[-2\cos 2t + 3\sin 2t] + \frac{3\cos 3t - 2\sin 3t}{26}.$$

4.51 A differential equation. Let us solve the initial-value problem

(1) $$y(0-) = 0,\ y'(0-) = 1, \quad \text{and} \quad y'' - y' - 6y = \cos 2t.$$

From (1), 3.9—11, and 3.28 it follows that

$$D^2y - Dy - 6y = \frac{D^2}{D^2+4} + D = \frac{D^3+D^2+4D}{D^2+4}.$$

Solving for y:

$$y = \frac{D^3+D^2+4D}{(D-3)(D+2)(D^2+4)}.$$

From 4.44 we see that

(2) $$y = \frac{0}{(0-3)(0+2)(0+4)} + \frac{\{\varphi_1 \mid D-3\}\, D}{D-3}$$

$$+ \frac{\{\varphi_2 \mid D+2\}\, D}{D+2} + \mathscr{R}[\varphi_3 \mid D^2+4],$$

where

(3) $$\varphi_1(z) = \frac{z^3+z^2+4z}{(z)(z+2)(z^2+4)}, \quad \varphi_2(z) = \frac{z^3+z^2+4z}{(z-3)(z)(z^2+4)},$$

and

(4) $$\varphi_3(z) = \frac{z^3+z^2+4z}{(z-3)(z+2)(z)}.$$

From 4.31 and (3) it now follows that

(5) $$\{\varphi_1 \mid D-3\} = \varphi_1(3) = \frac{16}{65} \quad \text{and} \quad \{\varphi_2 \mid D+2\} = \varphi_2(-2) = -\frac{3}{20}.$$

From 4.48 and (4) we get

$$(6) \quad [\varphi_3 \mid D^2 + 4] = [\varphi_3 \mid (D - 0)^2 + 2^2]$$

$$= e^0 \varphi_3(0 + 2i) \left(\frac{\exp 2it}{2i}\right) = \frac{1}{52}(-5 + i)\, e^{2it}.$$

Substituting (5) and (6) into (2), we obtain

$$y = \frac{16}{65} e^{3t} - \frac{3}{20} e^{-2t} - \frac{5 \cos 2t + \sin 2t}{52}.$$

Exercises

4.52.0 Solve for $Y(\,)$ the system (1)—(3) in 4.49.

Answer: $Y(t) = 1 - 2 \cos t + \cos 2t$.

4.52.1 Proceed as in 3.34 to solve the following classical problems:

(.2) $y(0-) = -1,\ y'(0-) = 2$ with $y'' + 2y' + y = t$.

Answer: $t - 2 + 2te^{-t} + e^{-t}$.

(.3) $y(0-) = 1,\ y'(0-) = -1$ with $y'' - 3y' + 2y = 4t + e^{3t}$.

Answer: $2t + 3 + 2^{-1}e^{3t} - 2e^{2t} - 2^{-1}e^{t}$.

(.4) $y(0-) = y'(0-) = 0$ with $y'' - 2y' + 5y = \sin 3t$.

Answer: $\frac{6}{52} \cos 3t - \frac{4}{52} \sin 3t - \frac{6}{52} e^t \cos 2t + \frac{9}{52} e^t \sin 2t$.

(.5) $y(0-) = A,\ y'(0-) = B$ with $y'' + y' = \cos t$.

Answer: $(A + B) + e^{-t}[2^{-1} - B] + 2^{-1}(\sin t - \cos t)$.

4.52.6 Find a $\mathcal{K}$-function $y(\,)$ such that $y(0-) = 1$ and

$$y'(t) + 2y(t) + \int_0^t y(u)\, du = \sin t \qquad \text{(all } t > 0).$$

Answer: $e^{-t} - (3/2)\, te^{-t} + 2^{-1} \sin t$.

4.52.7 Given a $\mathcal{K}$-function $G(\,)$ having no jumps on $[0,\infty)$, such that $G'(\,)$ is a $\mathcal{K}$-function, and such that $G(0-)=0$. Find a $\mathcal{K}$-function $y(\,)$ satisfying the equation

$$\int_0^t (e^{-u} - \sin u)\, y(t-u)\, du = G(t) \qquad \text{(all } t > 0).$$

Answer: $G' + 2G - \int_0^t G(u)\, du + 4 \int_0^t e^{t-u} G(u)\, du.$

4.52.8 Find the $\mathcal{K}$-function $y(\,)$ such that

$$\int_0^t (t-u) \cos(t-u)\, y(u)\, du = t^2/2.$$

Hint: use 4.52.10.

Answer: $-t^2/2 - 3 + 2(e^t + e^{-t})$.

4.52.9 Find the two $\mathcal{K}$-functions $f(\,)$ and $g(\,)$ such that

$$f = \frac{D+2}{(D+1)\,D^2} \quad \text{and} \quad g = \frac{D}{(D-2)^3\,(D+5)\,(D+7)}.$$

Answers: $f = t^2 - t + 1 - e^{-t}$,

$$g = (-7^{-3})\, 2^{-1} e^{-5t} + (9^{-3})\, 2^{-1} e^{-7t}$$
$$+ \frac{1}{63}\,(e^{2t})\, \frac{t^2}{2} - \frac{16}{7^2 9^2}\,(e^{2t})\, t + \frac{193}{7^3 9^3}\,(e^{2t}).$$

4.52.10 Derive the equation

$$\frac{(D^2-\beta^2)\,D}{(D^2+\beta^2)^2} = \langle t \cos \beta t \rangle.$$

Hint: apply D to both sides of 3.26 and use 3.8.0.

4.52.11 Proceed as in 3.34 to solve the equation

$$y^{(3)} - y' = 1$$

subject to the conditions

$$y(0-) = 1, \qquad y'(0-) = y''(0-) = 0.$$

Answer: $y = 1 - t + (e^t + e^{-t})/2$.

4.52.12 Proceeding as in 3.34, solve for x the system

$$x' - x - 2y = t,$$
$$-2x + y' - y = t$$

subject to the conditions: $y(0-) = 4, \quad x(0-) = 2.$

Answer: $x = (28/9)\,e^{3t} - e^{-t} - (1/9) - (t/3).$

4.52.13 Find a $\mathcal{K}$-function $y()$ having no jumps on $[0, \infty)$, such that $y'()$ is a $\mathcal{K}$-function, and such that

$$y(t) = \cos t + 2 \int_0^t \sin(t-u)\, y'(u)\, du \qquad \text{(all } t > 0).$$

Hint: use 3.8.3.

Answer: $y = e^t + (1 - 2a)\, t e^t$, where $a = y(0-)$.

4.52.14 Solve the equation

$$\frac{\partial^2}{\partial t^2} y + 2 \frac{\partial}{\partial t} y + y = t$$

subject to the conditions: $y(0-) = 1, \quad y'(0-) = 2.$

Answer: $y = t - 2 + 4te^{-t} + 3e^{-t}.$

Chapter 3

§ 5. Further Applications

This section is a direct continuation of § 3 (Elementary applications); it consists mostly of worked-out examples of the sort of problems that usually illustrate Laplace transform techniques.

We shall begin with a review of the translation property and some of its consequences. Recall that

$$\mathsf{T}_\alpha(\tau) = \begin{cases} 0 & (\tau \leq \alpha) \\ 1 & (\tau > \alpha). \end{cases} \tag{5.0}$$

Until further notice, suppose that $\alpha \geq 0$; we have $\mathsf{T}_\alpha = \langle \mathsf{T}_\alpha(t) \rangle$ (the canonical operator of the function $\mathsf{T}_\alpha(\,)$), and the equation

$$\mathsf{T}_\alpha \cdot \varphi(\tau) = \varphi(\tau - \alpha) \qquad (-\infty < \tau < \infty) \tag{5.1}$$

holds for any test-function $\varphi(\,)$: see 2.13. Further,

$$\boxed{\mathsf{T}_\alpha \mathsf{T}_\lambda = \mathsf{T}_{\alpha+\lambda}} \qquad (\alpha \geq 0 \text{ and } \lambda \geq 0\text{: see 2.14}); \tag{5.2}$$

consequently,

$$\mathsf{T}_\lambda^k = \mathsf{T}_{k\lambda} \qquad (\text{for } k = 1, 2, 3, \ldots). \tag{5.3}$$

Equation 5.3 holds for $k = 0$:

$$\boxed{\mathsf{T}_\lambda^0 \overset{\text{def}}{=} 1 = \mathsf{T}_0} \qquad (\text{see 2.17}). \tag{5.4}$$

Recall that

$$V1 = V = 1V \qquad \text{(for any operator } V\text{: see 2.18).} \tag{5.5}$$

If $h(\,)$ is a $\mathcal{K}$-function [1.36], then

$$\boxed{\mathsf{T}_\alpha\langle h(t)\rangle = \langle \mathsf{T}_\alpha(t)\, h(t-\alpha)\rangle} \qquad \text{(see 3.12).} \tag{5.6}$$

Since the canonical operator $\langle h(t)\rangle$ of the function $h(\,)$ is also denoted by h (see [2.1]), we can also write 5.6 as follows:

$$\mathsf{T}_\alpha h = \langle \mathsf{T}_\alpha(t)\, h(t-\alpha)\rangle.$$

The graph of the function $\{\mathsf{T}_\pi(t)\sin(t-\pi)\}(\,)$ has the following shape:

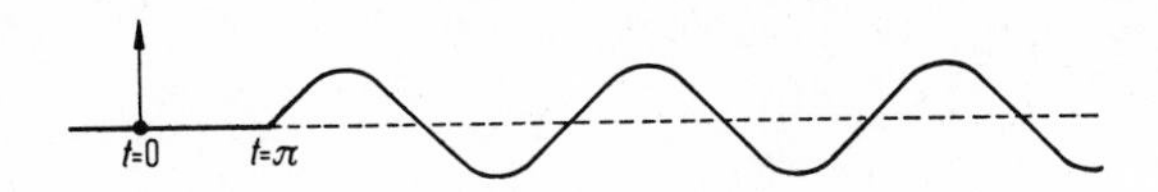

Clearly,

$$\{\sin t\}(\tau) + \mathsf{T}_\pi(\tau)\sin(\tau-\pi) = \begin{cases} \sin\tau & (0<\tau<\pi) \\ 0 & \text{(otherwise)}, \end{cases}$$

so that 5.6 gives

$$\langle \sin t\rangle + \mathsf{T}_\pi\langle \sin t\rangle = f, \tag{5.7}$$

where

$$f(\tau) = \begin{cases} \sin\tau & (0<\tau<\pi) \\ 0 & \text{(otherwise)}. \end{cases} \tag{5.8}$$

Equation 5.7 can also be written

$$f = (1+\mathsf{T}_\pi)\langle \sin t\rangle = \frac{(1+\mathsf{T}_\pi)\,D}{1+D^2} \qquad \text{(by 3.27).}$$

In other words:

(5.9)
$$\frac{1+\mathsf{T}_\pi}{1+D^2}\,D = \begin{cases}\sin t & (0 < t < \pi)\\ 0 & \text{(otherwise)};\end{cases}$$

the angular brackets have been omitted from the right-hand side.

5.10 **Theorem.** *Let $F(\,)$ be a $\mathcal{K}$-function whose derivative $F'(\,)$ is also a $\mathcal{K}$-function. If $F(\,)$ is continuous on the interval* $(0,\infty)$, *except possibly at $t = 0$ and at $t = \alpha$, then*

(5.11)
$$F = D^{-1}F' + F(0+) + [F(\alpha+) - F(\alpha-)]\,\mathsf{T}_\alpha.$$

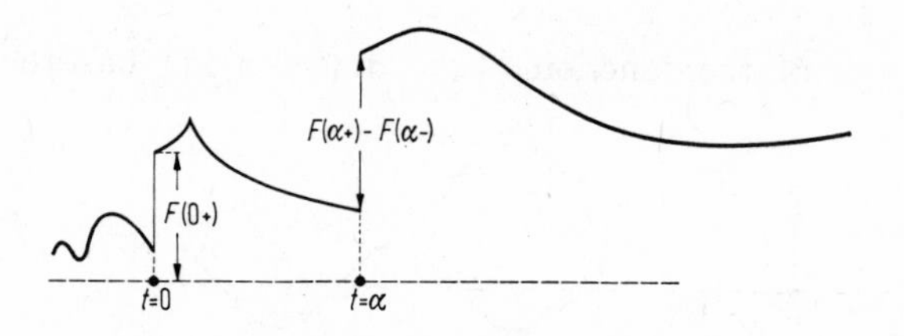

Proof. When $0 < \tau < \alpha$ the equation

(1)
$$\int_0^\tau F'(u)\,du = F(\tau) - F(0+)$$

is an immediate consequence of the Fundamental Theorem of Calculus (see the proof of 1.42). Since $\mathsf{T}_\alpha(\tau) = 0$ when $\tau \le \alpha$, Equation (1) gives

(2)
$$\int_0^\tau F'(u)\,du$$
$$= F(\tau) - F(0+) + [F(\alpha+) - F(\alpha-)]\,\mathsf{T}_\alpha(t)$$
(in case $0 < \tau < \alpha$).

Next, consider the case $\tau > \alpha$:

(3)
$$\int_0^\tau F'(u)\,du = \int_\alpha^\tau F'(u)\,du + \int_0^\alpha F'(u)\,du$$

(4)
$$= F(\tau) - F(\alpha+) + F(\alpha-) - F(0+)$$

(5)
$$= F(\tau) - [F(\alpha+) - F(\alpha-)]\,\mathsf{T}_\alpha(\tau) - F(0+):$$

the second equation is a consequence of the Fundamental Theorem of Calculus (note that $F(\,)$ is continuous inside the interval $(0, \alpha)$); the last equation comes from the fact that $\mathsf{T}_\alpha(\tau) = 1$ when $\tau > \alpha$. Combining (2) and (3)—(5):

$$\text{(6)} \qquad \int_0^\tau F'(u)\, du$$

$$= F(\tau) - [F(\alpha+) - F(\alpha-)]\, \mathsf{T}_\alpha(\tau) - F(0+)$$

$$(\text{all } \tau > 0).$$

From 2.32 it follows that

$$D^{-1}F' = \left\langle \int_0^t F'(u)\, du \right\rangle = F - [F(\alpha+) - F(\alpha-)]\, \mathsf{T}_\alpha - F(0+):$$

the second equation is from (6) and 2.66. This gives 5.11.

5.12.0 *Remarks*. In case $F(\,)$ is continuous at $t = \alpha$ and has no jump at $t = 0$, then $F(0-) = F(0+)$ and 5.11 becomes

$$F = D^{-1}F' + F(0-),$$

which implies the important formula 2.25 (which we called the Derivation Property). Formula 5.11 can easily be generalized as follows:

$$F = D^{-1}F' + F(0+) + \sum_{k=1}^{n} [F(\alpha_k+) - F(\alpha_k-)]\, \mathsf{T}_0(t - \alpha_k),$$

where α_k $(k = 1, 2, 3, \ldots, n)$ are the discontinuity points of the function $F(\,)$.

5.12.1 **The road ahead.** We shall illustrate how 5.11 can be used to find certain elementary operators. Our basic building block is the function $c(\mathsf{T}_\alpha - \mathsf{T}_\beta)(\,)$ defined by

$$\text{(5.13)} \qquad c(\mathsf{T}_\alpha - \mathsf{T}_\beta)(\tau) = \begin{cases} c & (\alpha < \tau \leq \beta) \\ 0 & (\text{otherwise}). \end{cases}$$

In case $c > 0$, $m > 0$, and $0 < \beta < a < b$ the graph of

$$c(1 - \mathsf{T}_\beta)() - m(\mathsf{T}_a - \mathsf{T}_b)()$$

has the following shape:

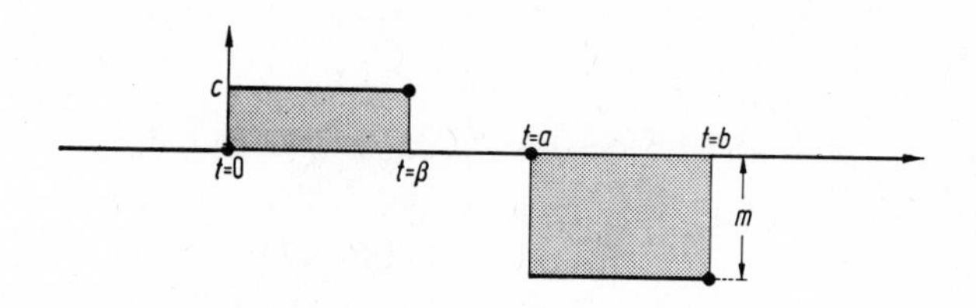

5.14 **For example,** consider the functions $F_\lambda()$ and $G_\lambda()$ defined by the following graphs:

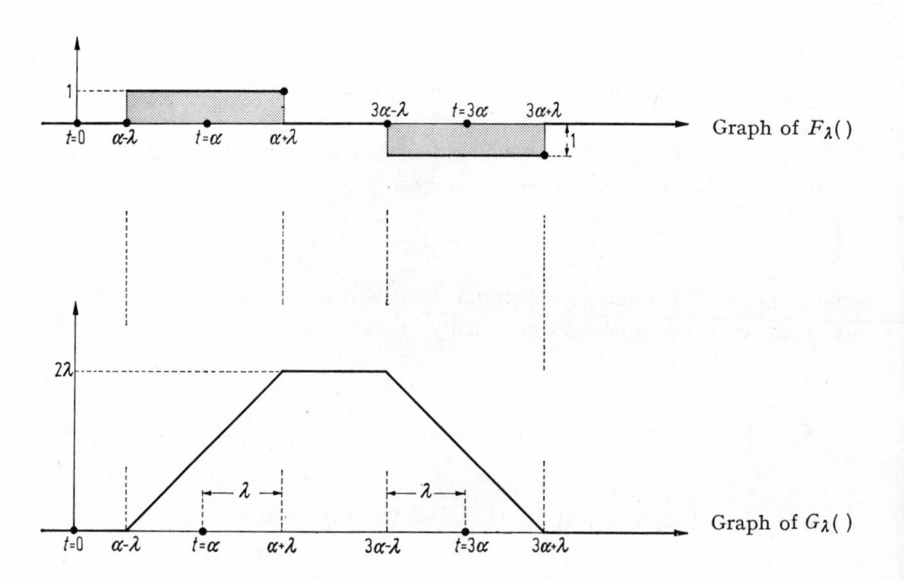

We suppose that $0 \leq \lambda \leq \alpha$; from the above graphs it is clear that

$$F_\lambda = \mathsf{T}_{\alpha-\lambda} - \mathsf{T}_{\alpha+\lambda} - [\mathsf{T}_{3\alpha-\lambda} - \mathsf{T}_{3\alpha+\lambda}], \tag{5.15}$$

and $G'_\lambda() = F_\lambda()$; we may therefore use 5.11 to conclude that

$$G_\lambda = D^{-1}(\mathsf{T}_{\alpha-\lambda} - \mathsf{T}_{\alpha+\lambda} - \mathsf{T}_{3\alpha-\lambda} + \mathsf{T}_{3\alpha+\lambda}). \tag{5.16}$$

In particular, for $\lambda = \alpha$ we have

(5.17) $$G_\alpha = D^{-1}(1 - \mathsf{T}_{2\alpha})^2 \qquad \text{(by 5.16 and 5.3)}.$$

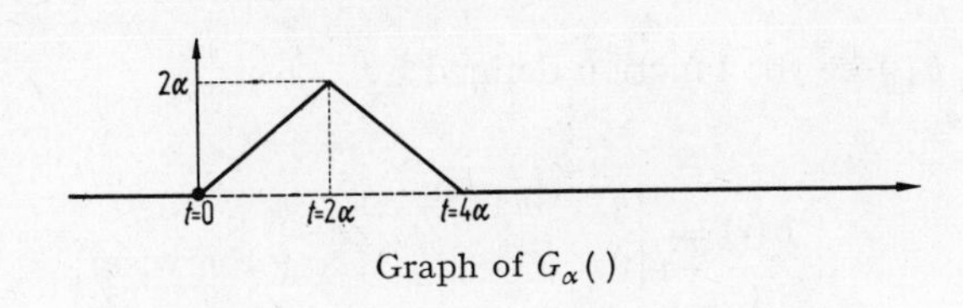

Graph of $G_\alpha(\,)$

5.18 Let $Y(\,)$ be the function described by the left-hand graph.

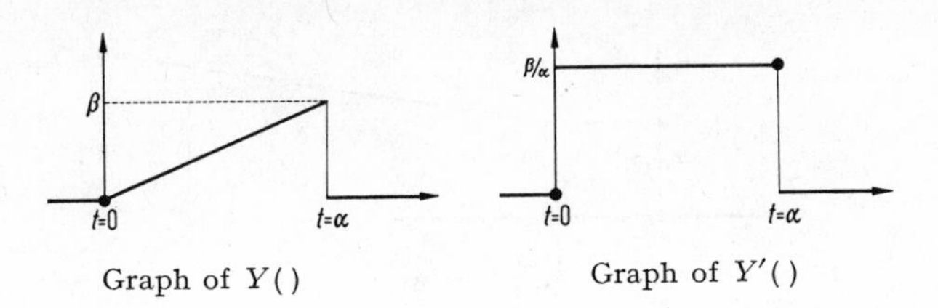

Graph of $Y(\,)$ Graph of $Y'(\,)$

Since $Y' = (\beta/\alpha)\,(1 - \mathsf{T}_\alpha)$ and $Y(\alpha+) - Y(\alpha-) = \beta$, we may use 5.11 with $F = Y$ to obtain

(5.19) $$Y = \frac{\beta}{\alpha} D^{-1}(1 - \mathsf{T}_\alpha) - \beta\,\mathsf{T}_\alpha.$$

The function $F(\,)$ defined by the graph

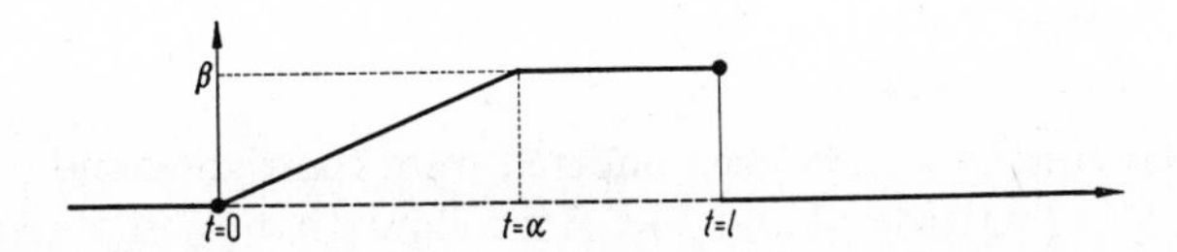

is such that $F' = (\beta/\alpha)\,(1 - \mathsf{T}_\alpha)$ and $F(l+) - F(l-) = -\beta$; Formula 5.11 gives

(5.20) $$F = \frac{\beta}{\alpha} D^{-1}(1 - \mathsf{T}_\alpha) - \beta\,\mathsf{T}_l.$$

The function $g(\,)$ described by the graph

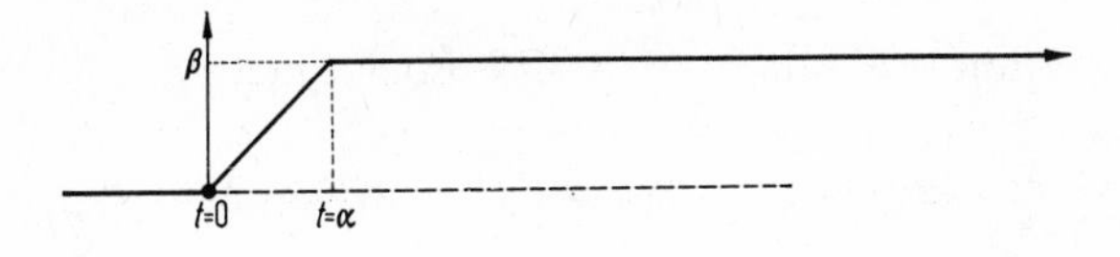

is such that $g' = (\beta/\alpha)(1 - \mathsf{T}_\alpha) = F'$; Formula 5.11 again gives

$$g = \frac{\beta}{\alpha} D^{-1}(1 - \mathsf{T}_\alpha) \qquad \text{(for } \alpha > 0\text{)}. \tag{5.21}$$

Finally, let $h(\,)$ be the function defined by

$$h(\tau) = \begin{cases} b + m(\tau - \alpha) & (\tau > \alpha) \\ 0 & \text{(otherwise)}; \end{cases}$$

in case $b > 0$ and $m > 0$ its graph has the following shape:

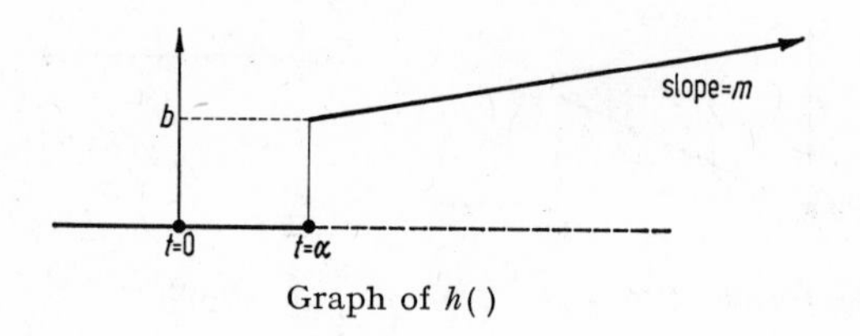

Graph of $h(\,)$

Clearly, $h' = m\mathsf{T}_\alpha$ and $h(\alpha+) - h(\alpha-) = b$; Formula 5.11 gives

$$h = mD^{-1}\mathsf{T}_\alpha + b\mathsf{T}_\alpha \qquad \text{(when } \alpha \geq 0\text{)}. \tag{5.22}$$

5.23 **Application.** Let us solve the initial-value problem

(1) $y(0-) = 1,\ y'(0-) = 0 \quad \text{with} \quad \frac{\partial^2}{\partial t^2} y + 4y = \begin{cases} 4t & (0 < t \leq 1) \\ 4 & (t > 1): \end{cases}$

the angular brackets have been omitted from the right-hand side. From [3.9] and 5.21 (with $\alpha = 1$ and $\beta = 4$) it follows that (1) implies the equation

(2) $$(D^2 + 4)\, y = D^2 + 4(1 - \mathsf{T}_1)\, D^{-1}.$$

Solving for y, we obtain

$$y = \frac{D^2}{D^2 + 4} + \frac{4(1 - \mathsf{T}_1)}{D(D^2 + 4)} = \frac{D^2}{D^2 + 4} + (1 - \mathsf{T}_1)\left(\frac{-D}{D^2 + 4} + \frac{1}{D}\right);$$

the last equation is obtained as in 4.17. Consequently,

$$y = \langle \cos 2t \rangle + (1 - \mathsf{T}_1)\langle -2^{-1} \sin 2t + t \rangle;$$

which, by 5.6, gives

(3) $y = \cos 2t + t - 2^{-1} \sin 2t - \mathsf{T}_1(t)\,[(t-1) - 2^{-1} \sin 2(t-1)]$.

Reversing our steps, we see that the equation

(4) $y(t) = \cos 2t + t - 2^{-1} \sin 2t - \mathsf{T}_1(t)\,[(t-1) - 2^{-1} \sin 2(t-1)]$

$(-\infty < t < \infty)$ implies $y(0-) = 1$, $y'(0-) = 0$, and (3); since (3) $\Rightarrow$ (2) and (2) $\Rightarrow$ (1) when $y(0-) = 1$ and $y'(0-) = 0$, it follows that (4) $\Rightarrow$ (1). We have solved our problem; however, a stronger assertion is readily obtained by noting that neither $y(\,)$ (as defined by (4)) nor $y'(\,)$ have jumps on $[0, \infty)$; since $y''(\,)$ is continuous we may infer from 2.55 and (1) that (4) implies

$$y(0+) = 1,\ \ y'(0+) = 0 \quad \text{with} \quad y''(t) + 4y(t) = \begin{cases} 4t & (0 < t \leq 1) \\ 4 & (t > 1). \end{cases}$$

5.24.0 **Procedure.** To solve a starting-value problem of the form

(5) $y(0+) = c_0,\ y'(0+) = c_1 \quad \text{with} \quad y'' + ay' + by = f,$

find the solution of the more general problem

$$y(0-) = c_0,\ y'(0-) = c_1 \quad \text{with} \quad \frac{\partial^2 y}{\partial t} + a\,\frac{\partial y}{\partial t} + by = f$$

and verify that it satisfies (5).

Exercises

5.24.1 Proceed as in 5.24.0 to solve the following starting-value problems.

(.2) $y(0+) = y'(0+) = 0 \quad \text{with} \quad y'' + y = \begin{cases} 4 & (0 < t \leq 2) \\ t + 2 & (t > 2). \end{cases}$

Answer: $4(1 - \cos t) + \mathsf{T}_2(t)\,[(t-2) - \sin(t-2)]$.

(.3) $y(0+) = 1,\ y'(0+) = 0 \quad \text{with} \quad y'' + y = \begin{cases} 3 & (0 < t \leq 4) \\ 2t - 5 & (t > 4). \end{cases}$

Answer: $3 - 2\cos t + 2\,\mathsf{T}_4(t)\,[t - 4 - \sin(t-4)]$.

$$(.4) \qquad y(0+) = y'(0+) = 0 \quad \text{with} \quad y'' + y = \begin{cases} 1 & (0 < t \leq 1) \\ -1 & (1 < t \leq 2) \\ 0 & (t > 2). \end{cases}$$

Hint: the graph of the right-hand side has the following shape:

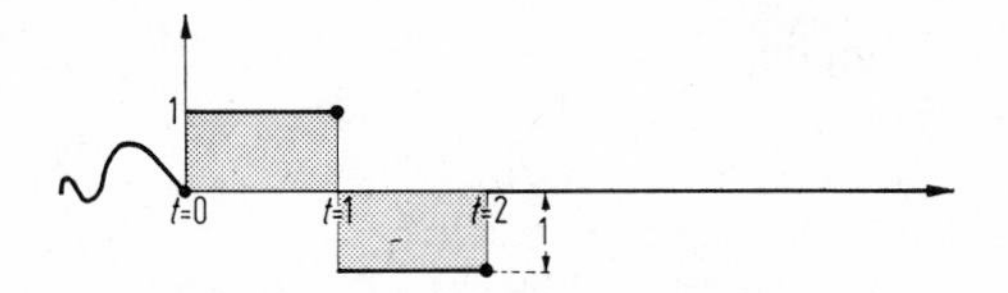

use 3.30 and observe that $1 - \cos t = 2\sin^2\left(\frac{t}{2}\right)$.

Answer: $2\sin^2\left(\frac{t}{2}\right) - 4\,\mathsf{T}_1(t)\sin^2\left(\frac{t-1}{2}\right) + 2\,\mathsf{T}_2(t)\sin^2\left(\frac{t-2}{2}\right)$.

5.24.5 Given $\alpha > 0$, find $\langle |t - \alpha| \rangle$; that is, find the canonical operator of the function $F(\ )$ defined by $F(t) = |t - \alpha|$.

Hint: use 5.11 and observe that $F' = -1 + 2\,\mathsf{T}_\alpha$.

Answer: $\langle |t - \alpha| \rangle = \alpha + \dfrac{2\,\mathsf{T}_\alpha - 1}{D}$.

Series of Translates

Suppose that $f(\ )$ is a $\mathcal{K}$-function such that $f(t) = 0$ for $t \leq 0$. If $\alpha \geq 0$ then

$$(5.24.6) \qquad f(t - \alpha) = 0 \quad \text{for} \quad t > \alpha,$$

so that

$$f(t - \alpha) = \mathsf{T}_\alpha(t)\, f(t - \alpha) \qquad (-\infty < t < \infty);$$

consequently, 5.6 implies that

$$(5.25) \qquad \boxed{\mathsf{T}_\alpha f = \langle f(t - \alpha) \rangle} \qquad (\text{if } \alpha \geq 0).$$

5.26 **Theorem.** *Suppose that* $x \geq 0$ *and* $a > 0$; *further, let c be a number. If* $f(\,)$ *is a* $\mathcal{K}$*-function such that* $f(t) = 0$ *for* $t \leq 0$, *and such that* $|f(\tau)| < \infty$ *for* $\tau \geq 0$, *then*

$$\boxed{\frac{\mathsf{T}_x f}{1 - c\,\mathsf{T}_a} = \left\langle \sum_{k=0}^{\infty} c^k f(t - ka - x) \right\rangle}. \tag{5.27}$$

Proof: see 5.27.7.

5.27.1 **The road ahead.** The above is the only important result that remains to be proved in this chapter; some of its applications are given in 5.28—38. The other sections (5.27.2—27.7) may be omitted on a first reading; in fact, the reader may prefer to study the less tiresome proof of the more informative theorem 11.58.1.

5.27.2 **Lemma.** Let $y(\,)$ be the function defined by

$$y(t) = \sum_{k=0}^{\infty} c^k f(t - ka - x) \qquad (-\infty < t < \infty); \tag{1}$$

from [2.1] it follows that

$$y = \langle y(t) \rangle = \left\langle \sum_{k=0}^{\infty} c^k f(t - ka - x) \right\rangle. \tag{2}$$

In fact, y is a perfect operator; this is because $y(\,)$ is a $\mathcal{K}$-function (see 5.27.3) and because of 1.39—40.

5.27.3 **Lemma.** To verify that $y(\,)$ is a $\mathcal{K}$-function, let n be any integer: obviously,

$$y(t) = \sum_{k=0}^{n-1} c^k f(t - ka - x) + \sum_{k=n}^{\infty} c^k f(t - ka - x). \tag{3}$$

If $t < na + x$, then $t < ka + x$ in the second summation (since $k \geq n$ there): consequently, $t - ka - x < 0$ in the second summation; our hypothesis ($f(\tau) = 0$ for $\tau \leq 0$) therefore gives

$$f(t - ka - k) = 0 \qquad \text{(when } t < na + x) \tag{4}$$

for all the values of k in the second summation. Consequently, every term in the second summation equals zero:

$$y(t) = \sum_{k=0}^{n-1} c^k f(t - ka - x) \qquad (\text{when } t < na + x). \tag{5.27.4}$$

Thus, in any interval with right end-point $t = na + x$, the function $y(\,)$ is the sum of finitely-many $\mathcal{K}$-functions: it is therefore also a $\mathcal{K}$-function.

5.27.5 **Lemma.** Setting $t = \tau - a$ in (1), we obtain

$$c y(\tau - a) = \sum_{k=0}^{\infty} c^{k+1} f(\tau - a - ka - x),$$

whence the equation

$$c y(\tau - a) = \sum_{m=1}^{\infty} c^m f(\tau - ma - x) \tag{5}$$

now follows by setting $m = k + 1$. From (1) we see that

$$y(\tau) = f(\tau - x) + \sum_{k=1}^{\infty} c^k f(\tau - ka - x). \tag{6}$$

Subtracting (5) from (6):

$$y(\tau) - c y(\tau - a) = f(\tau - x) \qquad (-\infty < \tau < \infty);$$

consequently,

$$\langle y(t)\rangle - c\langle y(t-a)\rangle = \langle f(t-x)\rangle \qquad (\text{by } 2.4\text{—}5),$$

so that

$$y - c\,\mathsf{T}_a y = \mathsf{T}_x f \qquad (\text{by } 5.25). \tag{7}$$

The equation (7) implies that

$$(1 - c\mathsf{T}_a)\, y = \mathsf{T}_x f \qquad (\text{by } 5.5). \tag{8}$$

5.27.6 **Lemma.** In case $f(\,)$ is the unit step function $\mathsf{T}_0(\,)$, we have $f = 1$ (by 5.4); Equation (8) becomes

$$(1 - c\mathsf{T}_a)\, y = \mathsf{T}_x.$$

In case $x = 0$, this gives

$$(1 - c\mathsf{T}_a) = 1 \qquad \text{(see 5.4)}. \tag{9}$$

Since y is a perfect operator (by 5.27.2), it now follows from (9) and [1.65] that $1 - c\,\mathsf{T}_a$ is an invertible operator.

5.27.7 *Proof of* 5.26. Since the operator $1 - c\,\mathsf{T}_a$ is invertible, Theorem 1.77 enables us to infer from (8) that

$$y = \frac{\mathsf{T}_x f}{1 - c\mathsf{T}_a};$$

in view of (2), we have proved 5.26.

5.28 **Applications.** Setting $a = \lambda$, $c = b$, $x = 0$, and $f(\,) = \mathsf{T}_0(\,)$ in 5.27, we obtain

$$\frac{1}{1 - b\,\mathsf{T}_\lambda} = \sum_{k=0}^{\infty} b^k \mathsf{T}_0(t - k\lambda). \tag{10}$$

In case $b > 0$, the graph of the function

$$\left\{\sum_{k=0}^{\infty} b^k\, \mathsf{T}_0(t - k\lambda)\right\}(\,)$$

has the following shape:

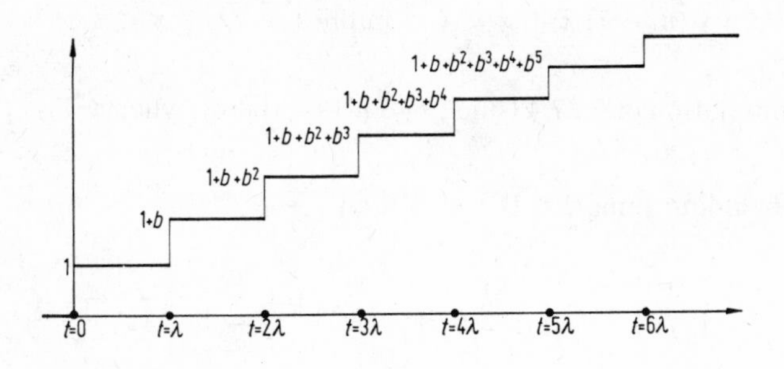

5.29 **Theorem.** *If $h(\,)$ is a $\mathcal{K}$-function, then the equation*

$$\boxed{\frac{\mathsf{T}_x}{1 - c\,\mathsf{T}_\lambda}\langle h(t)\rangle = \left\langle \sum_{k=0}^{\infty} \mathsf{T}_{k\lambda+x}(t)\, c^k\, h(t - k\lambda - x)\right\rangle} \tag{5.30}$$

holds for any numbers c, $x \geq 0$, and $\lambda > 0$. Moreover, if

$$(n-1)\,\lambda + x < t < n\lambda + x \qquad (n = 1, 2, 3, \ldots),$$

then

$$\sum_{k=0}^{\infty} \mathsf{T}_{k\lambda+x}(t)\, c^k\, h(t - k\lambda - x) = \sum_{k=0}^{n-1} c^k h(t - k\lambda - x). \tag{5.30.1}$$

Proof. Let $f(\,)$ be the function defined by

$$f(\tau) = \mathsf{T}_0(\tau)\, h(\tau)$$

whenever τ is a continuity-point of $h(\,)$, and by $f(\tau) = 0$ otherwise. This function $f(\,)$ satisfies the hypothesis of 5.26 and is such that

$$f(\,) = \{h(t)\}(\,) \qquad \text{(see [2.0])};$$

consequently, $f = \langle h(t)\rangle$; substituting into 5.27, we obtain 5.30 by noting that

$$\mathsf{T}_0(t - k\lambda - x) = \mathsf{T}_{k\lambda+x}(t). \tag{11}$$

On the other hand, the conclusion 5.30.1 comes from 5.27.4 and by observing that our hypothesis

$$(n-1)\,\lambda + x < t \quad \text{implies} \quad k\lambda + x < t$$

in the summation of 5.27.4 (since $k \leq n - 1$ there), whence $\mathsf{T}_{k\lambda+x}(t) = 1$.

5.31 **Concluding remarks.** If $b \neq 0$ then

$$\frac{1}{\mathsf{T}_\lambda - b} = \frac{1}{-b\,(1 - b^{-1}\mathsf{T}_\lambda)} = -\,b^{-1}\sum_{k=0}^{\infty} b^{-k}\,\mathsf{T}_{k\lambda}(t):$$

the last equation is from 5.29. On the other hand, 2.19 gives

$$\frac{1}{\mathsf{T}_\lambda - b} = [\![\mathsf{T}_{-\lambda}]\!] \qquad \text{(if } b = 0\text{)}.$$

Consequently, **the operator $\mathsf{T}_\lambda - b$ is invertible for any number** b (whether b be a complex or a real number).

Further theorems and additional facts are given in 11.9—59.

Periodic Functions

5.32 Let $G(\,)$ be a $\mathcal{K}$-function vanishing on the negative axis and having period $= \lambda > 0$; that is, $G(\tau + \lambda) = G(\tau)$ for all $\tau > 0$. It will be convenient to denote by $G^\lambda(\,)$ the function defined by

$$G^\lambda(\tau) = \begin{cases} G(\tau) & (0 < \tau \leq \lambda) \\ 0 & \text{(otherwise)}; \end{cases} \tag{5.33}$$

from our hypothesis $G(x) = 0$ for $x \leq 0$ it follows easily that

$$G^\lambda(t) = G(t) - G(t - \lambda) \qquad (-\infty < t < \infty),$$

so that

$$G^\lambda = G - \langle G(t - \lambda)\rangle,$$

and 5.25 now gives

$$G^\lambda = G - \mathsf{T}_\lambda G:$$

solving for G, we may use 5.31 to obtain

$$\boxed{G = \frac{G^\lambda}{1 - \mathsf{T}_\lambda}} \qquad \big(\text{when } \lambda = \text{the period of } G(\,)\big). \tag{5.34}$$

For example, let G_0 be the $\mathcal{K}$-function of period $= 2\pi$ such that

$$G_0^\lambda(\tau) = \begin{cases} \sin\tau & (0 < \tau \leq \pi) \\ 0 & \text{(otherwise)}; \end{cases}$$

it is often called the *half-wave rectification* of the sine function.

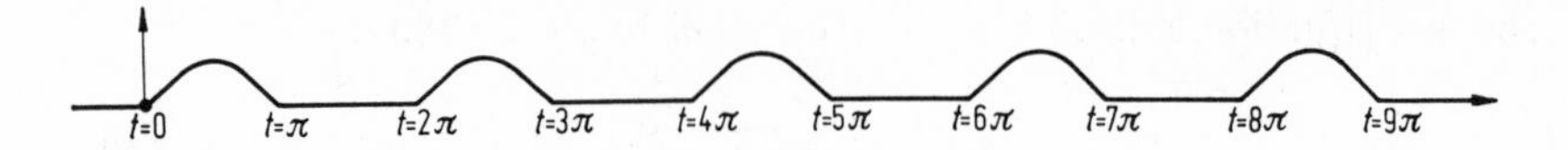

From 5.34 and 5.8—9 it follows immediately that

$$G_0 = \frac{(1 + \mathsf{T}_\pi)\, D}{(1 + D^2)\,(1 - \mathsf{T}_{2\pi})} = \frac{D}{(1 + D^2)\,(1 - \mathsf{T}_\pi)}:$$

the second equation is from 5.3 and 1.73.

5.35 *Remarks.* If $h(\,)$ is a $\mathcal{K}$-function vanishing outside the interval $(0, \lambda)$, it is easily verified that $h/(1 - \mathsf{T}_\lambda)$ is the canonical operator of a function of period $= \lambda$ (this is a converse of 5.34).

For other consequences of 5.34, see the table of formulas on pp. 343 to 344.

5.36 **The square-wave function** is the $\mathcal{K}$-function $G(\,)$ of period $= 2\alpha$ defined by

$$G(\tau) = \begin{cases} 1 & (0 < \tau \leq \alpha) \\ -1 & (\alpha < \tau \leq 2\alpha); \end{cases}$$

its graph has the following shape:

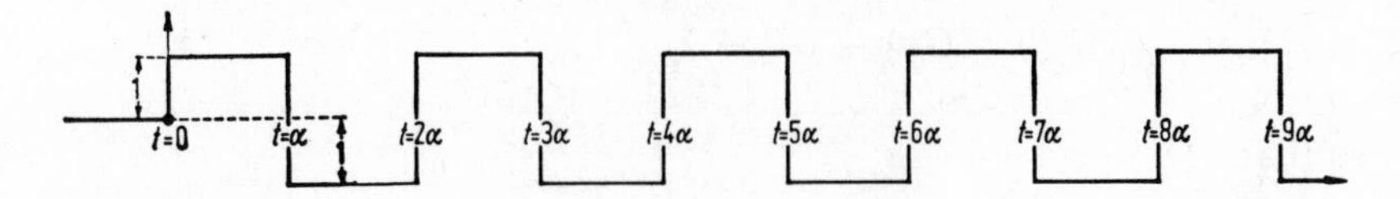

Setting $\lambda = 2\alpha$ in 5.34, we obtain

$$(1) \qquad G = \frac{G^{2\alpha}}{1 - \mathsf{T}_{2\alpha}} = \frac{G^{2\alpha}}{1 - (\mathsf{T}_\alpha)^2} = \frac{G^{2\alpha}}{(1 - \mathsf{T}_\alpha)\,(1 + \mathsf{T}_\alpha)} \qquad \text{(by 5.3)},$$

where $G^{2\alpha}(\,)$ is a function described by the following graph:

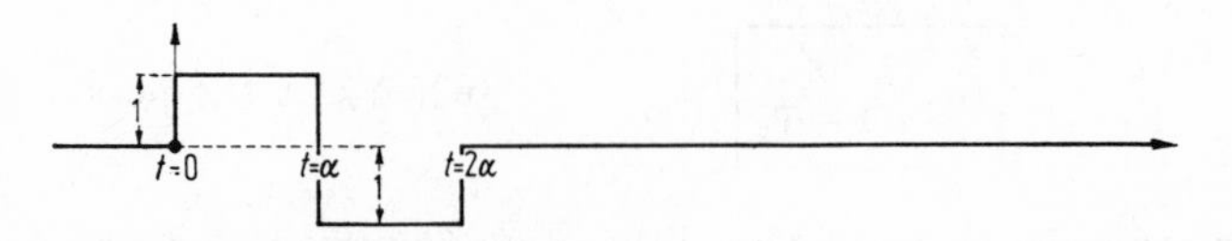

As in 5.12.1 we see that $G^{2\alpha} = 1 - \mathsf{T}_\alpha - (\mathsf{T}_\alpha - \mathsf{T}_{2\alpha})$; consequently, 5.3 gives $G^{2\alpha} = (1 - \mathsf{T}_\alpha)^2$. Substituting into (1):

$$(5.37) \qquad G = \frac{(1 - \mathsf{T}_\alpha)\,(1 - \mathsf{T}_\alpha)}{(1 - \mathsf{T}_\alpha)\,(1 + \mathsf{T}_\alpha)} = \frac{1 - \mathsf{T}_\alpha}{1 + \mathsf{T}_\alpha} \qquad \text{(by 1.73)}.$$

5.38 **The square-wave as a forcing function.** As in 5.36, let $G()$ be the square-wave function: let us solve the initial-value problem

$$(2) \qquad i(0-) = 0 \quad \text{with} \quad Li' + ri = G.$$

Proceeding as in 3.34, we consider instead the more general problem

$$(3) \qquad i(0-) = 0 \quad \text{with} \quad L\frac{\partial i}{\partial t} + Ri = G.$$

If i is an operator satisfying (3), then $LDi + Ri = G$ (by [3.7]); we may use 5.37 to write

$$LDi + Ri = \frac{1 - \mathsf{T}_\alpha}{1 + \mathsf{T}_\alpha} = 1 - \frac{2\,\mathsf{T}_\alpha}{1 + \mathsf{T}_\alpha}.$$

Solving for i, we obtain

$$(4) \qquad i = \frac{1}{LD + R} - 2\frac{\mathsf{T}_\alpha}{1 + \mathsf{T}_\alpha}\left(\frac{1}{LD + R}\right).$$

Note that

$$(5) \qquad \frac{1}{L}\left(\frac{1}{D + R/L}\right) = \frac{1}{L}\left\langle\frac{e^{-Rt/L} - 1}{-R/L}\right\rangle = \left\langle\frac{1 - e^{-Rt/L}}{R}\right\rangle:$$

the second equation is from 3.24 (with $a = -R/L$). The equations (4)—(5) imply that

$$i = \left\langle\frac{1 - e^{-Rt/L}}{R}\right\rangle - 2\frac{\mathsf{T}_\alpha}{1 + \mathsf{T}_\alpha}\left\langle\frac{1 - e^{-Rt/L}}{R}\right\rangle,$$

so that 5.30 gives

$$i = \frac{1 - e^{-Rt/L}}{R} - 2\sum_{k=0}^{\infty} \mathsf{T}_{k\alpha+\alpha}(t)\,(-1)^k\,\frac{1 - e^{-R(t-k\alpha-\alpha)/L}}{R}:$$

we have omitted the angular brackets on the right-hand side. In particular,

$$i(t) = \frac{1 - e^{-Rt/L}}{R} \qquad \text{for } 0 < t < \alpha,$$

and

$$i(t) = \frac{1 - e^{-Rt/L} - 2\,(1 - e^{-R(t-\alpha)/L})}{R} \qquad \text{for } \alpha < t < 2\alpha.$$

In general:

$$i(t) = \frac{1 - e^{-Rt/L}}{R} - 2 \sum_{k=0}^{n-1} (-1)^k \left(\frac{1 - e^{-R(t-k\alpha-\alpha)/L}}{R}\right)$$

for $n\alpha < t < (n+1)\alpha$ and $n = 1, 2, 3, \ldots$: see 5.30.1.

Advantages of Operational Calculus

Except for non-standard problems (such as 2.56, 4.13.2, and the problem immediately preceding 4.0), all of the problems in the preceding chapters could have been solved by classical methods. Since the problems in Chapter 3 cannot be formulated in terms of classical analysis, this is a good place to make some comparisons.

Classical methods require finding a particular solution and a "general solution" involving coefficients to be determined by the initial conditions; this implies solving systems of linear equations that can be quite cumbersome — moreover, discontinuous functions (in cases such as 5.23) greatly complicate the application of classical methods. On the contrary, operational calculus automatically makes the initial values appear as coefficients in operator equations; further, decomposition theorems (such as 4.23) provide a means of avoiding systems of linear equations.

Use of Laplace Transform Tables

5.39 If $G(\,)$ is a function, its usual Laplace transform $\hat{g}(\,)$ is defined by

$$\hat{g}(s) = \int_0^\infty e^{-su}\, G(u)\, du.$$

Let $G(\,)$ be a function whose usual Laplace transform $\hat{g}(\,)$ satisfies an equality of the form

$$\hat{g}(s) = f_1(s)\, f_2(e^{-s\alpha}); \tag{1}$$

more precisely, we suppose the existence of two rational functions $f_k(\,)$ $(k = 1, 2)$ and a positive number α such that the equation (1) holds for all sufficiently large values of s; under these circumstances, the operator $\langle G(t)\rangle$ is given by

$$\langle G(t)\rangle = D[f_1(D)]\, f_2(\mathsf{T}_\alpha). \tag{2}$$

Recall our notational convention $G = \langle G(t)\rangle$ (see 2.1). For example the Laplace transform $\hat{g}(\,)$ of the square-wave function $G(\,)$ satisfies the equation

$$\hat{g}(s) = s^{-1} \tanh s\alpha;$$

thus, $f_1(s) = 1/s$ and

$$f_2(e^{-s\alpha}) = \frac{e^{s\alpha/2} - e^{-s\alpha/2}}{e^{s\alpha/2} + e^{-s\alpha/2}} = \frac{1 - e^{-s\alpha}}{1 + e^{-s\alpha}};$$

from (1)—(2) it now follows that

$$G = DD^{-1} \frac{1 - \mathsf{T}_\alpha}{1 + \mathsf{T}_\alpha} \qquad \text{(compare with 5.37)}.$$

On the other hand, the usual Laplace transform of the function defined by the right-hand side of the relation

$$(D + \alpha)^{-m} = \alpha^{-m} - \frac{e^{-\alpha t}}{\alpha^m}\left[1 + \alpha t + \cdots + \frac{(\alpha t)^{m-1}}{(m-1)!}\right]$$

(proved above 4.32.1) is the function $\hat{g}(\,)$ defined by $\hat{g}(s) = s^{-1}(s + \alpha)^{-m}$: the corresponding relation is not found in standard Laplace transform tables.

Difference Equations

5.40 Given a $\mathcal{K}$-function $f(\,)$ and two different numbers a and b, let us find a $\mathcal{K}$-function such $y(\,)$ that $y(t) = 0$ for $t \leq 0$ and

$$y(t) + (a + b)\, y(t - 1) + ab\, y(t - 2) = f(t) \qquad (-\infty < t < \infty);$$

the problem requires $f(t) = 0$ for $t \leq 0$. If $y(\,)$ is such a function, it follows from 2.4—5 that

$$\langle y(t)\rangle + (a + b)\,\langle y(t - 1)\rangle + ab\langle y(t - 2)\rangle = \langle f(t)\rangle;$$

we may therefore use [2.1] and 5.25 to write

$$y + (a+b)\,\mathsf{T}_1 y + ab\,\mathsf{T}_2 y = f;$$

but $\mathsf{T}_2 = \mathsf{T}_1^2$ (by 5.3), whence

$$(1 + a\,\mathsf{T}_1)\,(1 + b\,\mathsf{T}_1)\,y = f,$$

which (by 5.29, 1.71, and 1.77) implies the equation

$$y = \frac{1}{(1 + a\,\mathsf{T}_1)\,(1 + b\,\mathsf{T}_1)}\,f.$$

Decomposing into two parts (as in 4.14—17), we obtain

$$y = \frac{1}{a-b}\left[\frac{a}{1 + a\,\mathsf{T}_1} - \frac{b}{1 + b\,\mathsf{T}_1}\right] f = \frac{1}{a-b}\left[a\,\frac{f}{1 + a\,\mathsf{T}_1} - b\,\frac{f}{1 + b\,\mathsf{T}_1}\right]$$

$$= \frac{1}{a-b}\left[a \sum_{k=0}^{\infty} (-a)^k f(t-k) - b \sum_{k=0}^{\infty} (-b)^k f(t-k)\right]:$$

the last equation is from 5.27. Consequently,

$$y(t) = \frac{1}{a-b} \sum_{k=0}^{\infty} [(-a)^k a - b(-b)^k]\, f(t-k) \qquad (-\infty < t < \infty).$$

The case $a = b$ is treated in 11.57.

Bibliographical Comments

Oliver Heaviside (1850—1925) was highly successful in applying his operational calculus to problems of importance in electrodynamics and electrical engineering; his work effectively challenged mathematicians to attempt justifying his operational calculus (see the first paragraph of the classical textbook by Laurent Schwartz [S 1] "Théorie des distributions").

5.41 Z. Hennyey introduced the term "entering function" in his book entitled "Linear electric circuits" (Reading, Mass.: Addison-Wesley 1962); his book contains a sketch of one of the more recent attempts at constructing a mathematical framework to justify HEAVISIDE's operational calculus based on the notion of linear operator (it should be said that his approach is quite different from the one adopted in the present book).

5.42.0 J. D. WESTON [W 1—4] uses the word "perfect operator" in a sense which is somewhat different from the one used in this book: his "perfect operators" are restricted to Laplace transformable test-functions with support in the positive axis. Perfect operators (in our sense of the word) are roughly what I. I. RJABCEV [R 1—2] calls "Mikusiński operators" (his continuity condition is un-necessary) — except that RJABCEV's operators are restricted to test-functions that vanish on the negative axis.

Notes and Further Comments

5.42.1 As far as I know, the two decomposition theorems (4.23 and 4.39) have not appeared in print; I formulated them in the belief that they are more efficient than other systematic procedures (such as the method of residues, which seems best adapted to problems arising from partial differential equations).

5.42.2 In spite of the differences indicated in 5.42.0, the basic mathematical backbone of this book (such as the notion of perfect operator and the corresponding properties) owes much to the work of WESTON and RJABCEV. Needless to say, the present book has been deeply influenced by the ideas of JAN MIKUSIŃSKI and LAURENT SCHWARTZ.

5.42.3 *The calculations in* 4.7 *are simpler than the ones required by the usual Laplace transformation* (5.39): this is because the usual Laplace transformation gives rise to *four* linear factors in the denominator of (8)—*instead of three.* In fact, the approach adopted in this book requires exactly the same calculations as the ones required by the Carson-Laplace transformation (adopted in [V 1]).

5.42.4 The problems 3.38—39 are not Laplace transformable, but if this fact is overlooked, the answers obtained will turn out to be correct: this situation is typical in operational calculus; it points to the existence of an approach (such as the one adopted in this book) which justifies the calculus but discards the Laplace transform approach.

5.42.5 Since the Laplace transformation maps functions of the positive variable t into analytic functions of a complex variable, it requires a double notation (for example, small letters for object-functions and capitals for functions of a complex variable): such double notation is unnecessary in the present approach to operational calculus (and also in MIKUSIŃSKI's [M 1]).

5.42.6 V. A. DITKIN [D 1] and L. BERG [B 2] have extended the Laplace transformation so that it can handle all the problems that we have dealt with (except the non-standard problems, such as 2.56); however, its very generality gives it an even greater notational disadvantage than the Laplace transformation.

5.42.7 CARSON ("Electric circuit theory and the operational calculus", New York: McGraw-Hill 1926) was the first to utilize the Laplace transformation as a substitute for HEAVISIDE's operational calculus; but his theory was mathematically deficient, and almost as questionable as HEAVISIDE's. In fact, PAUL LÉVY [Bull. Sci. Math. **50**, 174 (1926)] found HEAVISIDE's approach *more* convincing than CARSON's; a year later, G. DOETSCH's article provided the first convincing reasons for utilizing the Laplace transformation to attain a peaceful conscience (at the cost of restrictions that are often un-necessary).

Exercises

5.43.0 The sawtooth function is the $\mathcal{K}$-function $F(\,)$ of period $= \alpha$ defined by

$$F(t) = \begin{cases} 0 & (t \leq 0) \\ t/\alpha & (0 < t \leq \alpha): \end{cases}$$

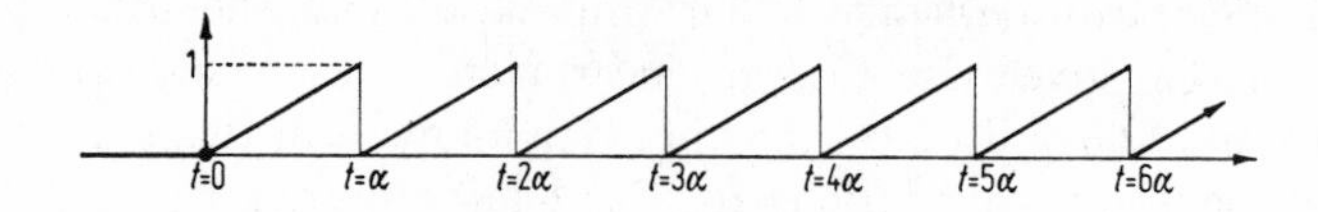

Use 5.18 and 5.34 to find the operator F.

Answer: $F = \dfrac{1}{\alpha D} - \dfrac{\mathsf{T}_\alpha}{1 - \mathsf{T}_\alpha}$.

5.43.1 Let $G(\,)$ be the $\mathcal{K}$-function of period $= 2$ defined by

$$G(t) = t - 2(t-1)\,\mathsf{T}_1(t) + (t-2)\,\mathsf{T}_2(t) \qquad (0 < t \leq 2):$$

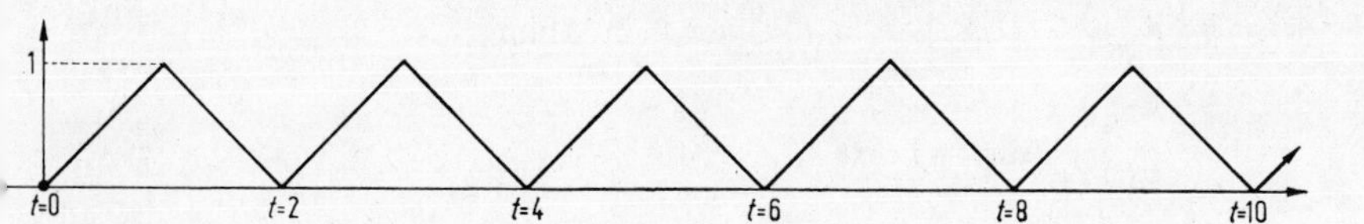

Find the operator G.

Answer: $G = (1 - \mathsf{T}_1)\, D^{-1} (1 + \mathsf{T}_1)^{-1}$.

5.43.2 Find the canonical operator of the $\mathcal{K}$-function defined by $F(t) = |\sin t|$:

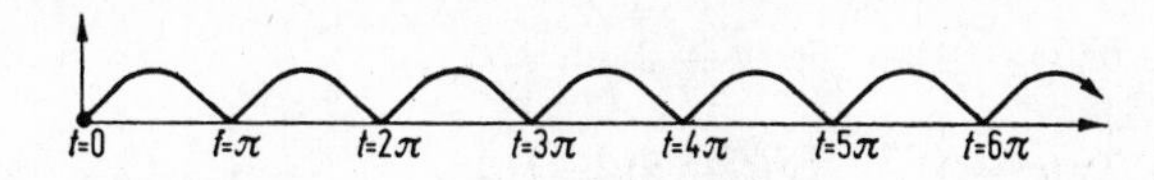

Answer: $\dfrac{D(1 + \mathsf{T}_\pi)}{(D^2 + 1)\,(1 - \mathsf{T}_\pi)} = \langle |\sin t| \rangle$.

5.43.3 Suppose that $0 \leq x \leq \alpha$; draw a rough sketch of the function $H_x(\,)$ such that

$$H_x = \frac{\mathsf{T}_{\alpha - x} - \mathsf{T}_{\alpha + x}}{(1 + \mathsf{T}_{2\alpha})\, D};$$

further, verify that $H_x(\,)$ has period $= 4\alpha$.

5.43.4 Solve the initial-value problem

$$y(0-) = y'(0-) = 0,$$

with

$$\frac{\partial^2 y}{\partial t^2} + 3\,\frac{\partial y}{\partial t} + 2y = \begin{cases} \sin t & (0 < t < \pi) \\ 0 & (\text{otherwise}). \end{cases}$$

Answer: $y(t) = F(t) + \mathsf{T}_\pi(t)\, F(t - \pi)$, where

$$F(t) = (1/2)\, \mathrm{e}^{-t} - (1/5)\, \mathrm{e}^{-2t} - (3/10) \cos t + (1/10) \sin t.$$

5.43.5 Proceed as in 5.24.0 to solve the following starting-value problem:

$$y(0+) = y'(0+) = 0 \quad \text{with} \quad y'' + y = \begin{cases} |\sin t| & (t > 0) \\ 0 & (\text{otherwise}). \end{cases}$$

Hint: use 5.43.2, 3.32, and 5.30.1.

Answer: if $(n-1)\pi + \pi < t < n\pi + \pi$, then

$$y(t) = \frac{\sin t - t\cos t}{2}$$

$$+ \sum_{k=0}^{n-1} (t - k\pi - \pi)\cos(t - k\pi) - \sin(t - k\pi).$$

5.43.6 Let $f(\,)$ be a $\mathcal{K}$-function such that $f(t) = 0$ for $t \leq 0$; solve the two difference equations

$$y(t) - 5y(t-1) + 6y(t-2) = f(t)$$

$$y(t) + 2y(t-1) - 3y(t-2) = f(t) \qquad (-\infty < t < \infty).$$

Hint: proceed as in 5.40.

Answers: $\sum_{k=0}^{\infty} [3^{k+1} - 2^{k+1}]\, f(t-k)$ and $\sum_{k=0}^{\infty} \frac{1-(-3)^{k+1}}{4} f(t-k)$.

5.43.7 Find the operator of the function $F(\,)$ described by the following graph:

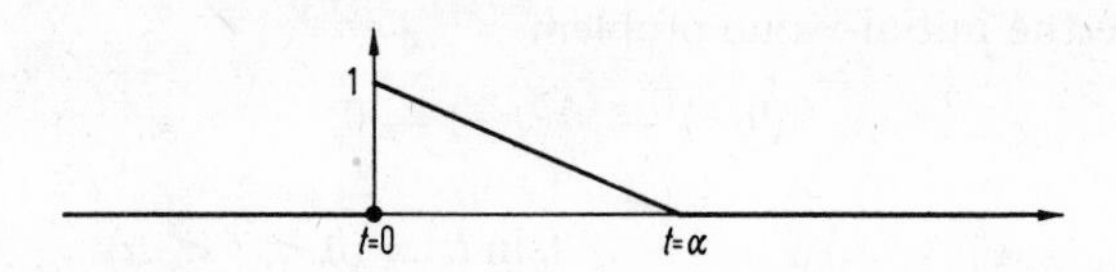

Hint: use 5.11 and note that $F' = -\alpha^{-1}(1 - \mathsf{T}_\alpha)$.

Answer: $F = -\alpha^{-1}D^{-1}(1 - \mathsf{T}_\alpha) + 1$.

5.43.8 Find the operator of the $\mathcal{K}$-function of period $= \alpha$ defined by the following graph:

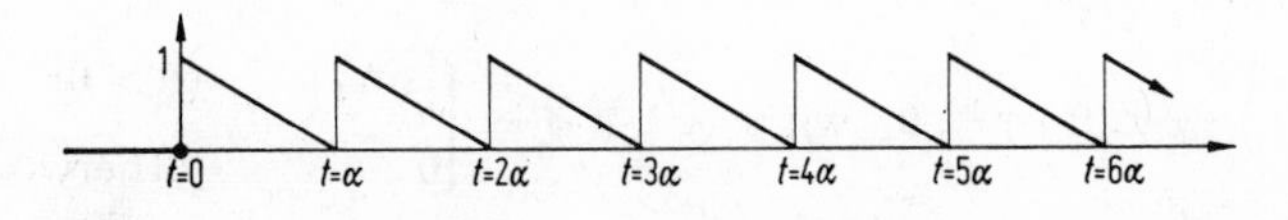

Answer: $(\mathsf{T}_\alpha - 1 + \alpha D)/\alpha D(1 - \mathsf{T}_\alpha)$.

§ 6. Calculus of Operators

In this section we shall integrate operators and show how certain properties of functions (value at a point, magnitude) can be carried over to operators. The practical usefulness of our definitions will be illustrated by many worked problems; besides justifying the definitions in this manner, we shall verify that they are valid extensions of familiar concepts. For example, we shall see that the operator $3D\mathsf{T}_2$ vanishes outside the point $t = 2$ and has magnitude $= 3$; in the words of OLIVER HEAVISIDE, it is "*of amount* 3 *wholly concentrated at* $t = 2$. *An impulsive function, so to speak*" (Electromagnetic Theory, ii, § 249). In fact, the operator $3D\mathsf{T}_2$ is the mathematical counterpart of an impulse of magnitude $= 3$ applied at $t = 2$ (see 6.41—42); further, $3D^2\mathsf{T}_2$ is a concentrated couple of moment $= -3$ applied at $t = 2$ (see 6.82). This section is subdivided as follows:

Convergence
Value of an operator at a point,
Integral of an operator,
Function-operators,
Impulse and dipole,
Distributions and operators*,
Two examples,
Multiplication by t,
Static deflection of beams.

We begin by reviewing some of the basic definitions. An operator V is a rule assigning to each test-function $\varphi(\,)$ a test-function $V \cdot \varphi(\,)$. If V_1 and V_2 are operators, the operator $V_1 V_2$ is defined by

$$(6.0) \qquad V_1 V_2 \cdot \varphi(\,) = V_1 \cdot (V_2 \cdot \varphi)(\,) \qquad \text{for every test-function } \varphi(\,).$$

If a and b are numbers, then the operator $a V_1 - b V_2$ is determined by the equation

$$(6.1) \qquad [a V_1 - b V_2] \cdot \varphi(\,) = a [V_1 \cdot \varphi](\,) - b [V_2 \cdot \varphi](\,)$$

* This item has the practical importance of a footnote.

for every test-function $\varphi(\,)$: see 1.56, 1.58, and 1.48. Suppose that $\varepsilon > 0$: combining 6.1 with 5.1, we find that the equation

$$\frac{1}{\varepsilon}(\mathsf{T}_0 - \mathsf{T}_\varepsilon) \cdot \varphi(\tau) = \frac{1}{\varepsilon}\varphi(\tau) - \frac{1}{\varepsilon}\varphi(\tau - \varepsilon) \tag{6.2}$$

holds for $-\infty < \tau < \infty$ and for any test-function $\varphi(\,)$.

6.3 Finally, recall that $V_1 = V_2$ if (and only if)

$$V_1 \cdot \psi(\,) = V_2 \cdot \psi(\,) \qquad \text{for every test-function } \psi(\,).$$

Convergence

6.4.0 **Definition.** Suppose that $-\infty \leq a \leq \infty$. If V_x is an operator depending on the real parameter x, we shall denote by

$$\lim_{x\to a} V_x$$

the mapping defined for any test-function $\varphi(\,)$ by the equation

$$\left[\lim_{x\to a} V_x\right] \cdot \varphi(\tau) = \lim_{x\to a} V_x \cdot \varphi(\tau) \qquad (-\infty < \tau < \infty).$$

6.4.1 **Remarks.** In view of its importance, it might be worth-while to discuss the preceding definition. Consider a test-function $\varphi_x(\,)$, and let E be the set of all the real numbers τ such that

$$\left|\lim_{x\to a} \varphi_x(\tau)\right| < \infty;$$

further, let us denote by

$$\lim_{x\to a} [\varphi_x(\,)] \tag{6.5}$$

the function that assigns to each number τ in E the number

$$\lim_{x\to a} \varphi_x(\tau)\,.$$

Thus, $\lim_{x \to a} V_x$ *is the mapping that assigns to each test-functions* $\varphi(\,)$ *the function* $\lim_{x \to a} [V_x \cdot \varphi(\,)]$. In other words, the operator $\lim_{x \to a} V_x$ is defined by the equation

$$\boxed{\left[\lim_{x \to a} V_x\right] \cdot \varphi(\,) = \lim_{x \to a} [V_x \cdot \varphi(\,)]}. \tag{6.6}$$

If the function on the right-hand side of 6.6 is defined everywhere whenever $\varphi(\,)$ is a test-function, then $\lim_{x \to a} V_x$ is a perfect operator (we shall not use this fact; it is a consequence of unpublished discoveries by HARRIS SHULTZ).

Let us prove that

$$\boxed{D = \lim_{\varepsilon \to 0+} \frac{1}{\varepsilon} (\mathsf{T}_0 - \mathsf{T}_\varepsilon)}: \tag{6.7}$$

as usual, $\varepsilon \to 0+$ means that ε approaches zero through positive values. To prove 6.7, let $\varphi(\,)$ be any test-function; if τ is a real number, then

$$\varphi'(\tau) = \lim_{\varepsilon \to 0+} \frac{\varphi(\tau) - \varphi(\tau - \varepsilon)}{\tau - (\tau - \varepsilon)}$$

$$= \lim_{\varepsilon \to 0+} \frac{1}{\varepsilon} \varphi(\tau) - \frac{1}{\varepsilon} \varphi(\tau - \varepsilon)$$

$$= \lim_{\varepsilon \to 0+} \frac{1}{\varepsilon} (\mathsf{T}_0 - \mathsf{T}_\varepsilon) \cdot \varphi(\tau) \qquad \text{by 6.2;}$$

since τ is any real number, we may apply Definition 6.6:

$$\varphi'(\,) = \lim_{\varepsilon \to 0+} \left[\frac{1}{\varepsilon} (\mathsf{T}_0 - \mathsf{T}_\varepsilon) \cdot \varphi(\,)\right].$$

Definitions 0.7 and 6.6 therefore imply that

$$D \cdot \varphi(\,) = \left[\lim_{\varepsilon \to 0+} \frac{1}{\varepsilon} (\mathsf{T}_0 - \mathsf{T}_\varepsilon)\right] \cdot \varphi(\,). \tag{1}$$

Since $\varphi(\,)$ is an arbitrary test-function, Conclusion 6.7 is immediate from (1) and [6.3].

6.8 **Right-multiplication Property.** *If W is an operator, then*

$$\left(\lim_{x\to a} V_x\right) W = \lim_{x\to a} V_x W. \tag{6.9}$$

Proof. Let $\psi()$ be any test-function; we have

$$\left[\left(\lim_{x\to a} V_x\right) W\right] \cdot \psi() = \left(\lim_{x\to a} V_x\right) \cdot (W \cdot \psi)() \qquad \text{by [6.0]}$$

$$= \lim_{x\to a} [V_x \cdot (W \cdot \psi)()] \qquad \text{by [6.6]}$$

$$= \lim_{x\to a} [V_x W \cdot \psi()] \qquad \text{by [6.0]}$$

$$= \left[\lim_{x\to a} V_x W\right] \cdot \psi() \qquad \text{by [6.6]}$$

in view of [6.3], Conclusion 6.9 is at hand.

6.10 *Remarks.* Let V_n $(n = 0, 1, 2, \ldots)$be a sequence of perfect operators such that the function

$$\lim_{n\to\infty} [V_n \cdot \varphi()]$$

is defined everywhere whenever $\varphi()$ is a test-function; in other words, we suppose that the sequence $V_n \cdot \varphi(\tau)$ $(n = 0, 1, 2, \ldots)$ converges for every real number τ and every test-function $\varphi()$: under these circumstances, there exists a unique perfect operator V such that

$$V = \lim_{n\to\infty} V_n$$

(unpublished result proved by HARRIS SHULTZ). Thus, the equation

$$V \cdot \varphi() = \lim_{n\to\infty} [V_n \cdot \varphi()] \qquad (\text{for every test-function } \varphi())$$

defines a perfect operator V *if the right-hand side is defined everywhere whenever $\varphi()$ is a test-function.*

6.10.1 **Theorem.** *Suppose that $F()$ and $H_n()$ $(n = 0, 1, 2, \ldots)$ are $\mathcal{K}$-functions such that*

$$F(t) = \lim_{n\to\infty} H_n(t) \qquad (0 < t < \infty).$$

If there exists a continuous function $f(\ )$ *such that the inequality*

$$|H_n(t)| \leq f(t) \qquad (0 \leq t < \infty)$$

holds for each $n = 0, 1, 2, \ldots$, *then*

$$F = \lim_{n\to\infty} H_n$$

Proof. Set $H_x(\ ) = \{f(t)\}(\ )$ in 15.31.

6.10.2 **Examples.** The limits

$$\lim_{x\to\infty} \cos xt$$

and

$$\lim_{x\to\infty} x \sin xt$$

do not exist, yet the corresponding operators converge:

$$\lim_{x\to\infty} \langle \cos xt\rangle = 0 \qquad \text{(see 6.12.1)},$$

$$\lim_{x\to\infty} \langle x \sin xt\rangle = D \qquad \text{(see 6.12.2)}.$$

6.11 **Another example.** Take $\lambda > 0$ and let c_k $(k = 0, 1, 2, \ldots)$ be any sequence of numbers; let us verify that the equation

$$\sum_{k=0}^{\infty} c_k D^k \mathsf{T}_{k\lambda} \overset{\text{def}}{=} \lim_{n\to\infty} \sum_{k=0}^{n} c_k D^k \mathsf{T}_{k\lambda}$$

defines a perfect operator. To that effect, set

$$V_n = \sum_{k=0}^{n} c_k D^k \mathsf{T}_{k\lambda},$$

and observe that the equations

$$V_n \cdot \varphi(t) = \sum_{k=0}^{n} c_k [D^k \mathsf{T}_{k\lambda} \cdot \varphi](t)$$

$$= \sum_{k=0}^{n} c_k \varphi^{(k)}(t - k\lambda) \qquad \text{(by 5.1)}$$

hold for any test-function $\varphi(\ )$. In view of 6.10 it will suffice to prove the convergence of the series

$$\sum_{k=0}^{\infty} c_k \varphi^{(k)}(t - k\lambda). \tag{2}$$

Let α be one of the points such that $\varphi(\tau) = 0$ for $\tau < \alpha$: if $k > (t - \alpha)/\lambda$ then $t - k\lambda < \alpha$, whence

$$\varphi(t - k\lambda) = 0 = \varphi^{(k)}(t - k\lambda) \qquad \text{whenever } k > (t - \alpha)/\lambda. \tag{3}$$

Consequently, if $n > m > (t - \alpha)/\lambda$ then

$$V_n \cdot \varphi(t) = \sum_{k=0}^{m} c_k \varphi^{(k)}(t - k\lambda) + \sum_{k=m+1}^{n} c_k \varphi^{(k)}(t - k\lambda) = V_m \cdot \varphi(t); \tag{4}$$

the last equation comes from (3) and the fact that $k > (t - \alpha)/\lambda$ in the last summation. Letting $n \to \infty$ in (4), we see that the series (2) converges to the number $V_m \cdot \varphi(t)$.

Exercises

6.12.0 Prove the following equation:

$$\lim_{x\to\infty} \left\langle \frac{\sin xt}{x} \right\rangle = 0.$$

Hints: Note that

$$\left[\lim_{x\to\infty} \left\langle \frac{\sin xt}{x} \right\rangle\right] \cdot \varphi(\tau) = \lim_{x\to\infty} \left\langle \frac{\sin xt}{x} \right\rangle \cdot \varphi(\tau) = 0:$$

the first equation is from [6.4.0]; the last equality comes from

$$\left\langle \frac{\sin xt}{x} \right\rangle \cdot \varphi(\tau) = \int_0^{\infty} \varphi'(\tau - u) \frac{\sin xu}{x}\, du \qquad \text{(see 2.2)}$$

and

$$\left| \int_0^{\infty} \varphi'(\tau - u) \frac{\sin xu}{x}\, du \right| \leq \frac{1}{x} \int_0^{\infty} |\varphi'(\tau - u)|\, du.$$

6.12.1 Prove that

$$\lim_{x\to\infty} \langle \cos xt \rangle = 0.$$

Hints. From 3.8.1 it follows that

$$D\left\langle \frac{\sin x t}{x} \right\rangle = \langle \cos xt \rangle:$$

apply 6.8 and 6.12.0.

6.12.2. Prove that

$$\lim_{x\to\infty} \langle x \sin xt \rangle = D$$

Hints. From 3.8.1 it follows that

$$\left\langle \frac{d}{dt} \cos xt \right\rangle = D\langle \cos xt \rangle - D;$$

thus,

$$-x \langle \sin xt \rangle = D\langle \cos xt \rangle - D;$$

next, apply 6.8 and 6.12.1 to obtain

$$-\lim_{x\to\infty} \langle \sin xt \rangle = 0 - D.$$

6.12.3 Prove that

$$\lim_{x\to\infty} \frac{x^2 D}{D^2 + x^2} = D.$$

Hint: use 6.12.2 and 3.27.

6.12.4 Prove that

$$\lim_{x\to\infty} \mathsf{T}_x = 0.$$

Hint: use [6.4.0] and 5.1.

6.12.5 Suppose that $\alpha > 0$: if V is an operator such that $\mathsf{T}_\alpha V = V$, prove that $V = 0$.

Hint: we have $\mathsf{T}_\alpha^2 V = \mathsf{T}_\alpha V$, whence $\mathsf{T}_{2\alpha} V = V$; in fact, $\mathsf{T}_{k\alpha} V = V$ (by an easy induction proof); finally, use 6.12.4.

6.12.6 Prove that the equation

$$\lim_{x\to\infty} f(x)\,\mathsf{T}_x = 0$$

holds for any function $f(\,)$.

Value of an Operator at a Point

Let V be any operator. In 8.21 we shall define a function $\{V\}(\,)$ assigning to any real number τ a number $\{V\}(\tau)$ (which may be the number ∞); this function $\{V\}(\,)$ has the following property:

(6.13) *if the equality* $V = D^n [\![f]\!]$ *holds for some integer* $n \geq 1$ *and for some entering function* $f(\,)$, *then* $\{V\}(\tau) = f^{(n)}(\tau)$ *at each point* τ *such that* $|f^{(n)}(\tau)| < \infty$.

6.14.0 *Remarks.* Recall that $D^n [\![f]\!]$ is the product of the operator D^n with the operator $[\![f]\!]$ of the function $f(\,)$ (see [1.14]); as usual, $f^{(n)}(\tau)$ is the value at the point τ of the nth order derivative of $f(\,)$. Property 6.13 (to be proved in 8.25) can be re-phrased as follows: if $V = D^n [\![f]\!]$, then the equality $\{V\}(\tau) = f^{(n)}(\tau)$ holds at each point τ at which the function $f(\,)$ is n times differentiable. It would be still simpler to express the conclusion of 6.13 by saying that the equation $\{V\}(\tau) = f^{(n)}(\tau)$ holds whenever its right-hand side has a meaning. The number $\{V\}(\tau)$ will be called *the value at* τ *of the operator* V.

6.14.1 From 6.13 we see that the equation

$$\{D^n [\![f]\!]\}(\tau) = f^{(n)}(\tau) \qquad (n = 1, 2, 3, \ldots)$$

holds whenever the right-hand side has a meaning. In particular, if $F(\,)$ is a $\mathcal{K}$-function, then $F = [\![f]\!]$ with $f(\,) = \{F(t)\}(\,)$ (see [2.1]); consequently,

$$\boxed{\{D^n F\}(\tau) = F^{(n)}(\tau)} \qquad (\text{all } \tau > 0) \tag{6.14.2}$$

— provided that $F^{(n)}(\tau)$ has a meaning; in particular, the equation

$$\{DF\}(\tau) = F'(\tau) = \frac{d}{d\tau} F(\tau) \qquad \text{(all } \tau > 0) \tag{6.14.3}$$

holds whenever the right-hand side has a meaning.

6.14.4 **Theorem.** *Let V_1 and V_2 be operators. If a and b are numbers, then*

$$\{aV_1 + bV_2\}(\tau) = a\{V_1\}(\tau) + b\{V_2\}(\tau) \tag{6.14.5}$$

at each point τ where both $\{V_1\}(\tau) \neq \infty$ and $\{V_2\}(\tau) \neq \infty$.

Proof: see 8.26.

6.14.6 **Theorem.** *If V is the operator of an entering function $g(\,)$, then $\{V\}(\tau) = g(\tau)$ at each point τ where $g(\,)$ is continuous*:

$$\{[\![g]\!]\}(\tau) = g(\tau). \tag{6.15}$$

Proof. Our hypothesis $V = [\![g]\!]$ implies

$$V = D[\![\mathsf{T}_0 * g]\!] \qquad \text{(by 1.29)}. \tag{1}$$

Let $f(\,)$ be the function $\mathsf{T}_0 * g(\,)$; from 0.26 we see that

$$f(\tau) = \int_{-\infty}^{\tau} g(u)\,du \qquad (-\infty < \tau < \infty). \tag{2}$$

Let τ be a point where $g(\,)$ is continuous: Equation (2) implies

$$f'(\tau) = \frac{d}{d\tau} \int_{-\infty}^{\tau} g(u)\,du = g(\tau). \tag{3}$$

Since $f(\,)$ is the function $\mathsf{T}_0 * g(\,)$, Equation (1) gives $V = D[\![f]\!]$: we can apply 6.13 (with $n = 1$) to obtain

$$\{V\}(\tau) = f^{(1)}(\tau) = f'(\tau) = g(\tau):$$

the last equation is from (3). Our conclusion $\{V\}(\tau) = g(\tau)$ is at hand.

6.16 *Remark.* In particular, if the equality $V = \langle F(t) \rangle$ holds for some $\mathcal{K}$-function $F(\,)$, then $V = [\![\{F(t)\}]\!]$ (see [2.1]), so that 6.15 gives

$$(4) \qquad \{V\}(\tau) = \{F(t)\}(\tau) = \begin{cases} 0 & (\tau < 0) \\ F(\tau) & (\tau > 0) \end{cases}:$$

the last equation holds only at the points τ where the function $F(\,)$ is continuous. Recall that $\langle F(t) \rangle$ is the canonical operator of the function $F(\,)$; since we often write F instead of $\langle F(t) \rangle$, Equation (4) implies that

$$(6.17) \qquad \boxed{\{F\}(\tau) = \begin{cases} 0 & (\tau < 0) \\ F(\tau) & (\tau > 0) \end{cases}}$$

– provided that the function $F(\,)$ is continuous at the point τ. In other words, the *value at τ of the operator* F equals the *value at τ of the function* $F(\,)$ (whenever τ is a continuity-point of the function $F(\,)$ in the interval $(0, \infty)$: see 0.32). In any case

$$\{F\}(\,) = \{F(t)\}(\,) \qquad \text{(see 1.47).}$$

Let us consider a still more particular case: if the equation $V = c$ holds for some number c, then $V = \langle c \rangle$ (see [2.16]), whence

$$(6.18) \qquad V = c \quad \text{implies} \quad \{V\}(\tau) = \{c\}(\tau) = \begin{cases} 0 & (\tau < 0) \\ c & (\tau > 0). \end{cases}$$

6.19 **An important null-function.** Let c be a number, and suppose that $\alpha \geq 0$: observe that $[\![c\mathsf{T}_\alpha]\!] = c\mathsf{T}_\alpha$ (by 2.11). If $V = cD^n\mathsf{T}_\alpha$ then

$$V = D^n[\![f]\!] \text{ with } f(\,) = c\mathsf{T}_\alpha(\,);$$

on the other hand, since

$$(5) \qquad |f^{(n)}(\tau)| = |c\mathsf{T}^{(n)}(\tau)| = \left|\frac{d^n}{d\tau^n}\, c\mathsf{T}_\alpha(\tau)\right| = 0 \qquad (\text{for } \tau \neq \alpha),$$

we may apply 6.13 to obtain

$$\boxed{\{cD^n \mathsf{T}_\alpha\}(\tau) = 0 \qquad \text{for each } \tau \neq \alpha} \tag{6.20}$$

in case $\alpha \geq 0$ and $n = 1, 2, 3, \ldots$; Equation (5) comes from the fact that the nth derivative of a constant is zero.

In particular,

$$\{D\mathsf{T}_\alpha\}(\tau) = 0 \qquad \text{for each } \tau \neq \alpha \tag{6.20.1}$$

and

$$\{D\}(\tau) = 0 \qquad \text{for each } \tau \neq 0. \tag{6.20.2}$$

Integral of an Operator

6.21 **Definition.** Suppose that $-\infty < \tau < \infty$. If A is an operator, we denote by

$$\int_{-\infty}^{\tau} A$$

the value at τ of the operator $D^{-1}A$.

6.22 *Remarks.* The operator $D^{-1}A$ is the product of the two operators D^{-1} and A. In consequence of the preceding definition, we have

$$\boxed{\int_{-\infty}^{\tau} A = \{D^{-1}A\}(\tau)} \qquad (-\infty < \tau < \infty). \tag{6.23}$$

We set

$$\int_{-\infty}^{\infty} A \overset{\text{de}}{=} \lim_{\tau \to \infty} \int_{-\infty}^{\tau} A \tag{6.24}$$

and

$$\frac{\partial A}{\partial t} \overset{\text{def}}{=} D[A - \{A\}(0-)] \qquad \text{(see 3.7)}. \tag{6.25}$$

Let c be a number, and suppose that $\alpha \geq 0$: the equations

$$\int_{-\infty}^{\tau} c\,D\mathsf{T}_\alpha = \{c\,\mathsf{T}_\alpha\}(\tau) = c\,\mathsf{T}_\alpha(\tau) = \begin{cases} 0 & (\tau < \alpha) \\ c & (\tau > \alpha) \end{cases} \tag{6.26}$$

are from [6.23] (with $A = cD\mathsf{T}_\alpha$), 6.15, and [5.0]. From [6.24] and 6.26 it obviously follows that

$$\text{(6.27)} \qquad \boxed{\int_{-\infty}^{\infty} cD\mathsf{T}_\alpha = c} \qquad \text{(whenever } \alpha \geq 0\text{)}.$$

A second example: using [6.23—24] and 6.20 it is easily verified that

$$\text{(6.28)} \qquad \boxed{\int_{-\infty}^{\infty} cD^n\mathsf{T}_\alpha = 0} \qquad \text{(whenever } n \geq 2 \text{ and } \alpha \geq 0\text{)}.$$

Finally, let us examine the case where A is the operator of an entering function $f()$: we have $A = [\![f]\!]$, which implies

$$A = D[\![\mathsf{T}_0 * f]\!] \qquad \text{(by 1.29)},$$

whence $D^{-1}A = [\![\mathsf{T}_0 * f]\!]$; from 6.15 (with $g = \mathsf{T}_0 * g$) it now follows that

$$\text{(6)} \qquad \{D^{-1}A\}(\tau) = \mathsf{T}_0 * f(\tau)$$

at each point τ where $\mathsf{T}_0 * f()$ is continuous; since $\mathsf{T}_0 * f()$ is continuous (2.61), Equation (6) holds for any real value of τ. The conclusion

$$\int_{-\infty}^{\tau} A = \{D^{-1}A\}(\tau) = \mathsf{T}_0 * f(\tau) = \int_{-\infty}^{\tau} f(u)\,\mathrm{d}u \qquad (-\infty < \tau < \infty)$$

is immediate from [6.23], (6), and 0.26. Thus,

$$\text{(6.29)} \qquad \int_{-\infty}^{\tau} [\![f]\!] = \int_{-\infty}^{\tau} f(u)\,\mathrm{d}u \qquad (-\infty < \tau < \infty).$$

In particular, if F is the canonical operator of a $\mathcal{K}$-function $F()$, we have

$$F = [\![\{F(t)\}]\!] \qquad \text{(see [2.1])};$$

we can therefore set $f = \{F(t)\}()$ in 6.29 to obtain

$$\text{(6.29.1)} \qquad \boxed{\int_{-\infty}^{\tau} F = \int_{0}^{\tau} F(u)\,\mathrm{d}u} \qquad (-\infty < \tau < \infty).$$

6.30 **An electrical problem.** The circuit

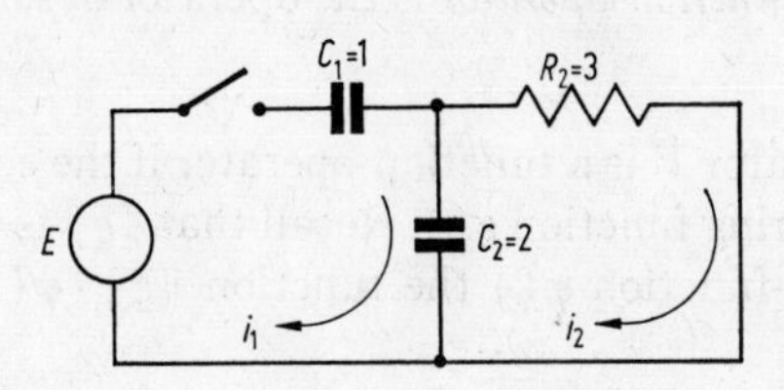

is initially at rest: there is no voltage and no charge at the time $t = 0-$. The system of equations

$$\left(\frac{1}{C_1} + \frac{1}{C_2}\right) i_1 - \frac{1}{C_2}\, i_2 = \frac{\partial E}{\partial t}\,, \tag{1}$$

$$-\frac{1}{C_2}\, i_1 + R_2 \frac{\partial}{\partial t}\, i_2 + \frac{1}{C_2}\, i_2 = 0 \tag{2}$$

governs the currents i_1 and i_2. Suppose that $E = 7$: from 6.18 it follows that $\{E\}(t) = 0$ for $t < 0$ and $\{E\}(\tau) = 7$ for $\tau > 0$; since $\{E\}(0-) = 0$, Definition 6.25 gives $\partial E/\partial t = DE = 7D$. Let us solve the system (1)—(2) for i_1 (in case $C_1 = 1$, $C_2 = 2$, and $R_2 = 3$):

$$i_1 = \frac{\begin{vmatrix} 7D & -\frac{1}{2} \\ 0 & 3D + \frac{1}{2} \end{vmatrix}}{\begin{vmatrix} \frac{3}{2} & -\frac{1}{2} \\ -\frac{1}{2} & 3D + \frac{1}{2} \end{vmatrix}} = 7D\, \frac{6D+1}{9D+1}\,;$$

since $(6D + 1)/(9D + 1) = 2/3 + 3^{-1}/(1 + 9D)$ (see 4.9), we obtain

$$i_1 = \frac{14}{3} D + 7\,(3^{-1})\, \frac{D}{1+9D} = \frac{14}{3} D + \frac{7}{27}\, e^{-t/9}: \tag{6.31}$$

The system (1)—(2) is usually expressed as a system of integral equations: see 7.15—20.

Function-operators

6.32 **Definition.** A *function-operator* is the operator of some entering function.

6.33 Thus, an operator V is a function-operator if the equation $V = [\![g]\!]$ holds for some entering function $g(\,)$. Recall that $[\![g]\!]$ is the operator that assigns to any test-function $\varphi(\,)$ the function $[\![g]\!] \cdot \varphi(\,)$ defined by

$$(1) \qquad [\![g]\!] \cdot \varphi(t) = \int_{-\infty}^{\infty} \varphi'(t-u)\, g(u)\, du \qquad (-\infty < t < \infty):$$

see 1.17.

6.34 **Theorem.** *Let c be a number, let n be an integer ≥ 1, and suppose that $\alpha \geq 0$. If $cD^n\mathsf{T}_\alpha$ is a function-operator, then $c = 0$.*

Proof. By hypothesis, $cD^n\mathsf{T}_\alpha$ is the operator of some entering function $g(\,)$:

$$(2) \qquad cD^n\mathsf{T}_\alpha = [\![g]\!];$$

the equations

$$(3) \qquad 0 = \{cD^n\mathsf{T}_\alpha\}(\tau) = g(\tau) \qquad (\text{if } \tau \neq \alpha)$$

are from 6.20 and 6.15: more precisely, they hold at each point $\tau \neq \alpha$ where $g(\,)$ is continuous. Since $g(\,)$ is an entering function [0.23], it is continuous in each interval $[0, t]$ with the possible exception of finitely-many points: from (3) it therefore follows that $g(\tau) = 0$ except for finitely-many points; consequently, (1) implies that

$$[\![g]\!] \cdot \varphi(t) = 0 \qquad (-\infty < t < \infty).$$

Thus, $[\![g]\!]$ is the zero operator 0: from (2) we now see that

$$(4) \qquad cD^n\mathsf{T}_\alpha = 0:$$

right-multiplying by $[\![\mathsf{T}_{-\alpha}]\!]D^{-n}$ both sides of (4), we obtain our conclusion

$$c = cD^n\mathsf{T}_\alpha [\![\mathsf{T}_{-\alpha}]\!] D^{-n} = 0 \qquad \big(\text{by 2.15 and (4)}\big).$$

6.35 **Consequence:** *if $c \neq 0$, then $cD^n\mathsf{T}_\alpha$ is not a function-operator.*

6.36 From 5.5 we see that

$$(5) \qquad 1D^n\mathsf{T}_\alpha = D^n\mathsf{T}_\alpha;$$

since $1 \neq 0$, it follows from 6.35 that $D^n\mathsf{T}_\alpha$ *is not a function-operator.* In the particular case where $\alpha = 0$, we have $D^n\mathsf{T}_\alpha = D^n$ (by 5.4 and 5.5): therefore, D^n *is not a function-operator* ($n = 1, 2, 3, \ldots$).

6.36.1 Let m be an integer. If $m \leq 0$, it follows from 3.20 and 3.12 that

$$D^m \mathsf{T}_\alpha = \left\langle \mathsf{T}_\alpha(t) \, \frac{(t-\alpha)^{-m}}{(-m)!} \right\rangle :$$

therefore, $D^m \mathsf{T}_\alpha$ is a function-operator. From 6.36 it now follows that

$D^m \mathsf{T}_\alpha$ *is a function-operator* *if (and only if)* $m \leq 0$.

6.36.2 **The road ahead.** Suppose that n is an integer ≥ 1: as we shall now see, $cD^n \mathsf{T}_\alpha$ is the limit of a sequence of function-operators (recall that $cD^n \mathsf{T}_\alpha$ is a perfect operator, by 1.15, 1.12, and 1.8).

6.37 **Theorem.** *Any perfect operator is the limit of a sequence of function-operators.*

Proof. From 15.8 and 15.19 it follows that the equation

$$\delta_n(t) = \frac{(nt)^{-3/2}}{\sqrt{\pi}} \, \mathsf{T}_0(t) \exp\left(\frac{-1}{nt}\right) \qquad (-\infty < t < \infty)$$

defines a sequence δ_n ($n = 1, 2, 3, \ldots$) of test-functions such that

$$\text{(6)} \qquad \boxed{D = \lim_{n \to \infty} [\![\delta_n]\!]} \, .$$

If V is any perfect operator, we can right-multiply by $D^{-1}V$ both sides of the equation (6) and use 6.9 to obtain

$$\text{(7)} \qquad V = \lim_{n \to \infty} [\![\delta_n]\!] \, D^{-1} V .$$

Since $D^{-1}V$ is a perfect operator (by 1.8), the equations

$$\text{(8)} \qquad [\![\delta_n]\!] \, D^{-1} V = (D^{-1}V) \, [\![\delta_n]\!] = [\![(D^{-1}V) \cdot \delta_n]\!]$$

are from 1.9 and 1.19. Since $\delta_n(\,)$ is a test-function, and since $D^{-1}V$ is an operator, we see (from [0.4]) that

(9) $\qquad (D^{-1}V) \cdot \delta_n(\,)$ is a test-function.

Equation (8) shows that $[\![\delta_n]\!] \, D^{-1}V$ is the operator of the test-function (9); accordingly, $[\![\delta_n]\!] \, D^{-1}V$ is a function-operator: the conclusion is now immediate from (7).

6.38 **Analogies.** Let m be an integer. Setting $\alpha = 0$ in 6.36.1, we see that

$$\{D^m \text{ is a function-operator}\} \Leftrightarrow \{m \leq 0\};$$

this is analogous to the fact that

$$\left\{\left(\frac{1}{2}\right)^m \text{ is a whole number}\right\} \Leftrightarrow \{m \leq 0\}.$$

Property 6.37 resembles the fact that any real number is the limit of a sequence of rational numbers. For example, the real number $\sqrt{2}$ is the limit of a sequence of rational numbers; it also is the only solution y of the equation $y^2 = 2$ but it is not a rational number. Analogously, the perfect operator D is the limit of a sequence of function-operators (6), is not a function-operator (6.36), but is the only solution y of the equation $y * 1 = 1$ (see [8.8]). Thus, perfect operators are generalized functions in the same sense that real numbers are generalized rational numbers. From a different viewpoint, the perfect operator D is analogous to the complex number $\sqrt{-1}$ (see 6.44).

Impulse and Dipole

Until further notice, suppose that $\alpha \geq 0$. Let c be a number; right-multiplying by $c\mathsf{T}_\alpha$ both sides of 6.7, we may use 6.9 and 5.2 to obtain

$$c D\mathsf{T}_\alpha = \lim_{\substack{\varepsilon \to 0 \\ \varepsilon > 0}} \frac{c}{\varepsilon} (\mathsf{T}_\alpha - \mathsf{T}_{\alpha+\varepsilon}). \tag{6.39}$$

From 5.13 we see that

$$\left\{\frac{c}{\varepsilon} (\mathsf{T}_\alpha - \mathsf{T}_{\alpha+\varepsilon})\right\}(\tau) = \begin{cases} c/\varepsilon & (\alpha < \tau < \alpha + \varepsilon) \\ 0 & (\text{otherwise}). \end{cases} \tag{1}$$

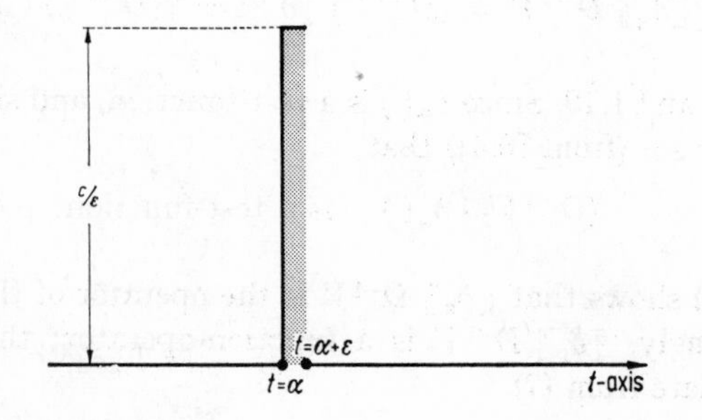

As ε approaches zero, the function (1) assumes the increasingly large value c/ε on the increasingly small interval of length with end-point $t = \alpha$. The equations

$$\int_{-\infty}^{\infty} \frac{c}{\varepsilon} (\mathsf{T}_\alpha - \mathsf{T}_{\alpha+\varepsilon}) = c = \int_{-\infty}^{\infty} c\, D\mathsf{T}_\alpha \tag{6.40}$$

are immediate consequences of (1), 6.29.1 and 6.27.

6.41 **Terminology.** An operator V will be said to be *of magnitude* $= b$ if (and only if)

$$\int_{-\infty}^{\infty} V = b \qquad \text{(see [6.24])}.$$

An operator V will be called an *impulse applied at* $t = \alpha$ if there exists a number m such that $V = m D\mathsf{T}_\alpha$.

6.42 The above terminology is justified by 6.39. If V is an impulse applied at $t = \alpha$, it follows from 6.40 that V is of magnitude $= c$ if (and only if) $V = c D\mathsf{T}_\alpha$. Recall that

$$\boxed{\{c D\mathsf{T}_\alpha\}(\tau) = 0 \quad \text{for all } \tau \neq \alpha} \qquad \text{(see 6.20)}. \tag{6.43}$$

In particular, the operator D is an impulse of magnitude $= 1$ applied at $t = 0$.

6.44 **Return to the electrical problem.** Set $F_1 = (7/27)\, e^{-t/9}$ and $c_1 = 14/3$; the equation

$$i_1 = F_1 + c_1 D \tag{2}$$

is the solution (6.31) of the problem 6.30: the operator i_1 is the result of adding the function-operator F_1 to the impulse of magnitude $= c_1$ applied at $t = 0$.

If F is a function-operator such that the equation

$$i_1 = F + cD$$

holds for some number c, then

$$F + cD = F_1 + c_1 D, \tag{3}$$

whence $F - F_1 = (c_1 - c)\, D$; consequently, $(c_1 - c) D$ is a function-operator: the conclusion $c_1 = c$ is now immediate from 6.34; substituting our conclusion $c_1 = c$ into (3), we obtain

$$F = F_1 \quad \text{and} \quad c_1 D = cD.$$

We have just proved that *the decomposition* (2) *is unique*: there is only one function-operator F and only one impulse I applied at $t=0$ such that $i_1 = F + I$. In analogy with the complex number situation, we could say that $(14/3)D$ is the *impulse part* of the operator i_1: it is the operator I such that $i_1 - I$ is a function-operator.

Our newly acquired terminology can be applied to conclude that a constant voltage applied at $t=0$ in the circuit of Problem 6.30 causes a current i_1 whose impulse part is $\neq 0$.

6.45 If $\delta_1(\,)$ is a non-negative function of magnitude $=1$ that vanishes on the negative axis, then

$$D = \lim_{n\to\infty} \langle n\,\delta_1(nt)\rangle \qquad \text{(see 15.21)}.$$

Suppose that $\alpha \geq 0$; multiplying by T_α both sides of the above equation, we may use 6.9 to obtain

$$(6.46) \qquad D\mathsf{T}_\alpha = \lim_{n\to\infty} \mathsf{T}_\alpha \langle n\,\delta_1(nt)\rangle \qquad \text{(when } \alpha \geq 0).$$

For example,

$$n\,\delta_1(nt) = n\exp(-nt),$$

$$n\,\delta_1(nt) = \left(\frac{2n}{\pi}\right)\frac{1}{1+n^2t^2},$$

$$n\,\delta_1(nt) = \frac{2}{\sqrt{\pi}}\exp(-n^2t^2).$$

Let us examine the case where $\delta_1(\,)$ is the function described by the following graph:

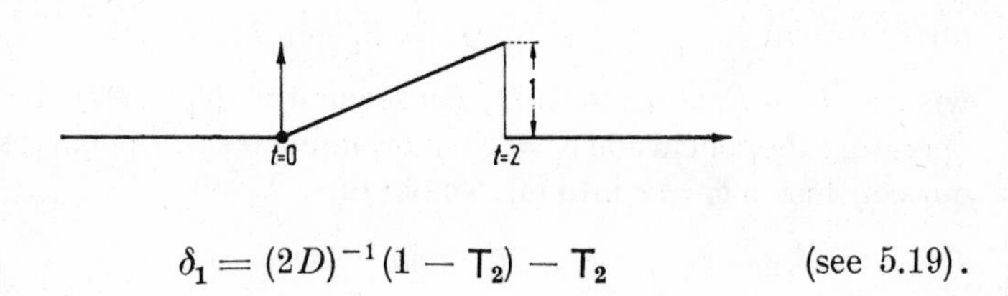

$$\delta_1 = (2D)^{-1}(1-\mathsf{T}_2) - \mathsf{T}_2 \qquad \text{(see 5.19)}.$$

Let $\delta_n(\,)$ be the function $\{n\,\delta_1(nt)\}(\,)$: its graph has the following shape:

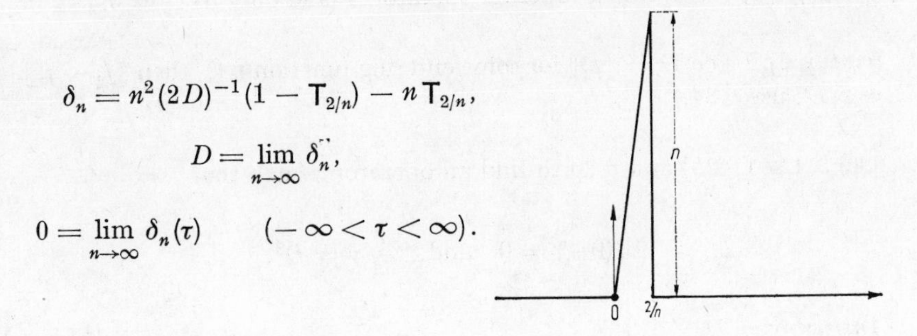

$$\delta_n = n^2(2D)^{-1}(1 - \mathsf{T}_{2/n}) - n\,\mathsf{T}_{2/n},$$

$$D = \lim_{n\to\infty} \delta_n^{\cdot\cdot},$$

$$0 = \lim_{n\to\infty} \delta_n(\tau) \qquad (-\infty < \tau < \infty).$$

The first operator-equation is from 5.19.

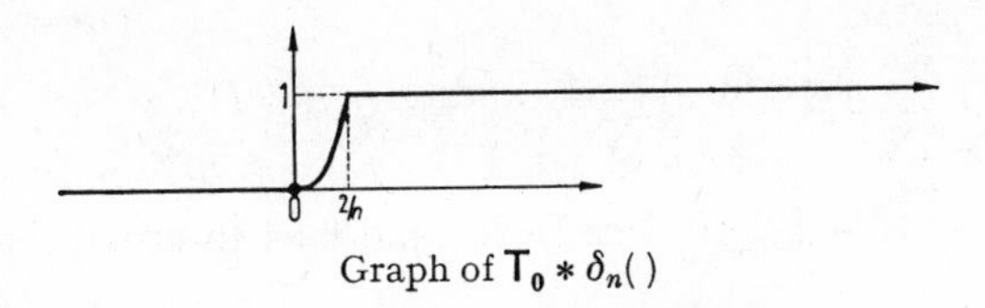

Graph of $\mathsf{T}_0 * \delta_n(\,)$

6.47 **Dipole.** Right-multiplying by D both sides of 6.39, we may use 6.9 to obtain

$$D^2\mathsf{T}_\alpha = \lim_{\substack{\varepsilon\to 0\\ \varepsilon>0}} \left(\frac{1}{\varepsilon}\,D\mathsf{T}_\alpha - \frac{1}{\varepsilon}\,D\mathsf{T}_{\alpha+\varepsilon}\right): \tag{6.48}$$

this could be translated into words by saying that the operator $D^2\mathsf{T}_\alpha$ is the limiting case (as $\varepsilon \to 0$) of two impulses: one (of magnitude $= 1/\varepsilon$) applied at $t = \alpha$, the other (of magnitude $= -1/\varepsilon$) applied at $t = \alpha + \varepsilon$. In the terminology of physics, $D^2\mathsf{T}_\alpha$ is a *dipole* applied at $t = \alpha$. For more information about $D^2\mathsf{T}_\alpha$, see 6.80—83.

Exercises

6.49.0 Find two operators i_1 and i_2 such that

$$3i_1 - i_2 = 14D \quad \text{and} \quad 6Di_2 + i_2 - i_1 = 0.$$

Answer: $i_1 = 7\left[\frac{2}{3}\,D + \frac{1}{27}\,\mathrm{e}^{-t/9}\right]$ and $i_2 = \frac{7}{9}\,\mathrm{e}^{-t/9}$.

6.49.1 Prove 6.28.

6.49.2 Given a function-operator $[\![f_1]\!]$ and a number c_1; prove that the operator $[\![f_1]\!] + c_1 D$ is a function-operator if (and only if) $c_1 = 0$.

Hint: if $[\![f_1]\!] + c_1 D = [\![f_2]\!]$ for some entering function $f_2()$, then $[\![f_2 - f_1]\!] = c_1 D$; use 6.34.

6.49.3 Use [6.25] and 6.20 to find an operator y such that

$$\{y\}(0-) = 0 \quad \text{and} \quad \frac{\partial}{\partial t} y = D^2.$$

Answer: $y = D$.

6.49.4 Use [6.25] and 6.20 to prove that there is no operator y such that

$$\{y\}(0-) = 1 \quad \text{and} \quad \frac{\partial}{\partial t} y = D^2.$$

6.49.5 Suppose that $\alpha \geq 0$. Use [6.25] and 6.54 to prove that

$$\frac{\partial}{\partial t} \delta(t - \alpha) = D\delta(t - \alpha) = D^2 \mathsf{T}_\alpha.$$

Distributions and Operators

The following §§ 6.50—54 are intended for the reader who is somewhat acquainted with the theory of distributions. Suppose that $f()$ is a distribution whose support is bounded to the left; it can be shown that the convolution $f * \psi()$ is a test-function whenever $\psi()$ is a test-function.

6.50 **Definition.** If $f()$ is a distribution with left-bounded support, we denote by $[\![f]\!]$ the mapping that assigns to each test-function $\varphi()$ the test-function $f * \varphi'()$:

(6.51) $$[\![f]\!] \cdot \varphi() = f * \varphi'() \qquad \text{for every test-function } \varphi().$$

6.52 In view of 0.16 and 0.35, Definition 6.51 is equivalent to [1.17] in the particular case where $f()$ is a function. It can be shown that $[\![f]\!]$ is a perfect operator; in fact, the correspondence $f() \mapsto [\![f]\!]$ is a one-to-one mapping of $\mathscr{D}'_+$ onto the algebra of perfect operators ($\mathscr{D}'_+$ is the space of all distributions with left-bounded supports; the surjectivity has been proved by HARRIS SHULTZ). Let α be a real number; the *Dirac Delta* is a distribution $\delta(t-\alpha)(\ )$ such that the equation

$$(1) \qquad \delta(t-\alpha) * \psi(\tau) = \psi(\tau-\alpha) \qquad (-\infty < \tau < \infty)$$

holds for any test-function $\psi(\ )$. In view of 1.35, Equation (1) implies that

$$(2) \qquad \delta(t-\alpha) * \varphi'(\ ) = [\![\mathsf{T}_\alpha]\!] \cdot \varphi'(\ ) \qquad \text{for every test-function } \varphi(\ ).$$

Combining [6.51] with (2), we obtain

$$(3) \qquad [\![\delta(t-\alpha)]\!] \cdot \varphi(\ ) = [\![\mathsf{T}_\alpha]\!] \cdot \varphi'(\ ) = [\![\mathsf{T}_\alpha]\!] D \cdot \varphi(\ );$$

the last equation is from [0.7] and [6.0]. From 6.3, (3), and 1.9 it now follows that

$$(6.53) \qquad \boxed{[\![\delta(t-\alpha)]\!] = [\![\mathsf{T}_\alpha]\!] D = D[\![\mathsf{T}_\alpha]\!]} \qquad (-\infty < \alpha < \infty).$$

In case $\alpha \geq 0$ we shall write $\delta(t-\alpha)$ instead of $[\![\delta(t-\alpha)]\!]$; from 2.11 we therefore have

$$(6.54) \qquad \boxed{\delta(t-\alpha) = D\mathsf{T}_\alpha} \qquad (\text{when } \alpha \geq 0).$$

Two Examples

6.54.1 Consider a particle of mass m attached to a weightless spring having spring constant $= b$; the particle slides on a smooth horizontal surface. When the particle is subjected to a force f, NEWTON'S law of motion asserts that the equation

$$(1) \qquad m \frac{\partial^2 y}{\partial t^2} + by = f$$

governs the displacement y from the equilibrium position.

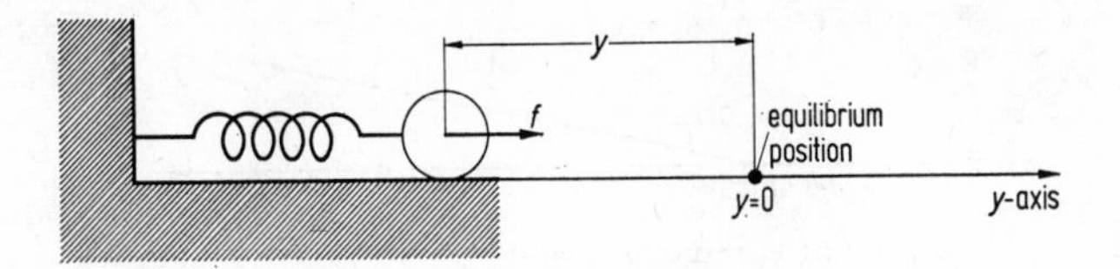

In view of [3.9], Equation (1) is equivalent to

$$m\left(D^2y - y(0-)D^2 - y'(0-)D\right) + by = f;$$

solving for y:

$$(2)\qquad y = \frac{y(0-)D^2 + y'(0-)D}{D^2+c^2} + \frac{m^{-1}f}{D^2+c^2} \qquad \left(\text{with } c = \sqrt{\frac{b}{m}}\right).$$

We shall consider two particular cases. First, suppose that the particle is initially at the equilibrium position with velocity v (that is, $y(0-) = 0$ and $y'(0-) = v$); further, let the particle be subjected to an impulse of magnitude $= -mv$ applied at $t = \alpha$: in view of 6.42, this means that $f = -mvD\mathsf{T}_\alpha$; substituting into (2), we obtain

$$y = \frac{v}{c}\langle \sin ct\rangle - v\mathsf{T}_\alpha \frac{D}{D^2+c^2} \qquad \text{by 3.25}$$

$$= \frac{v}{c}\{\langle \sin ct\rangle - \mathsf{T}_\alpha\langle \sin ct\rangle\} \qquad \text{by 3.25}$$

$$= v\sqrt{\frac{m}{b}}\left\{\sin\left[t\sqrt{\frac{b}{m}}\right] - \mathsf{T}_\alpha(t)\sin\left[(t-\alpha)\sqrt{\frac{b}{m}}\right]\right\} \qquad \text{by 3.12.}$$

In case $\alpha = 0$, we see that $y = 0$: *the motion is brought to a stop* by the impulse applied at $t = \alpha = 0$ having magnitude $= -mv$ (compare with the non-classical problem 2.50).

Last particular case: $b = 0 = y(0-) = y'(0-)$ and $f = 8\delta(t-\alpha)$ with $\alpha \geq 0$. This means that there is no spring, the particle is at rest initially at the origin and is subjected at the time $t = \alpha$ to an impulse of magnitude $= 8$ (see 6.54). Consequently, $f = 8D\mathsf{T}_\alpha$, and Equation (2) becomes

$$y = \frac{m^{-1}\,8D\mathsf{T}_\alpha}{D^2} = \frac{8}{m}\mathsf{T}_\alpha D^{-1} = \frac{8}{m}\mathsf{T}_\alpha\langle t\rangle = \frac{8}{m}\mathsf{T}_\alpha(t)\,(t-\alpha):$$

the last three equations are from 3.18 and 3.12.

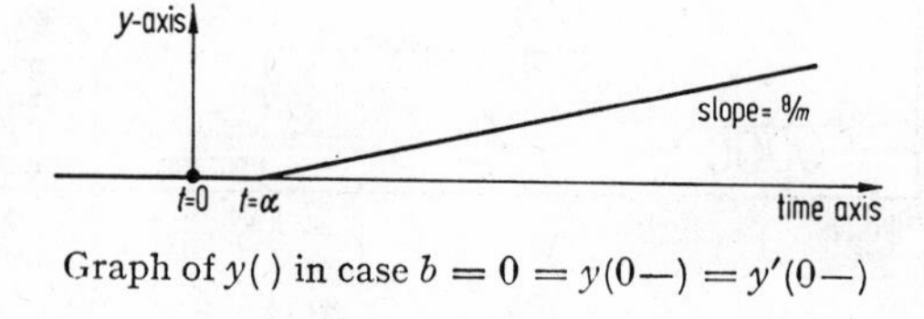

Graph of $y(\,)$ in case $b = 0 = y(0-) = y'(0-)$

6.55 **Second example.** The equations

$$i_1 = \left(\frac{1}{R_1} + C_1 \frac{\partial}{\partial t}\right) E_1 = -\left(\frac{1}{R_2} + C_2 \frac{\partial}{\partial t}\right) E_2, \tag{3}$$

$$i_2 = -i_1, \tag{4}$$

$$E_1 = E_2 \tag{5}$$

govern the currents (i_1, i_2) and the voltages $(E_1(), E_2())$ in the circuit

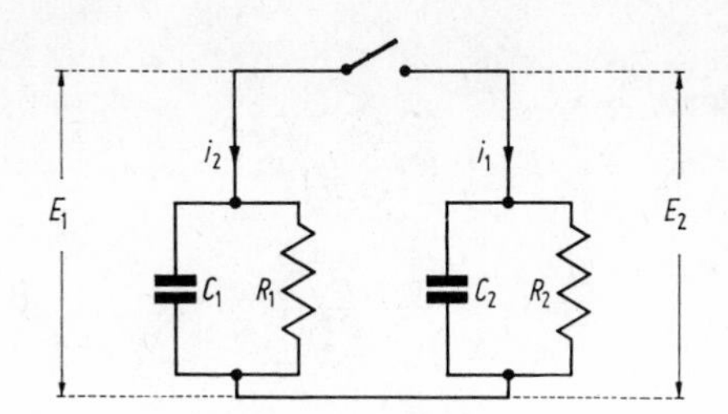

Closing the switch at the time $t = 0$, we connect the initially charged capacitor C_1 to the initially uncharged capacitor C_2:

$$E_1(0-) = v \quad \text{and} \quad E_2(0-) = 0. \tag{6}$$

Thus, it is possible to have simultaneously the two relations $E_1 = E_2$ and $E_1(0-) \neq E_2(0-)$; in this connection, it is important to note that both $E_1()$ and $E_2()$ are supposed to be $\mathcal{K}$-functions (this hypothesis will be removed in 7.15.1); as can be seen from 2.2, the equation (5) does *not* imply anything about the behavior of these functions on the left-hand side of $t = 0$: in fact, the equation (5) does not contradict the implication

$$E_1(0-) \neq E_2(0-) \qquad \text{(which comes from (6))}.$$

From (3), (6) and [3.7] we obtain

$$\left(\frac{1}{R_1} + C_1 D\right) E_1 - C_1 v D = -\left(\frac{1}{R_2} + C_2 D\right) E_2;$$

combining this with (5), we can solve for E_1:

$$E_1 = \left(\frac{C_1 v}{C}\right) \frac{D}{D + R^{-1} C^{-1}} = \left(\frac{C_1 v}{C}\right) \mathrm{e}^{-t/RC}, \tag{7}$$

where

$$C \stackrel{\text{def}}{=} C_1 + C_2$$

and

$$R \stackrel{\text{def}}{=} \left(\frac{1}{R_1} + \frac{1}{R_2}\right)^{-1}.$$

From (7), 3.21 and (6) we see that the voltage $E_1()$ has a jump at $t = 0$:

$$E_1(0+) - E_1(0-) = \frac{C_1 v}{C} - v.$$

Let us find i_2:

$$\begin{aligned} i_2 &= \frac{1}{R_2} E_2 + C_2 \frac{\partial}{\partial t} E_2 && \text{by (4)–(3)} \\ &= \frac{1}{R_2} E_2 + C_2 D E_2 && \text{by [3.7] and (6)} \\ &= \left(\frac{1}{R_2} + C_2 D\right) E_1 && \text{by (5)}. \end{aligned}$$

From (7) it now follows that

$$i_2 = \frac{C_1 v}{C}\left[\frac{1}{R_2}\left(\frac{D}{D + R^{-1}C^{-1}}\right) + C_2\left(D - \frac{D R^{-1} C^{-1}}{D + R^{-1} C^{-1}}\right)\right];$$

consequently,

$$i_2 = \frac{C_1 v C_2}{C}\,\delta(t) + \frac{C_1 v C_2}{C}\left[\frac{1}{R_2 C_2} - \frac{1}{RC}\right] e^{-t/RC}:$$

recall that $\delta(t) = D$ (see 6.54 and 5.4—5).

Exercises

6.56.0 Solve the problem

$$y(0-) = c_0 \quad \text{with} \quad \frac{\partial y}{\partial t} - a y = \delta(t - \alpha) \qquad (\text{in case } \alpha > 0).$$

Hint: recall that $\delta(t - \alpha) = D\mathsf{T}_\alpha$.

Answer: $y(t) = \mathsf{T}_\alpha(t)\, e^{at - a\alpha} + c_0 e^{at} \qquad (-\infty < t < \infty).$

6.56.1 Solve the preceding problem in case $\alpha = 0$.

Answer: $y(t) = \mathsf{T}_0(t)\, e^{at} + c_0 e^{at} \qquad (-\infty < t < \infty)$.

6.56.2 Find a family of $\mathcal{K}$-functions $y()$ such that

$$\frac{\partial y}{\partial t} = \delta(t).$$

Answer: $y(t) = c + \mathsf{T}_0(t) \qquad (-\infty < t < \infty)$.

6.56.3 Suppose that $\alpha > 0$; solve the three following problems:

$$y(0-) = c_0 \quad \text{with} \quad \frac{\partial y}{\partial t} + y = 2t + \delta(t-\alpha),$$

$$y(0-) = c_0,\ y'(0-) = c_1 \quad \text{with} \quad \frac{\partial^2 y}{\partial t^2} = \sin t + \delta(t-\alpha),$$

$$y(0-) = y'(0-) = 0, \quad \frac{\partial^2 y}{\partial t^2} + 2\frac{\partial y}{\partial t} + 2y = \delta(t-\alpha).$$

Answers: $2t - 2 + \mathsf{T}_\alpha(t)\, e^{t-\alpha} + (2 + c_0)\, e^{-t}$,

$$c_0 + (c_1 + 1)\, t - \sin t + (t-\alpha)\, \mathsf{T}_\alpha(t),$$

$$[e^{-t+\alpha} \sin(t-\alpha)]\, \mathsf{T}_\alpha(t).$$

6.56.4 Solve for i the problem

$$i(0-) = 0 \quad \text{with} \quad R\frac{\partial i}{\partial t} + C^{-1} i = \frac{\partial E}{\partial t},$$

when $E(t) = R$ for $0 < t \leq \alpha$ but $E(t) = 0$ otherwise. Find the jump at $t = 0$ of the function $i()$. *Hint*: use 5.13.

Answer: $i(t) = \mathsf{T}_0(t)\, e^{-t/RC} - \mathsf{T}_\alpha(t)\, e^{-(t-\alpha)/RC} \qquad (-\infty < t < \infty)$,

and

$$i(0+) - i(0-) = 1.$$

Multiplication by t

Let T be the mapping that assigns to each test-function $\varphi()$ the function $T \cdot \varphi()$ defined by

$$T \cdot \varphi(\tau) = \tau \varphi(\tau) \qquad (-\infty < \tau < \infty): \tag{6.57}$$

a moment's thought will show that T is an operator. When V is an operator, we set

(6.58) $$\boldsymbol{t}V \stackrel{\text{def}}{=} TV - VD^{-1}TD.$$

The operator $\boldsymbol{t}V$ should not be confused with the product of the operators $t = D^{-1}$ and V. Let a and b be numbers; if V_1 and V_2 are perfect operators it is easily verified that

(6.59) $$\boldsymbol{t}(aV_1 + bV_2) = a\boldsymbol{t}V_1 + b\boldsymbol{t}V_2.$$

6.60 **Theorem.** *If A is a perfect operator, then*

(6.61) $$D[\boldsymbol{t}A] = A + \boldsymbol{t}[DA].$$

Proof. Since

$$[DT] \cdot \varphi(\tau) = \frac{d}{d\tau}[\tau\varphi(\tau)] = \varphi(\tau) + \tau\varphi'(\tau),$$

the equation $[DT] \cdot \varphi(\,) = \varphi(\,) + T \cdot \varphi'(\,)$ is valid for any test-function $\varphi(\,)$; consequently,

(1) $$DT = 1 + TD.$$

Observe that

$$D[\boldsymbol{t}A] = DTA - DAD^{-1}TD \qquad \text{by [6.58]}$$

$$= (1 + TD)A - DAD^{-1}TD \qquad \text{by (1)}$$

$$= A + T(DA) - (DA)D^{-1}TD.$$

Conclusion 6.61 is now immediate from [6.58].

6.62 **Theorem.** *If $f(\,)$ is a $\mathcal{K}$-function, then $\boldsymbol{t}f = \langle tf(t)\rangle$.*

Proof. Let $\varphi(\,)$ be any test-function; from [6.58] we see that

(2) $$\boldsymbol{t}f \cdot \varphi(\,) = T \cdot (f \cdot \varphi)(\,) - f \cdot \psi(\,),$$

where

(3) $$\psi(\,) = (D^{-1}TD \cdot \varphi)(\,) = D^{-1} \cdot (T \cdot \varphi')(\,).$$

If $-\infty < \tau < \infty$ then

$$(4) \qquad t f \cdot \varphi(\tau) = \tau (f \cdot \varphi)(\tau) - (f \cdot \psi)(\tau) \qquad \text{by (2) and [6.57]}$$

$$(5) \qquad = \tau (f \cdot \varphi)(\tau) - \int_0^\infty \psi'(\tau - u) f(u)\, du \qquad \text{by 2.2.}$$

The equations

$$(6) \qquad \psi'(x) = D \cdot \psi(x) = D \cdot (D^{-1} \cdot [T \cdot \varphi'])(x) = [T \cdot \varphi'](x) = x\varphi'(x)$$

are from [0.7], (3), and [6.57]. Combining (4)—(6), we obtain

$$(7) \qquad t f \cdot \varphi(\tau) = \tau (f \cdot \varphi)(\tau) - \int_0^\infty (\tau - u)\, \varphi'(\tau - u)\, f(u)\, du$$

$$= \tau (f \cdot \varphi)(\tau) - \tau (f \cdot \varphi)(\tau) + \int_0^\infty \varphi'(\tau - u)\, u f(u)\, du$$

$$= 0 + \langle t f(t) \rangle \cdot \varphi(\tau):$$

the last two equations are from 2.2. This concludes the proof of 6.62.

6.63 *Remarks.* If $f(\,)$ is a $\mathcal{K}$-function, it follows from 6.62 and 6.17 that the equation

$$\{tf\}(\tau) = \tau f(\tau)$$

holds at each point $\tau > 0$ where $f(\,)$ is continuous: here $\{tf\}(\tau)$ is the value at τ of the operator tf (see 6.14.0). Suppose $\alpha \geq 0$: from 6.62 we have

$$t\mathsf{T}_\alpha = \langle t\mathsf{T}_\alpha(t) \rangle = \langle \mathsf{T}_\alpha(t)\,(t - \alpha) \rangle + \alpha \langle \mathsf{T}_\alpha(t) \rangle,$$

so that 3.12 and 2.1 give

$$(8) \qquad t\mathsf{T}_\alpha = \mathsf{T}_\alpha \langle t \rangle + \alpha \mathsf{T}_\alpha = D^{-1}\mathsf{T}_\alpha + \alpha \mathsf{T}_\alpha.$$

Left-multiplying by D both sides of (8):

$$(9) \qquad D[t\mathsf{T}_\alpha] = \mathsf{T}_\alpha + \alpha D \mathsf{T}_\alpha.$$

On the other hand, 6.60 gives

(10) $$D(\boldsymbol{t}\mathsf{T}_\alpha) = \mathsf{T}_\alpha + \boldsymbol{t}(D\mathsf{T}_\alpha).$$

Combining (10) with (9):

$$\mathsf{T}_\alpha + \boldsymbol{t}(D\mathsf{T}_\alpha) = \mathsf{T}_\alpha + \alpha D\mathsf{T}_\alpha;$$

solving for $\boldsymbol{t}(D\mathsf{T}_\alpha)$ this last equation:

(6.64) $$\boxed{\boldsymbol{t}[D\mathsf{T}_\alpha] = \alpha D\mathsf{T}_\alpha}.$$

Left-multiplying by D both sides of 6.64, we obtain

$$\alpha D^2\mathsf{T}_\alpha = D[\boldsymbol{t}(D\mathsf{T}_\alpha)] = D\mathsf{T}_\alpha + \boldsymbol{t}(D^2\mathsf{T}_\alpha) \qquad \text{(by 6.60)};$$

solving for $\boldsymbol{t}(D^2\mathsf{T}_\alpha)$:

(6.65) $$\boxed{\boldsymbol{t}(D^2\mathsf{T}_\alpha) = \alpha D^2\mathsf{T}_\alpha - D\mathsf{T}_\alpha}.$$

We can use the notation introduced in 6.54 to write 6.64—65 in the following form:

(6.66) $$\boldsymbol{t}\delta(t-\alpha) = \alpha\delta(t-\alpha),$$

(6.67) $$\boldsymbol{t}[D\delta(t-\alpha)] = \alpha D\delta(t-\alpha) - \delta(t-\alpha).$$

Static Deflection of Beams

As we shall see presently, the operators $D\mathsf{T}_\alpha$ and $D^2\mathsf{T}_\alpha$ occur in practical problems connected with the loading of beams. Recall that $D\mathsf{T}_\alpha$ is the operator of the Dirac Delta:

(6.68) $$\delta(t-\alpha) = D\mathsf{T}_\alpha \qquad \text{(see 6.54)}.$$

In the above equation — and until further notice — we suppose $\alpha \geq 0$. As we saw in 6.57—67, the correspondence $V \mapsto \boldsymbol{t}V$ extends to operators the correspondence $f \mapsto tf(t)$ that assigns to any $\mathcal{K}$-function $f(\,)$ the result of multiplying it by the independent variable t; we shall define the *moment* of an operator V as the magnitude of the operator $\boldsymbol{t}V$ (see [6.78]).

First, a few recalls and comments. An operator A is said to be **of magnitude** $= a$ if (and only if)

(6.69) $$a = \int_{-\infty}^{\infty} A \qquad \text{(see 6.24—23).}$$

From [6.24] and 6.20 we see that

(6.70) $$\int_{-\infty}^{\infty} D^2 \mathsf{T}_\alpha = 0 .$$

Also, recall that

(6.71) $$\int_{-\infty}^{\infty} c D \mathsf{T}_\alpha = c \qquad \text{(see 6.27).}$$

6.72 Consider a uniform beam of unit flexural rigidity whose equilibrium position lies along the t-axis, with end-points at $t = 0$ and at $t = l$. If the beam is subjected to a load f (which need not be a function-operator), then the upwards deflection y satisfies the equation

(6.73) $$\frac{\partial^4 y}{\partial t^4} = -f .$$

In view of [3.11], Equation 6.73 is equivalent to

(6.74) $$y = \frac{-f}{D^4} + y(0-) + \frac{y'(0-)}{D} + \frac{y''(0-)}{D^2} + \frac{y^{(3)}(0-)}{D^3} .$$

6.75 We suppose throughout that $0 \leq \alpha < \alpha + \varepsilon \leq l$. When c is a number, the operator $c(\mathsf{T}_\alpha - \mathsf{T}_{\alpha+\varepsilon})$ can be described as a *uniform load* applied to the interval $(\alpha, \alpha + \varepsilon)$: recall that $c(\mathsf{T}_\alpha - \mathsf{T}_{\alpha+\varepsilon})$ is the canonical operator of the $\mathcal{K}$-function $F()$ defined by

(1) $$F(\tau) = \begin{cases} c & (\alpha < \tau \leq \alpha + \varepsilon) \\ 0 & \text{(otherwise):} \end{cases} \qquad \text{see 3.4.}$$

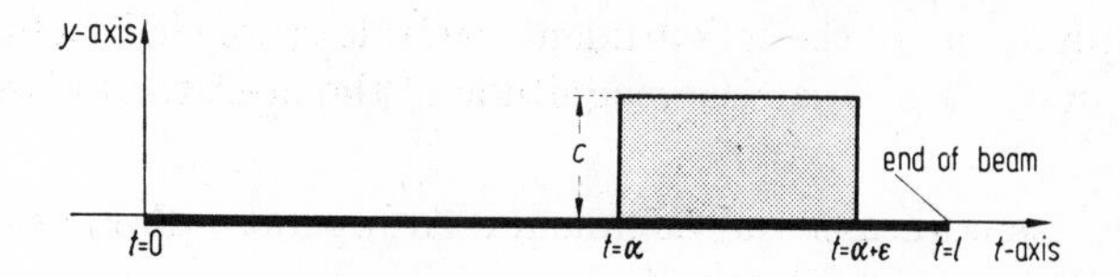

Since

$$\int_{-\infty}^{\infty} c(\mathsf{T}_\alpha - \mathsf{T}_{\alpha+\varepsilon}) = c\varepsilon \tag{6.76}$$ (by 6.29.1 and (1)).

we can say that the equation

$$\frac{\partial^4}{\partial t^4} y = -c(\mathsf{T}_\alpha - \mathsf{T}_{\alpha+\varepsilon})$$

governs the deflection y of a beam subjected to a uniform load of magnitude $= c\varepsilon$ applied to the interval $(\alpha, \alpha + \varepsilon)$.

6.77 **Force applied at a point.** Consider a uniform load f_ε applied to an interval $(\alpha, \alpha + \varepsilon)$: there exists a number c such that $f_\varepsilon = c(\mathsf{T}_\alpha - \mathsf{T}_{\alpha+\varepsilon})$. Moreover, suppose that f_ε is of magnitude $= a$: it then follows from 6.76 that $c\varepsilon = a$, whence

$$f_\varepsilon = \frac{a}{\varepsilon}(\mathsf{T}_\alpha - \mathsf{T}_{\alpha+\varepsilon}). \tag{2}$$

Note that

$$\lim_{\varepsilon \to 0} f_\varepsilon = \lim_{\varepsilon \to 0} \frac{a}{\varepsilon}(\mathsf{T}_\alpha - \mathsf{T}_{\alpha+\varepsilon}) = aD\mathsf{T}_\alpha: \tag{3}$$

the last equation is from 6.39. The load $aD\mathsf{T}_\alpha$ can be described as a *force applied at* $t = \alpha$; from 6.71 we see that it is of magnitude $= a$. Equations (2)—(3) imply that the force $aD\mathsf{T}_\alpha$ (applied at $t = \alpha$) is the limit (as $\varepsilon \to 0$) of a uniform load of magnitude $= a$ applied to the interval $(\alpha, \alpha + \varepsilon)$.

6.78 **Definitions.** Let V be an operator; the integral

$$\int_{-\infty}^{\infty} \boldsymbol{t} V$$

will be called the "*moment*" of V (here $\boldsymbol{t} V$ is the operator that was defined in 6.58; the integral was defined in 6.23). The expression mom(V) will be used as an abbreviation of "*moment of* V":

$$\mathrm{mom}(V) \overset{\mathrm{def}}{=} \int_{-\infty}^{\infty} \boldsymbol{t} V.$$

6.78.1 **Definition.** An operator V *is of moment* $= m$ if (and only if)

$$\operatorname{mom}(V) = m.$$

6.78.2 **Consequently,**

$$\operatorname{mom}(V) = m \quad \text{if (and only if)} \quad \int_{-\infty}^{\infty} tV = m.$$

6.79 **One more definition:** an operator V is called a *couple applied to the interval* $(\alpha, \alpha + \varepsilon)$ if (and only if) there exists a number c such that $V = Dc(\mathsf{T}_\alpha - \mathsf{T}_{\alpha+\varepsilon})$. Until further notice, suppose that $V = Dc(\mathsf{T}_\alpha - \mathsf{T}_{\alpha+\varepsilon})$; thus,

$$V = cD\mathsf{T}_\alpha - cD\mathsf{T}_{\alpha+\varepsilon}; \tag{4}$$

we can use the terminology of 6.77 to describe V as the load resulting from the following two forces: one force of magnitude $= c$ applied at $t = \alpha$, the other force of magnitude $= -c$ applied at $t = \alpha + \varepsilon$. Observe:

$$\int_{-\infty}^{\infty} tV = \int_{-\infty}^{\infty} ct(D\mathsf{T}_\alpha) + \int_{-\infty}^{\infty} -ct(D\mathsf{T}_{\alpha+\varepsilon}) \qquad \text{(by (4) and 6.59)} \tag{5}$$

$$= c\int_{-\infty}^{\infty} t(D\mathsf{T}_\alpha) - c\int_{-\infty}^{\infty} t(D\mathsf{T}_{\alpha+\varepsilon}) \tag{6}$$

$$= c\int_{-\infty}^{\infty} \alpha D\mathsf{T}_\alpha - c\int_{-\infty}^{\infty} (\alpha + \varepsilon)\, D\mathsf{T}_{\alpha+\varepsilon} \qquad \text{(by 6.64)} \tag{7}$$

$$= c\alpha - c(\alpha + \varepsilon) = -c\varepsilon \qquad \text{(by 6.71)}. \tag{8}$$

We may now use Definition 6.78 to say that V is **of moment** $= -c\varepsilon$. The situation can be depicted as follows:

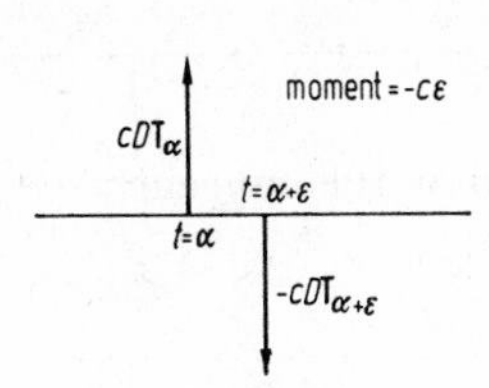

The load $cD\mathsf{T}_\alpha - cD\mathsf{T}_{\alpha+\varepsilon}$ in case $c > 0$

6.80 **Concentrated couple.** Let the beam be subjected to a couple V of moment $= -m$ applied to the interval $(\alpha, \alpha+\varepsilon)$: there exists a number c such that $V = cD\mathsf{T}_\alpha - cD\mathsf{T}_{\alpha+\varepsilon}$; since

$$-c\varepsilon = \operatorname{mom}(V) = -m \qquad \bigl(\text{by } (5)-(8)\bigr),$$

we have $c = m/\varepsilon$ and

$$V = \frac{m}{\varepsilon} D\mathsf{T}_\alpha - \frac{m}{\varepsilon} D\mathsf{T}_{\alpha+\varepsilon}.$$

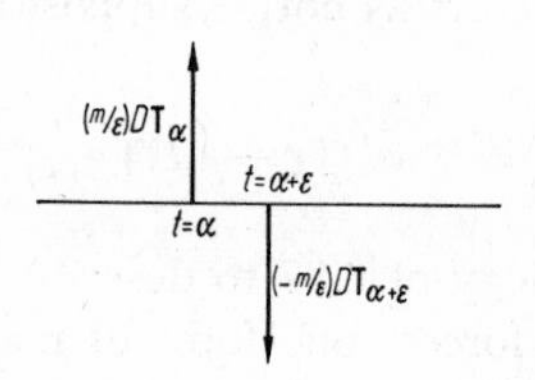

Therefore,

$$\lim_{\varepsilon\to 0} V = m\left[\lim_{\varepsilon\to 0}\left(\frac{1}{\varepsilon} D\mathsf{T}_\alpha - \frac{1}{\varepsilon} D\mathsf{T}_{\alpha+\varepsilon}\right)\right] = mD^2\mathsf{T}_\alpha:$$

the last equation is from 6.48. Thus, *$mD^2\mathsf{T}_\alpha$ is the limit (as $\varepsilon \to 0$) of a couple of moment $= -m$ applied to the interval $(\alpha, \alpha+\varepsilon)$.* In the terminology of physics, $mD^2\mathsf{T}_\alpha$ is a *concentrated couple applied at $t = \alpha$.* The equations

$$\int_{-\infty}^{\infty} \boldsymbol{t}(mD^2\mathsf{T}_\alpha) = \int_{-\infty}^{\infty} m\,\boldsymbol{t}(D^2\mathsf{T}_\alpha) = m\int_{-\infty}^{\infty} -D\mathsf{T}_\alpha + m\int_{-\infty}^{\infty} \alpha D^2\mathsf{T}_\alpha$$

are from 6.59 and 6.65; they imply that

$$(6.81) \qquad \boxed{\int_{-\infty}^{\infty} \boldsymbol{t}(mD^2\mathsf{T}_\alpha) = -m} \qquad (\text{by } 6.71-70).$$

Equation 6.81 states that the concentrated couple $mD^2\mathsf{T}_\alpha$ is of moment $= -m$; in particular,

$$\boxed{\operatorname{mom}(D^2) = -1}.$$

6.82 **Remark.** *If V is a concentrated couple of moment $= c$ applied at $t = \alpha$, then $V = -cD^2\mathsf{T}_\alpha$.*

This fact is an immediate consequence of 6.81 and the following definition: an operator V is called a *concentrated couple applied at* $t = \alpha$ if (and only if) there exists a number m such that $V = mD^2\mathsf{T}_\alpha$; consequently, if V is of moment $= c$, then

$$c = \operatorname{mom}(mD^2\mathsf{T}_\alpha) = -m \qquad \text{(by 6.81)},$$

which implies the conclusion $V = -cD^2\mathsf{T}_\alpha$.

6.83 We saw in 6.80 that the concentrated couple $D^2\mathsf{T}_\alpha$ is the limit of a couple of moment $= -1$; we now propose to verify that $D^2\mathsf{T}_\alpha$ is also the limit of a pair of uniform loads. The equation $\delta_1 = D^{-1}(1 - \mathsf{T}_1)^2$ defines a non-negative function $\delta_1(\,)$ of unit magnitude: its graph is sketched on the left-hand side below:

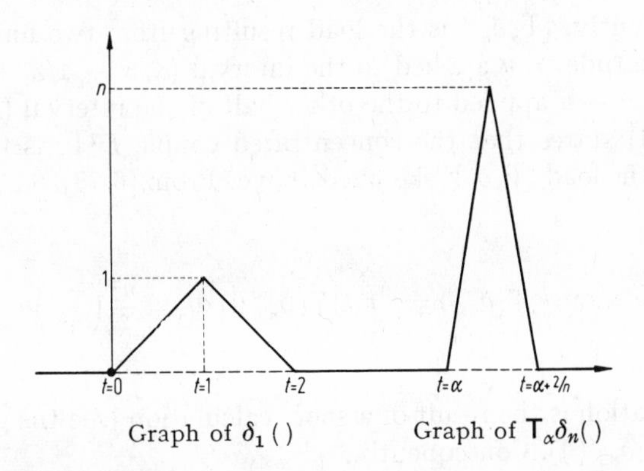

Graph of $\delta_1(\,)$ Graph of $\mathsf{T}_\alpha\delta_n(\,)$

From 6.46 it follows that

$$D\mathsf{T}_\alpha = \lim_{n\to\infty} \mathsf{T}_\alpha\delta_n, \qquad \text{where } \delta_n(\,) = \{n\delta_1(nt)\}(\,);$$

right-multiplying by D both sides, we can use 6.9 to obtain

$$D^2\mathsf{T}_\alpha = \lim_{n\to\infty} D\mathsf{T}_\alpha\delta_n = \lim_{n\to\infty} [\mathsf{T}_\alpha\delta_n]' : \tag{1}$$

the second equation is from the Derivation Property (3.8.1): recall that $\alpha \geq 0$, which implies $[\mathsf{T}_\alpha\delta_n](0-) = 0$. From the above graph of $\mathsf{T}_\alpha\delta_n(\,)$ it is easily seen that $[\mathsf{T}_\alpha\delta_n]'(\,)$ has the following graph

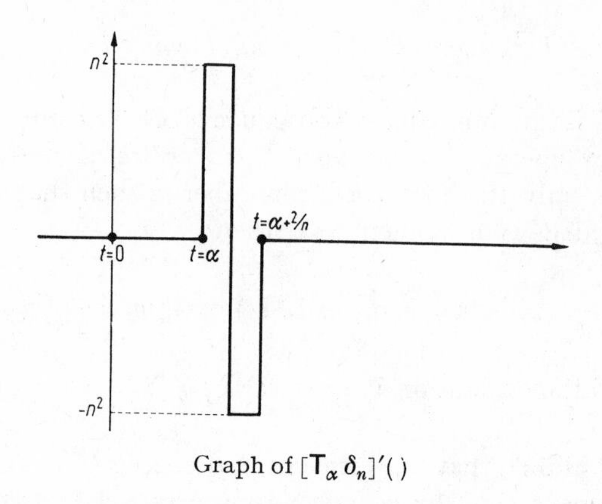

Graph of $[\mathsf{T}_\alpha \delta_n]'(\,)$

In fact,

$$[\mathsf{T}_\alpha \delta_n]' = n^2 [\mathsf{T}_\alpha - \mathsf{T}_{\alpha+1/n} - (\mathsf{T}_{\alpha+1/n} - \mathsf{T}_{\alpha+2/n})].$$

Consequently, $[\mathsf{T}_\alpha \delta_n]'$ is the load resulting from two uniform loads: one of magnitude $= n$ applied to the interval $(\alpha, \alpha + 1/n)$, the other of magnitude $= -n$ applied to the other half of the interval $(\alpha, \alpha + 2/n)$. Equation (1) states that the concentrated couple $D^2\mathsf{T}_\alpha$ is the limit (as $n \to \infty$) of the load $[\mathsf{T}_\alpha \delta_n]'$ sketched above. From [6.78], 6.29.1, and 6.62 we see that

$$\text{(2)} \qquad \operatorname{mom}([\mathsf{T}_\alpha \delta_n]') = \int_{-\infty}^{\infty} t [\mathsf{T}_\alpha \delta_n]'(t)\, dt = \frac{n^2}{2} \left(\frac{-2}{n^2}\right):$$

the last equation is the result of a short calculation (see the graph of the function $[\mathsf{T}_\alpha \delta_n]'(\,)$). Consequently,

$$\text{(3)} \qquad \operatorname{mom}([\mathsf{T}_\alpha \delta_n]') = -1 = \operatorname{mom}(D^2\mathsf{T}_\alpha) \qquad \text{(by (2) and 6.81)}.$$

The preceding discussion is based on a specific choice of the function $\delta_1(\,)$; let $\delta_1(\,)$ now be *any* non-negative function of magnitude $= 1$: as in (1), we may use 6.46 and 6.9 to obtain

$$D^2\mathsf{T}_\alpha = \lim_{n\to\infty} [\mathsf{T}_\alpha \delta_n]'.$$

The shape of the graph of $\delta_1(\,)$ is not important; for example, we could have

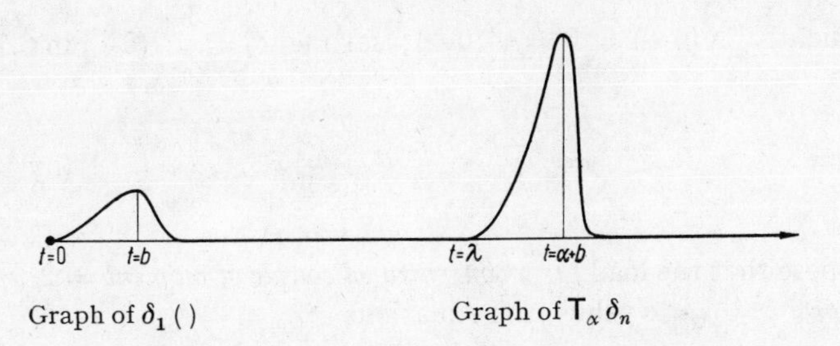

Graph of δ_1 ( ) Graph of $\mathsf{T}_\alpha\,\delta_n$

which gives the following kind of graph for the function $[\mathsf{T}_\alpha\delta_n]'(\,)$:

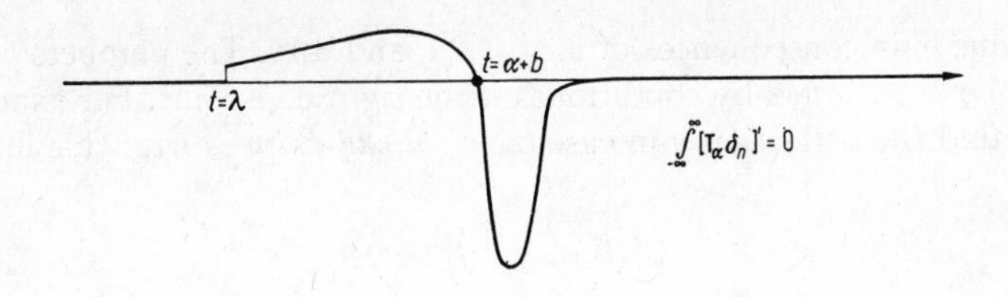

6.84 **Static deflection of a beam.** Let a uniform beam be subjected to the following combination of loads:

(1) a concentrated couple of moment $= c$ applied at $t = \gamma$,
(2) a force of magnitude $= b$ applied at $t = \beta$,
(3) a uniform load of magnitude $= A$ applied to the interval $(\alpha, \alpha + \varepsilon)$;

in this case, the deflection upwards y satisfies the equation

$$\text{(6.85)} \qquad \frac{\partial^4 y}{\partial t^4} = -\left[-cD^2\mathsf{T}_\gamma + bD\mathsf{T}_\beta + \frac{A}{\varepsilon}\left(\mathsf{T}_\alpha - \mathsf{T}_{\alpha+\varepsilon}\right)\right]:$$

see 6.82 and 6.72.

6.86 **Cantilever beam.** Suppose that the beam is of length l and is clamped horizontally at $t = 0$:

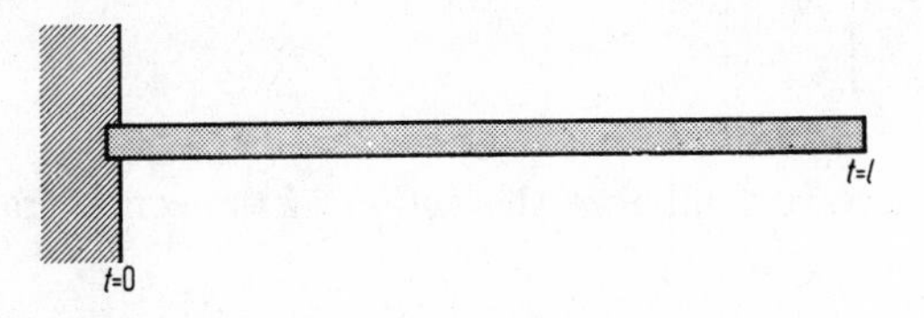

Consequently, $y(0-) = 0 = y'(0-)$; setting $c_k = y^{(k)}(0-)$ in 6.74, we obtain

$$y = \frac{-f}{D^4} + \frac{c_2}{2} t^2 + \frac{c_3}{6} t^3 \quad \text{(by 3.20)}. \tag{4}$$

Suppose that the load f *is a concentrated couple of moment* $= 2$ *applied at the point* $t = \alpha < l$; thus, the equations

$$\frac{-f}{D^4} = -\frac{-2D^2\mathsf{T}_\alpha}{D^4} = \mathsf{T}_\alpha \frac{2}{D^2} = \mathsf{T}_\alpha\langle t^2\rangle = \mathsf{T}_\alpha(t)\,(t-\alpha)^2 \tag{5}$$

are immediate consequences of 6.82, 3.20, and 3.12. The numbers c_1 and c_2 can be determined by conditions on the right end-point. For example, let us find the deflection y in case *the right end-point is free*: this implies that

$$y''(l) = y^{(3)}(l) = 0. \tag{6}$$

To find y, we first combine (4) and (5):

$$y = \frac{c_2}{2} t^2 + \frac{c_3}{6} t^3 + \mathsf{T}_\alpha(t)\,(t-\alpha)^2, \tag{7}$$

which implies that

$$y''(l) = c_2 + c_3 l + 2 \quad (\text{since } \alpha < l),$$

and $y^{(3)}(l) = c_3 = 0$. Consequently, (7) gives the positive deflection

$$y = -t^2 + \mathsf{T}_\alpha(t)\,(t-\alpha)^2 = \begin{cases} -t^2 & (t \le \alpha) \\ -2\alpha t + \alpha^2 & (t > \alpha). \end{cases}$$

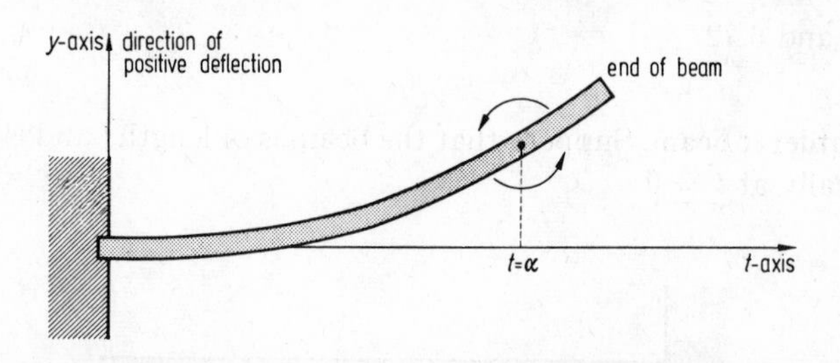

The two arrows symbolize the action of the concentrated couple at $t = \alpha$.

Exercises

6.87.0 A uniform cantilever beam is of length $2a$, and its right end-point is free. If the beam has weight w per unit length and is subjected to a force of magnitude $= wa$ applied at $t = a$, find the upwards deflection.

Hint: the equation is

$$\frac{\partial^4 y}{\partial t^4} = w[-(1-\mathsf{T}_{2a}) \quad a\,\delta(t-a)].$$

Answer: If $0 < t < a$, then

$$-w^{-1} y(t) = \frac{3a^2t^2}{2} - \frac{at^3}{2} + \frac{t^4}{4!} + \mathsf{T}_a(t)\frac{a(t-a)^3}{6}.$$

6.87.1 Let the beam in 6.87.0 be subjected to two loads: one uniform load of magnitude $= c$ applied to the first half of the beam, and a force of magnitude $= m$ applied at $t = \lambda < 2a$. Find the deflection upwards.

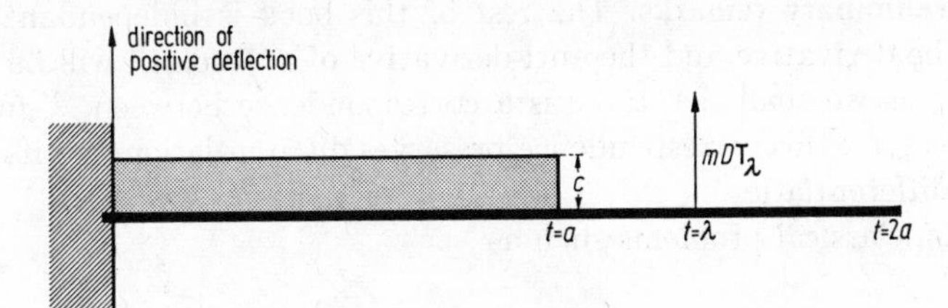

Answer: $\frac{c}{24}[\mathsf{T}_a(t)(t-a)^4 - t^4] - m\mathsf{T}_\lambda(t)\frac{(t-\lambda)^3}{6} - \frac{m\lambda t^2}{2} + \frac{mt^3}{6}$.

6.87.2 Determine the upwards deflection y of a simply supported beam of length l subjected to a force of magnitude $= 1$ applied at $t = \alpha < l$.

Hint: the end-conditions are

$$0 = y(0-) = y''(0-) = y(l) = y''(l).$$

Answer: $y = -\frac{(t-\alpha)^3}{3!}\mathsf{T}_\alpha(t) - \frac{(\alpha - l)\,\alpha(\alpha - 2l)\,t}{6l} - \frac{(\alpha - l)\,t^3}{6l}$.

6.87.3 Given three numbers c_0, c_1, and c_2; solve the problem

$$y(0-) = c_0, \quad y'(0-) = c_1, \quad y''(0-) = c_2$$

with

$$\frac{\partial^3 y}{\partial t^3} = 120t^2 + 2\delta(t-\alpha) - 5D^2\delta(t-\lambda) \qquad \text{and} \qquad 0 \le \alpha < \lambda.$$

Hint: recall 6.68.

Answer: $y = c_0 + c_1 t + (c_2/2)t^2 + 2t^5 + (t-\alpha)^2\mathsf{T}_\alpha(t) - 5\,\mathsf{T}_\lambda(t)$.

§ 7. Vectors

Motivation. The equations governing electric circuits involve not only derivatives, but also anti-derivatives; since (as we saw in 6.30 and 6.55) the currents in such circuits need not be function-operators, there arises the need to give a meaning to derivatives and anti-derivatives of electric currents that are not function-operators. The theoretical material presented in this § 7 will enable us to formulate such problems in the form customary in electrical engineering practice, and to solve them in the case of **non-zero initial values.**

Another motivation: in 6.55 the initial data combines with the existing machinery at our disposal to force us to assume* that the voltages are function-operators; although this assumption turned out to be consistent with the problem at hand, we shall presently be able to dispense with such *a priori* un-necessary assumptions on the solutions.

7.0 **Preliminary remarks.** The rest of this book is independant of this § 7. The derivative and the anti-derivative of a "vector" will be defined in 7.6; as we shall see, there is a correspondence between $\mathcal{K}$-functions and vectors: this correspondence preserves differentiation and its inverse (anti-differentiation).

Non-classical problems such as

$$\left(\frac{\partial^2}{\partial t^2} + 3\frac{\partial}{\partial t} + 2\right) y = 2D + 3D^2$$

will be solved in full generality: see 7.24—36.

First, recall that

$$\delta(t-\alpha) = D\mathsf{T}_\alpha \qquad \text{(for } \alpha \geq 0\text{):} \tag{7.1}$$

see 6.54. Note the particular case $\alpha = 0$:

$$\delta(t) = D; \tag{7.1.1}$$

as far as we are concerned, these two equations serve to introduce customary and convenient alternative notations for the operators $D\mathsf{T}_\alpha$ and D.

* As with the Laplace transform machinery: it forces us to assume that the solutions satisfy growth restrictions; such *a priori* assumptions are un-necessary and are needed only to make the machinery applicable.

7.2 **Definitions.** If $f()$ is a function and if V is an operator, then $(f(), V)$ denotes the ordered pair whose first element is $f()$ and whose second element is V: we set

$$\frac{d}{dt}(f(), V) \stackrel{\text{def}}{=} -f(0)D + DV \tag{7.3}$$

and

$$\int (f(), V) \stackrel{\text{def}}{=} \int_{-\infty}^{0} f(u)\, du + D^{-1}V. \tag{7.4}$$

7.5 **Examples.** If $f()$ is a $\mathcal{K}$-function such that $f(0) = f(0-)$, then [7.3] gives

$$\frac{d}{dt}(f(), f) = -f(0-)D + Df = \frac{\partial}{\partial t} f;$$

see [3.7] and recall that f is the canonical operator $\langle f(t)\rangle$ of the function $f()$.

7.5.1 In particular, if $f'()$ is a $\mathcal{K}$-function and if

$$f(\tau-) = f(\tau) = f(\tau+) \qquad \text{for each } \tau \geq 0,$$

then

$$\frac{d}{dt}(f(), f) = \left\langle \frac{d}{dt} f(t) \right\rangle:$$

this comes from 7.5 and 3.8.0.

7.5.2 **Theorem.** *If $f()$ is an entering function, then*

$$\int (f(), f) = \left\langle \int_{-\infty}^{t} f(u)\, du \right\rangle. \tag{7.5.3}$$

Proof. Definition 7.4 gives

$$\int (f(), f) = D^{-1}f + \int_{-\infty}^{0} f(u)\, du:$$

Conclusion 7.5.3 is now immediate from 3.13.

7.6 **Definitions.** A "*vector*" is an ordered pair whose first element (called "*the initial-part*") is a function, and whose second element (called "*the operator-part*") is a perfect operator. If $\boldsymbol{y}$ is a vector, then its initial-part will be denoted by $\boldsymbol{y}()$ and its operator-part will be denoted by y; consequently, $\boldsymbol{y}$ is the result of pairing the function $\boldsymbol{y}()$ with the operator y:

$$\boldsymbol{y} = (\boldsymbol{y}(), y) \tag{7.7}$$ (see [7.2])

and

$$\boxed{\frac{\mathrm{d}}{\mathrm{d}t}\boldsymbol{y} = -\boldsymbol{y}(0)D + Dy} \tag{7.8}$$ (by [7.3]);

further,

$$\int \boldsymbol{y} = \int_{-\infty}^{0} \boldsymbol{y}(u)\,\mathrm{d}u + D^{-1}y \tag{7.9}$$ (by [7.4]).

7.9.1 *Remarks.* To any $\mathcal{K}$-function $f()$ there corresponds the vector $\boldsymbol{f} = (f(), f)$ obtained by pairing the function with its canonical operator. If $f(0) = f(0-)$, then

$$\frac{\mathrm{d}}{\mathrm{d}t}\boldsymbol{f} = \frac{\partial}{\partial t} f \qquad \text{(see 7.5)}.$$

If $f'()$ is a $\mathcal{K}$-function and if $f()$ has no jumps on $[0, \infty)$, then

$$\frac{\mathrm{d}}{\mathrm{d}t}\boldsymbol{f} = \left\langle \frac{\mathrm{d}}{\mathrm{d}t} f(t) \right\rangle \qquad \text{(see 7.5.1)}.$$

If $f()$ is an entering function, then

$$\int \boldsymbol{f} = \left\langle \int_{-\infty}^{t} f(u)\,\mathrm{d}u \right\rangle \qquad \text{(see 7.5.2)}.$$

7.9.2 **Example.** The equation

$$y(t) = 2 + \mathsf{T}_0(t) \qquad (-\infty < t < \infty)$$

defines a function $y()$ whose corresponding vector $\boldsymbol{y} = (y(), y)$ has operator-part $y = 3$; obviously,

$$\boldsymbol{y}(0) = 2 \quad \text{and} \quad \frac{\mathrm{d}}{\mathrm{d}t}\boldsymbol{y} = \frac{\partial}{\partial t} y = D = \delta(t).$$

7.9.3 **Anti-derivative of a vector.** Let $\boldsymbol{y}^{(-1)}()$ be the function defined by

$$\boldsymbol{y}^{(-1)}(\tau) = \int_{-\infty}^{\tau} \boldsymbol{y}(u)\, \mathrm{d}u \qquad \text{(for each } \tau \leq 0); \tag{7.10}$$

consequently, [7.9] becomes

$$\boxed{\int \boldsymbol{y} = \boldsymbol{y}^{(-1)}(0) + D^{-1}y}. \tag{7.11}$$

Finally, let $\boldsymbol{y}^{(-1)}$ be the vector obtained by pairing the function $\boldsymbol{y}^{(-1)}()$ with the operator 7.11:

$$\boldsymbol{y}^{(-1)} \stackrel{\text{def}}{=} (\boldsymbol{y}^{(-1)}(), \int \boldsymbol{y}). \tag{7.12}$$

7.13 **Theorem.** *If* $\boldsymbol{A}$ *is a vector, then*

$$\boxed{\frac{\mathrm{d}}{\mathrm{d}t} \boldsymbol{A}^{(-1)} = A}. \tag{7.14}$$

Proof. From [7.3] and [7.12] we have

$$\frac{\mathrm{d}}{\mathrm{d}t} \boldsymbol{A}^{(-1)} = -\boldsymbol{A}^{(-1)}(0)\, D + D \int \boldsymbol{A}$$

$$= -\boldsymbol{A}^{(-1)}(0)\, D + D[\boldsymbol{A}^{(-1)}(0) + D^{-1}A]:$$

the last equation is from [7.11]. Conclusion 7.14 is now at hand.

7.14.1 **Examples.** If $\boldsymbol{i}$ is a vector whose operator-part is equal to $3D + 2$, then [7.11] gives

$$\int \boldsymbol{i} = \boldsymbol{i}^{(-1)}(0) + 3 + 2t.$$

Any vector $\boldsymbol{i}_1$ whose operator-part i_1 is given by

$$i_1 = \left(\frac{14}{3}\right) \delta(t) + \left(\frac{7}{3}\right) \frac{D}{1 + 9D} \tag{7.14.2}$$

describes the current in Problem 6.30; the vector $\boldsymbol{i}_1^{(-1)}$ is the charge of the capacitor C_1: its operator-part is

(7.14.3)
$$\int \boldsymbol{i}_1 = \boldsymbol{i}_1^{(-1)}(0) + \frac{14}{3} + \left(\frac{7}{3}\right)\frac{1}{1+9D}$$
$$= \boldsymbol{i}_1^{(-1)}(0) + \frac{14}{3} + \frac{7}{3} - \left(\frac{7}{3}\right)e^{-t/9}.$$

The initial-part $\boldsymbol{i}_1^{(-1)}()$ of the vector $\boldsymbol{i}_1^{(-1)}$ can be considered as describing its past history.

7.14.4. Suppose that $\boldsymbol{y}_1$ and $\boldsymbol{y}_2$ are vectors; if c_1 and c_2 are numbers, then

$$c_1\boldsymbol{y}_1 + c_2\boldsymbol{y}_2 \overset{\text{def}}{=} (c_1 y_1() + c_2 y_2(), c_1 y_1 + c_2 y_2):$$

consequently, [7.11] gives

(7.14.5)
$$\int (\boldsymbol{y}_1 - \boldsymbol{y}_2) = \boldsymbol{y}_1^{(-1)}(0) - \boldsymbol{y}_2^{(-1)}(0) + D^{-1}(y_1 - y_2).$$

Electric Circuits

7.15 **First application.** The current law for the circuit

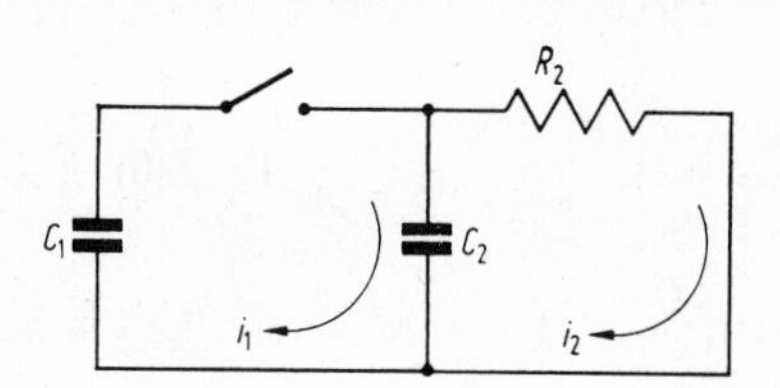

gives the following system of equations

(1)
$$\frac{1}{C_1}\int \boldsymbol{i}_1 + \frac{1}{C_2}\int (\boldsymbol{i}_1 - \boldsymbol{i}_2) = 0,$$

(2)
$$-\frac{1}{C_2}\int \boldsymbol{i}_1 + R_2 i_2 + \frac{1}{C_2}\int \boldsymbol{i}_2 = 0.$$

The number

$$\boldsymbol{i}_k^{(-1)}(0) = \int_{-\infty}^{0} \boldsymbol{i}_k(u)\, \mathrm{d}u \qquad \text{(see [7.10])}$$

is the initial charge on the capacitor C_k. We suppose that the initial charges on both capacitors equal $-7C_1$:

$$\text{(3)} \qquad \boldsymbol{i}_1^{(-1)}(0) = \boldsymbol{i}_2^{(-1)}(0) = -7C_1.$$

In view of 7.14.5, the system (1)—(2) becomes

$$(C_1^{-1} + C_2^{-1})(D^{-1}i_1 - 7C_1) - C_2^{-1}(D^{-1}i_2 - 7C_1) = 0,$$

$$-C_2^{-1}(D^{-1}i_1 - 7C_1) + R_2 i_2 + C_2^{-1}(D^{-1}i_2 - 7C_1) = 0;$$

effectuating the obvious cancellations, we obtain

$$(C_1^{-1} + C_2^{-1})\, i_1 - C_2^{-1} i_2 = 7D,$$

$$-C_2^{-1} i_1 + R_2 D i_2 + C_2^{-1} i_2 = 0:$$

this is precisely the system that we solved in 6.30 with the particular values $C_1 = 1$, $C_2 = 2$, and $R_2 = 3$. The operator i_1 is not a function-operator (in view of 6.49.2). The anti-derivative of $\boldsymbol{i}_1$ has been calculated in 7.14.2—14.3; in view of 7.14.3 and (3), the corresponding charge is

$$\int \boldsymbol{i}_1 = -7C_1 + 7 - \frac{7}{3}\, \mathrm{e}^{-t/9}.$$

7.15.1 In the electrical problem 6.55 we need not make the assumption that the voltages are function-operators; indeed, the problem 6.55 can be re-stated as follows:

$$i_1 = \frac{1}{R_1} E_1 + C_1 \frac{\mathrm{d}}{\mathrm{d}t} \boldsymbol{E}_1 = -\left(\frac{1}{R_1} E_2 + C_2 \frac{\mathrm{d}}{\mathrm{d}t} \boldsymbol{E}_2\right),$$

$$i_2 = -i_1, \qquad E_1 = E_2, \qquad \boldsymbol{E}_1(0) = v, \quad \text{and} \quad \boldsymbol{E}_2(0) = 0.$$

Solving this system of equations, we obtain the same answer as in 6.55.

7.15.2 **Another electric circuit.** The equations

$$2\int i_1 + \frac{d}{dt} i_1 - \frac{d}{dt} i_2 - \int i_2 = E, \tag{4}$$

$$2\int i_2 + \frac{d}{dt} i_2 - \frac{d}{dt} i_1 - \int i_1 = 0 \tag{5}$$

govern the current-vectors in a certain two-mesh circuit. To simplify matters, suppose that

$$i_1(0) = i_2(0) \tag{6}$$

and

$$i_1^{(-1)}(0) = i_2^{(-1)}(0) = 1 = E. \tag{7}$$

From [7.11] and [7.8] we see that the system (4)—(7) implies

$$2(1 + D^{-1} i_1) + D[i_1 - 0 - i_2] - (1 + D^{-1} i_2) = 1, \tag{8}$$

$$2(1 + D^{-1} i_2) + D[i_2 + 0 - i_1] - (1 + D^{-1} i_1) = 0: \tag{9}$$

the zero inside the two rectangular brackets stands for the number $i_1(0) - i_2(0)$, which equals zero by (6). Solving for i_2 the system (8)—(9):

$$i_2 = -\frac{\delta(t)}{2} - \left(\frac{1}{4}\sqrt{\frac{2}{3}}\right) \sin\left(t\sqrt{\frac{3}{2}}\right).$$

The corresponding charge is therefore

$$\int i_2 = \frac{1}{3} + \left(\frac{1}{6}\right) \cos\left(t\sqrt{\frac{3}{2}}\right). \tag{7.15.3}$$

7.16 **Simple electric circuits.** Let E be the voltage impressed on a circuit containing a resistance R, a capacitor C, and let L be an inductance; E can be *any* perfect operator (it need not be a function-operator).

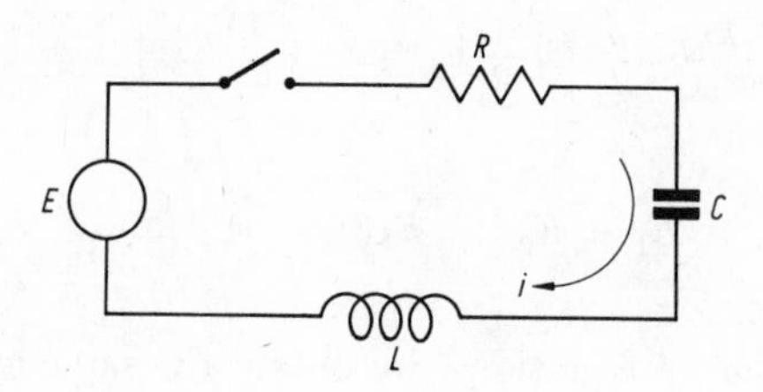

The equation

$$(7.17)\qquad L\frac{\mathrm{d}}{\mathrm{d}t}\boldsymbol{i} + Ri + \frac{1}{C}\int \boldsymbol{i} = E$$

governs the current vector $\boldsymbol{i}$. From [7.8] and [7.11] we can solve 7.17 to obtain

$$(7.18)\qquad i = \frac{E + \boldsymbol{i}(0)\,LD - C^{-1}\boldsymbol{i}^{(-1)}(0)}{LD + R + C^{-1}D^{-1}};$$

the number $\boldsymbol{i}(0)$ is the initial current; the number $\boldsymbol{i}^{(-1)}(0)$ is the initial charge of the capacitor.

7.19 **Short-circuit.** If $E = L = R = 0$, this means that the capacitor in the circuit

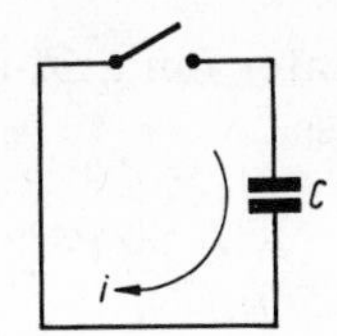

is short-circuited by closing the switch at $t = 0$. Equation 7.18 gives $i = -\boldsymbol{i}^{(-1)}(0)D$: this is an impulse applied at $t = 0$ whose magnitude $-\boldsymbol{i}^{(-1)}(0)$ equals the negative of the initial charge on the capacitor. We could also have reasoned directly by observing that 7.17 implies

$$(10)\qquad \int \boldsymbol{i} = 0;$$

In view of [7.11], Equation (10) means that

$$\boldsymbol{i}^{(-1)}(0) + D^{-1}i = 0:$$

the conclusion

$$i = -\boldsymbol{i}^{(-1)}(0)\;D = -\boldsymbol{i}^{(-1)}(0)\;\delta(t)$$

is now at hand.

7.20 **Miscellaneous situations.** Until further notice, let the capacitor have zero initial charge; this means that $\boldsymbol{i}^{(-1)}(0) = 0$ in Equation 7.18.

First, suppose that E is an impulse of magnitude $= b$ applied at $t = \alpha \geq 0$; in view of 6.41 and 7.1, this means that $E = b\,D\mathsf{T}_\alpha = b\,\delta(t-\alpha)$. We shall consider two particular cases. First, if $L \neq 0$ and $C^{-1} = 0 = \boldsymbol{i}(0)$, then 7.18 gives

$$i = b\mathsf{T}_\alpha \frac{D}{LD + R} = \frac{b}{L}\mathsf{T}_\alpha(t)\exp\left(-R(t-\alpha)/L\right):$$

the last equation is from 3.21 and 3.12. Our second particular case is $L = 0 = \boldsymbol{i}(0)$: Equation 7.18 becomes

$$i = b\mathsf{T}_\alpha D \frac{D}{RD + C^{-1}} = \frac{b}{R} \mathsf{T}_\alpha D \langle e^{-t/RC} \rangle \qquad \text{by 3.21}$$

$$= \frac{b}{R} \mathsf{T}_\alpha \left[\frac{-1}{RC} \langle e^{-t/RC} \rangle + D \right] \qquad \text{by 3.8.4;}$$

combining this with 3.12 and 7.1:

$$i = \frac{-b}{R^2 C} \mathsf{T}_\alpha(t)\, e^{-(t-\alpha)/RC} + \frac{b}{R}\, \delta(t - \alpha).$$

Finally, let E be the operator f of a $\mathcal{K}$-function $f(\,)$. If $C^{-1} = 0$ and $L \neq 0$, Equation 7.18 becomes

$$i = fD^{-1} \frac{D}{LD + R} + \boldsymbol{i}(0)\, e^{-tR/L} \qquad \text{(by 3.21)},$$

so that

$$i = L^{-1} \int_0^t f(t - u)\, e^{-uR/L}\, du + \boldsymbol{i}(0)\, e^{-tR/L}.$$

7.20.1 **Bibliographical comment.** The problems discussed in 7.15—20 can also be solved by means of the far-reaching theory developed in VÁCLAV DOLEŽAL's book "Dynamics of linear Systems" [P. Noordhoff, Groningen, and Academia, Prague, 1967].

Exercises

7.21.0 Closing the switch at $t = 0$, an impulsive voltage of magnitude $= 1$ is applied to the circuit

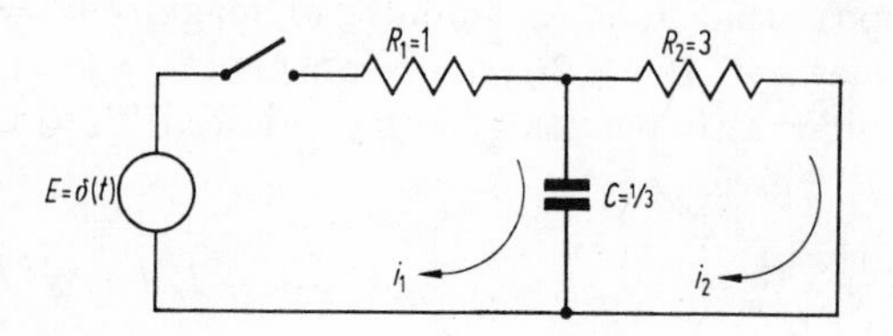

The current law gives

$$R_1 i_1 + \frac{1}{C}\int i_1 - \frac{1}{C}\int i_2 = \delta(t),$$

$$R_2 i_2 + \frac{1}{C}\int i_2 - \frac{1}{C}\int i_1 = 0.$$

Supposing that the capacitor has initial charge zero, solve this system for i_1 and i_2. *Hint*: use 7.11 and 7.1.1.

Answer: $i_1 = \delta(t) - 3\exp(-4t)$ and $i_2 = \exp(-4t)$.

7.21.1 The simple circuit described in 7.16 has $E = 0$, $L = 0$, and $R \neq 0$. If a is the initial charge on the capacitor, find the current and the corresponding charge.

Answers: $i = \frac{-a}{RC}\exp\left(\frac{-t}{RC}\right)$. The corresponding charge:

$$\int i = a\exp(-t/RC).$$

7.21.2 An impulse E of magnitude $= b$ is applied at $t = 0$ to the circuit described in 7.16. Moreover, suppose that $R = 0$ and let i_0 be the initial current. Find the current operator i in case there is no initial charge on the capacitor.

Answer: $i = \left(\frac{b}{L} + i_0\right)\cos\left(\frac{t}{\sqrt{LC}}\right)$.

7.21.3 Suppose that i is a vector such that

$$i^{(-1)}(0) = 1 \quad \text{and} \quad i = -\frac{\delta(t)}{2} - \left(\frac{1}{4}\sqrt{\frac{2}{3}}\right)\sin\left(t\sqrt{\frac{3}{2}}\right);$$

verify the that corresponding charge is given by the equation

$$\int i = \frac{1}{3} + \frac{1}{6}\cos\left(t\sqrt{\frac{3}{2}}\right).$$

7.21.4 Let $\boldsymbol{q}$ be a vector such that $\boldsymbol{q}(0) = 1$ and

$$q = \frac{1}{3} + \frac{D^2}{6D^2 + 9}:$$

Find $\frac{\mathrm{d}}{\mathrm{d}t}\boldsymbol{q}$.

Answer: $\frac{\mathrm{d}}{\mathrm{d}t}\boldsymbol{q} = -\frac{\delta(t)}{2} - \left(\frac{1}{4}\sqrt{\frac{2}{3}}\right)\sin\left(t\sqrt{\frac{3}{2}}\right)$.

7.21.5 The circuit described in 7.15 has $C_1 = 1$, $C_2 = 2$, and $R_2 = 3$. Find i_1 in case both capacitors have initially a charge equal to -9.

Answer: $i_1 = 6\delta(t) + \left(\frac{1}{3}\right)\exp\left(\frac{-t}{9}\right)$.

7.21.6 The simple electric circuit considered in 7.16 has $C^{-1} = 0$ and $L \neq 0$. If the initial current equals 5, find the current vector $\boldsymbol{i}$ in case

$$E = \sum_{k=0}^{\infty} \exp(k^4)\, D^k \mathsf{T}_k$$

Hint: E is a perfect operator (as we saw in 6.11).

Answers: $i = 5\exp(-Rt/L) + E(LD + R)^{-1}$, and

$$\boldsymbol{i}(t) = 5\exp(-Rt/L) \qquad \text{for each } t \leq 0.$$

7.21.7 If $\boldsymbol{B}$ is a vector, let $\boldsymbol{B}'$ be the result of pairing the derivative $\boldsymbol{B}'()$ of its initial-part $\boldsymbol{B}()$ with the operator $\mathrm{d}\boldsymbol{B}/\mathrm{d}t$:

$$\boldsymbol{B}' = \left(\boldsymbol{B}'(), \frac{\mathrm{d}}{\mathrm{d}t}\boldsymbol{B}\right).$$

Given a vector $\boldsymbol{A}$, use 7.12 and 7.13 to prove that $\boldsymbol{A}^{(-1)}$ is a vector $\boldsymbol{B}$ such that $\boldsymbol{B}' = \boldsymbol{A}$.

7.21.8 Let $\boldsymbol{y}$ be a vector whose initial-part $\boldsymbol{y}()$ is an entering function: use 7.21.7 to prove that $\boldsymbol{y}'$ is a vector $\boldsymbol{B}$ such that $\boldsymbol{B}^{(-1)} = \boldsymbol{y}$.

The General Case

7.22 Let y be the canonical operator of a function $y()$. Recall that

$$(0)\qquad \frac{\partial^k}{\partial t^k} y = D^k y - \sum_{s=0}^{k-1} y^{(s)}(0-)\, D^{k-s} \qquad (k = 1, 2, 3, \ldots).$$

Given two finite sequences

$$c_k (k = 0, 1, 2, \ldots, n-1) \quad \text{and} \quad a_k (k = 0, 1, 2, \ldots, n)$$

of numbers; let us denote by

$$(1)\qquad (\sum a_k D^k \,|||\, \sum c_k/D^k)$$

the impulse part of the operator

$$(2)\qquad (a_n D^n + \cdots + a_1 D + a_0)\left(c_0 + \frac{c_1}{D} + \cdots + \frac{c_{n-1}}{D^{n-1}}\right).$$

Thus, (1) *is the result of discarding all function-operators in the sum obtained by carrying out the multiplication* (2). In view of 6.36.1, this means that only the terms containing $D, D^2, \ldots, D^n$ are retained. For example,

$$(D^2 + a_1 D + a_0 \,|||\, c_0 + c_1/D) = (c_1 + a_1 c_0)\, D + c_0 D^2.$$

Let us verify that

$$(7.23)\qquad \boxed{(\sum a_k D^k \,|||\, \sum c_k/D^k) = \sum_{k=1}^{n} a_k \sum_{s=0}^{k-1} c_s D^{k-s}}.$$

To that effect, we carry out the multiplication indicated in (2):

$$\sum_{k=0}^{n} a_k \sum_{s=0}^{n-1} c_s D^k D^{-s} = a_0 \sum_{s=0}^{n-1} c_s D^{-s} + \sum_{k=1}^{n} a_k \left(\sum_{s=0}^{k-1} + \sum_{s=k}^{n-1}\right) c_s D^{k-s}$$

and discard the first sum on the right-hand side and the last sum inside the large parenthesis (the one with $s = k, k+1, \ldots, n-1$): in so doing, we discard all terms of the form bD^m with $m \leq 0$ and retain only the middle sum (the one consisting of terms of the form bD^m with $m = k - s \geq 1$). This concludes the proof of 7.23.

7.24 **Theorem.** *Let h be a given operator. If y is the canonical operator of a function $y()$ such that*

$$\left[a_n \frac{\partial^n}{\partial t^n} + a_{n-1} \frac{\partial^{n-1}}{\partial t^{n-1}} + \cdots + a_1 \frac{\partial}{\partial t} + a_0\right] y = h \tag{7.25}$$

and

$$y(0-) = c_0,\ y'(0-) = c_1,\ \ldots,\ y^{(n-1)}(0-) = c_{n-1}, \tag{7.26}$$

then

$$y = \frac{h + (\sum a_k D^k \,|||\, c_k/D^k)}{a_n D^n + a_{n-1} D^{n-1} + \cdots + a_1 D + a_0}. \tag{7.27}$$

Proof. From 7.25 and (0) it follows easily that

$$\sum_{k=0}^{n} a_k D^k y - \sum_{k=1}^{n} a_k \sum_{s=0}^{k-1} c_s D^{k-s} = h,$$

whence

$$\left(\sum_{k=0}^{n} a_k D^k\right) y = h + \sum_{k=1}^{n} a_k \sum_{s=0}^{k-1} c_s D^{k-s};$$

in view of 7.23, Conclusion 7.27 is at hand.

7.28 *Remark.* Consequently, if the problem 7.25—26 has a solution, then it has only one solution: that solution is given by 7.27.

7.29 **Terminology.** Let us call the given operator h the ***input.*** If we set

$$G = a_n D^n + a_{n-1} D^{n-1} + \cdots + a_1 D + a_0, \tag{7.30}$$

then 7.27 can be written

$$y = \frac{h}{G} + A_c, \quad \text{where} \quad A_c = \frac{(G \,|||\, \sum c_k/D^k)}{G}. \tag{7.31}$$

Let us accept the following fact without proof: the operator A_c is the canonical operator $\langle A_c(t)\rangle$ of an infinitely differentiable function $A_c()$ such that

$$A_c(0) = c_0,\ A_c'(0) = c_1, \ldots, A_c^{(n-1)}(0) = c_{n-1}. \tag{7.32}$$

Further, it can be shown that

(7.33) $$a_n A_c^{(n)}(t) + a_{n-1} A_c^{(n-1)}(t) + \cdots + a_1 A_c'(t) + a_0 A_c(t) = 0$$
$$(-\infty < t < \infty).$$

7.33.1 **A special case.** *If h/G is the canonical operator of a $\mathcal{K}$-function $F()$, then the equation*

(3) $$y(t) = A_c(t) + \mathsf{T}_0(t)\, F(t) \qquad (-\infty < t < \infty)$$

defines a function $y()$ satisfying the system 7.25—26.

7.33.2 **For example,** if the input h is the Dirac Delta operator $D = \delta(t)$, then

$$\frac{h}{G} = \frac{D}{G} = \frac{D}{a_n D^n + a_{n-1} D^{n-1} + \cdots + a_1 D + a_0};$$

since $n \geq 1$, it results from 4.23 that h/G is the canonical operator of a $\mathcal{K}$-function, called the **"Green's function"** of the problem 7.25—26 (the names "*transfer function*" and "*impulse response*" are also used).

7.34 **Remark.** The solution $y()$ defined by the equation (3) may have a jump at the point $t = 0$; it can also happen that several of the derivatives of $y()$ have jumps at the point $t = 0$.

7.35 **Example.** To find a function $y()$ such that

$$y(0-) = y'(0-) = c \quad \text{and} \quad \left(\frac{\partial^2}{\partial t^2} + 1\right) y = D^2.$$

Solving for y, we obtain

(4) $$y = \frac{c D^2 + c D}{D^2 + 1} + \frac{D^2}{D^2 + 1} = c \langle \cos t + \sin t \rangle + \langle \cos t \rangle.$$

Here h/G is the operator $\langle \cos t \rangle$: we have the situation discussed in 7.33.1; indeed, the equation

$$y(t) = c(\cos t + \sin t) + \mathsf{T}_0(t) \cos t \qquad (-\infty < t < \infty)$$

determines a solution of (4).

Exercises

7.36 Solve the problem

$$y(0-) = c_0, \quad y'(0-) = c_1 \quad \text{with} \quad \left(\frac{\partial^2}{\partial t^2} + 3\frac{\partial}{\partial t} + 2\right) y = 2D + 3D^2.$$

Hint: use 7.33.1 and 4.1.

Answer: If $-\infty < t < \infty$, then

$$y(t) = (2c_0 + c_1)\,e^{-t} - (c_0 + c_1)\,e^{-2t} + \mathsf{T}_0(t)\,(4e^{-2t} - e^{-t}).$$

7.37 Use 7.33.1 to solve the problem

$$y(0-) = y'(0-) = c \quad \text{with} \quad \left(\frac{\partial^2}{\partial t^2} - 1\right) y = D^2.$$

Hint: proceed as in 7.36.

Answer: $y(t) = c\,e^t + \mathsf{T}_0(t)\,(\cosh t) \qquad (-\infty < t < \infty).$

7.37.1 Use 7.33.1 to solve the problem

$$y(0-) = c \quad \text{with} \quad m\frac{\partial}{\partial t} y = \delta(t).$$

Answer: $y(t) = c + m^{-1}\,\mathsf{T}_0(t) \qquad (-\infty < t < \infty).$

§ 8. Non-integrable Functions

Set

$$\text{(8.0)} \qquad \mathsf{t}^{-n-1} \overset{\text{def}}{=} \frac{(-1)^n}{n!} D^{n+1}\langle \log t\rangle \qquad (n = 0, 1, 2, \ldots),$$

$$\text{(8.1)} \qquad \mathsf{t}^{-3/2} \overset{\text{def}}{=} D\langle -2t^{-1/2}\rangle,$$

and

$$\text{(8.2)} \qquad \mathsf{t}^{-5/2} \overset{\text{def}}{=} D^2\left\langle \frac{4}{3}\, t^{-1/2}\right\rangle.$$

Since $\{\log t\}(\ )$ and $\{t^{-1/2}\}(\ )$ are $\mathcal{K}$-functions, it follows from [2.1], 1.15, and 1.8 that each of the above equalities defines a perfect operator. Definition 2.1 does not assign an operator to the function $\{t^\alpha\}(\ )$ when $\alpha \leq -1$ (since this function is not a $\mathcal{K}$-function). From 6.14.2 it follows that the value of the operator t^α at any point $\tau > 0$ equals τ^α:

$$\boxed{\{\mathrm{t}^\alpha\}(\tau) = \tau^\alpha} \qquad \text{for each } \tau > 0 \tag{8.3}$$

— in case $\alpha = -3/2$, $\alpha = -5/2$, and $\alpha = -1, -2, -3, \ldots$ Let us verify 8.3 in the case $\alpha = -n - 1$:

$$\{\mathrm{t}^{-n-1}\}(\tau) = \left\{D^{n+1}\frac{(-1)^n}{n!}\langle\log t\rangle\right\}(\tau) \qquad \text{(by [8.0])}$$

$$= \frac{\mathrm{d}^{n+1}}{\mathrm{d}\tau^{n+1}}\left(\frac{(-1)^n}{n!}\log\tau\right) = \tau^{-n-1}:$$

the second equation is from 6.14.2.

8.4 **Application.** Let us find an operator y such that

$$D^2y + \sqrt{2}\,Dy + 2y = 2\sqrt{2}\,\langle\log t\rangle - 2\mathrm{t}^{-3}. \tag{1}$$

Since [8.0] gives $\mathrm{t}^{-3} = 2^{-1}D^3\langle\log t\rangle$, Equation (1) means that

$$y = \frac{(2\sqrt{2} - D^3)\langle\log t\rangle}{D^2 + \sqrt{2}\,D + 2} = (-D + \sqrt{2})\,\langle\log t\rangle: \tag{2}$$

the second equation is from 1.73. From (2) and [8.0] it now follows that

$$y = -\mathrm{t}^{-1} + \sqrt{2}\,\langle\log t\rangle. \tag{3}$$

8.5 Let a and b be numbers; from 6.14.5 we see that

$$\{a\mathrm{t}^{-n-1} + b\langle\log t\rangle\}(\tau) = a\,\{\mathrm{t}^{-n-1}\}(\tau) + b\,\{\langle\log t\rangle\}(\tau)$$

$$= a\tau^{-n-1} + b\log\tau \qquad \text{(all } \tau > 0):$$

the second equation is from 8.3 and 6.17.

8.6 From (3) and 8.5 it follows that

$$\{y\}(\tau) = -\tau^{-1} + \sqrt{2}\,\log\tau \qquad (\text{all } \tau > 0). \tag{4}$$

Consequently, the function $y()$ defined by $y(\tau) = \{y\}(\tau)$ for all $\tau > 0$ is infinitely differentiable on the interval $(0, \infty)$; we may therefore use 6.14.5 and 6.14.2 to write

$$\begin{aligned} y''(\tau) + \sqrt{2}\,y'(\tau) + 2y(\tau) &= \{D^2 y + \sqrt{2}\,Dy + 2y\}(\tau) \\ &= \{2\sqrt{2}\,\langle\log t\rangle - 2t^{-3}\}(\tau) \qquad (\text{by } (1)), \\ &= 2\sqrt{2}\,\log\tau - 2\tau^{-3} \qquad (\text{by } 8.5). \end{aligned}$$

In conclusion: the equation $y(\tau) = -\tau^{-1} + \sqrt{2}\,\log\tau$ (all $\tau > 0$) *defines a solution of the differential equation*

$$y''(\tau) + \sqrt{2}\,y'(\tau) + 2y(\tau) = 2\sqrt{2}\,\log\tau - 2\tau^{-3} \qquad (\text{all } \tau > 0).$$

Recall that this solution was obtained by solving the corresponding operator-equation (1) and writing down the function $y()$ whose values coincide with the values of the operator y (the number $\{y\}(\tau)$ is called *the value* at τ of the operator y: see 8.21).

8.7 In general, let $g()$ be a $\mathcal{K}$-function: the operator-equation

$$(a_n D^n + a_{n-1} D^{n-1} + \cdots + a_1 D + a_0 y)\, y = D^m \langle g(t)\rangle \tag{5}$$

implies the ordinary differential equation

$$a_n y^{(n)}(\tau) + a_{n-1} y^{(n-1)}(\tau) + \cdots + a_1 y'(\tau) + a_0 y(\tau) = g^{(m)}(\tau) \tag{6}$$

(all $\tau > 0$), where $y()$ is the function defined by $y(\tau) = \{y\}(\tau)$ for all $\tau > 0$. This fact can be established by an obvious extension of the reasoning in 8.5—6; it could be re-phrased as follows: *any solution y of the operator-equation* (5) *defines a function $y(\tau) = \{y\}(\tau)$ which satisfies the ordinary differential equation* (6). Note that $g^{(m)}()$ **does not have to be locally integrable.** Recall that

$$\{D^m\langle g(t)\rangle\}(\tau) = g^{(m)}(\tau) \qquad (\text{for each } \tau > 0).$$

— provided that the right-hand side exists: see 6.14.2.

Convolution of Operators

If $f()$ and $h()$ are entering functions, then $[\![f]\!]D^{-1}[\![h]\!]$ is the operator $[\![f * h]\!]$ of the function $f * h()$ (see 1.26); from 6.15 (with $V = [\![f]\!]D^{-1}[\![h]\!]$ and $g = f * h$) it therefore follows that

(1) $$\{[\![f]\!]D^{-1}[\![h]\!]\}(\tau) = f * h(\tau)$$

for all τ at which the function $f * h()$ is continuous. This suggests the following definition:

(8.8) $$\boxed{y * V \overset{\text{def}}{=} yD^{-1}V} \quad \text{(when } y \text{ and } V \text{ are any operators).}$$

In case y is the Dirac Delta operator D (see 7.1), Definition 8.8 implies that

(8.9) $$\boxed{D * V = V = V * D} \quad \text{(for any operator } V\text{).}$$

Suppose that $\alpha \geq 0$. If $f()$ is a $\mathcal{K}$-function, Definition 8.8 gives

(2) $$f * (D\mathsf{T}_\alpha) = \mathsf{T}_\alpha f.$$

From (2), [7.1], and 3.12 we obtain

$$f * \delta(t - \alpha) = \langle \mathsf{T}_\alpha(t)\, f(t - \alpha) \rangle.$$

Thus, if $f(t) = 0$ for $t \leq 0$:

(8.10) $$\boxed{f * \delta(t - \alpha) = \langle f(t - \alpha) \rangle}.$$

A more general convolution formula is given in 8.14.14.

8.10.1 Let $f()$ be a $\mathcal{K}$-function. From [8.8] and 3.15.1 it follows that the equation

(8.10.2) $$f * g = \{f(t)\} * \{g(t)\} = \left\langle \int_0^t f(t - u)\, g(u)\, du \right\rangle$$

holds for any $\mathcal{K}$-function $g()$: if at least one of the two functions $f()$ and $g()$ is regulated, we may use 2.61 and 6.17 to write

$$\{f * g\}(\tau) = \int_0^\tau f(\tau - u)\, g(u)\, du \qquad \text{(for each } \tau > 0\text{)}:$$

the right-hand side being the value at τ of the operator $f * g$.

Finite Part of a Divergent Integral

It is not hard to extend the definitions 8.0—2 by defining t^{-p} for *any* number $p \geq 1$. If $p \geq 1$ and if $f()$ is a $\mathcal{K}$-function, then the operator $f * \mathrm{t}^{-p}$ will be called **the finite part of the divergent integral**

$$\int_0^t \frac{f(u)}{(t-u)^p}\, du;$$

in symbols:

$$\text{(8.11)} \qquad \mathrm{FP} \int_0^t \frac{f(u)}{(t-u)^p}\, du \overset{\text{def}}{=} f * \mathrm{t}^{-p}.$$

For example, if n is an integer ≥ 0, we have

$$\text{(3)} \qquad \mathrm{FP} \int_0^t \frac{f(u)}{(t-u)^{n+1}}\, du = f * \left[D \, \frac{D^n(-1)^n}{n!} \langle \log t \rangle \right] \qquad \text{(by [8.0])},$$

$$\text{(4)} \qquad = \frac{1}{n!} f D^n (-1)^n \langle \log t \rangle \qquad \text{(by [8.9])}.$$

if $f()$ and $f'()$ are $\mathcal{K}$-functions such that $f()$ has no jumps on $[0, \infty)$, we can use (3)—(4) and 3.8.2 to obtain

$$\mathrm{FP} \int_0^t \frac{f(u)}{t-u}\, du = f' D^{-1} \langle \log t \rangle + f(0-) \langle \log t \rangle$$

$$= \int_0^t f'(t-u)\, (\log u)\, du + f(0-) \log t:$$

the last equation is from 3.15.

8.12 Application. Given a $\mathcal{K}$-function $h()$, let us solve the equation

$$\mathrm{FP}\int_0^t \frac{y(u)}{(t-u)^{5/2}}\,du = h.$$

In view of [8.11] and [8.2], this equation implies

$$h = y * t^{-5/2} = y * D^2\left\langle\frac{4}{3}t^{-1/2}\right\rangle, \tag{5}$$

whence

$$h = yD\left\langle\frac{4}{3}t^{-1/2}\right\rangle \qquad \text{(by [8.8])};$$

right-multiplying by $(3/4D)\,\langle t^{-1/2}\rangle$, we can use 3.16 to obtain

$$\frac{3}{4}\,D^{-1}\langle t^{-1/2}\rangle\,h = \pi D y. \tag{6}$$

The equations

$$y = \frac{3}{4\pi}\,2\langle t^{1/2}\rangle D^{-1}h = \frac{3}{2\pi}\int_0^t (t-u)^{1/2}\,h(u)\,du$$

are immediate from (6), 2.30, and 8.10.2.

8.12.1 **More information** on the subject of finite parts is found in the articles by T. K. Boehme ["Operational calculus and the finite part of divergent integrals", Transactions Amer. Math. Soc. **106**, 346—368 (1963)] and P. L. Butzer ["Singular integral equations of Volterra type and the finite part of divergent integrals", Arch. Rat. Mech. Anal. **3**, 194—205 (1959)].

8.13 An integral equation. Let λ be a given number, and consider the equation

$$f * y = \sin t + \lambda t. \tag{7}$$

Consequently, Equation (7) implies

$$f * y = \frac{D}{D^2+1} + \frac{\lambda}{D}; \tag{8}$$

if the operator f is invertible, Definition 8.8 gives

$$y = \frac{f^{-1}D^2}{D^2+1} + \lambda f^{-1}. \tag{9}$$

In particular, if (7) has the form

$$\langle t\rangle * y = \sin t + \lambda t,$$

then $f = \langle t\rangle = D^{-1}$ (see 3.18), so that (9) becomes

$$y = \frac{D^3}{D^2+1} + \lambda D = D - \frac{D}{D^2+1} + \lambda D.$$

Consequently,

$$y = (1+\lambda)\,\delta(t) - \sin t;$$

from 6.49.2 it follows that y is a function-operator if (and only if) $\lambda = -1$

Exercises

8.14.0 Find a particular solution of the equation $Dy - ay = \mathrm{t}^{-3/2}$, where a is a given number.

Answer: $y = -2t^{-1/2} - 2a \int\limits_0^t \frac{\mathrm{e}^{au}}{\sqrt{t-u}}\,du.$

8.14.1 Verify that

$$\mathrm{FP} \int\limits_0^t \frac{\mathrm{e}^{au}}{(t-u)^{3/2}}\,du = -2t^{-1/2} - 2a \int\limits_0^t \frac{\mathrm{e}^{au}}{\sqrt{t-u}}\,du.$$

8.14.2 Use the idea behind the definitions 8.0—2 to define the operator tan that corresponds to the function defined by the equation $f(t) = (\sin t)/(\cos t)$: this function is not a $\mathcal{K}$-function:

Answer: $\tan = D\langle -\log|\cos t|\rangle$.

8.14.3 Define the operator t^α when α is any non-integer < -1.

Answer: $\mathrm{t}^\alpha = D^n \left\langle \frac{t^{\alpha+n}}{(\alpha+1)(\alpha+2)\cdots(\alpha+n)} \right\rangle,$

where n is the smallest integer such that $\alpha + n > -1$.

8.14.4 Show that D is the only operator y such that $y * \langle t \rangle = \langle t \rangle$.

8.14.5 Given a $\mathcal{K}$-function $h(\,)$, solve the equation

$$\mathrm{FP} \int_0^t \frac{y(u)}{(t-u)^{3/2}}\, du = h.$$

Hint: use 8.11 and 3.16. *Answer*:

$$y = -\frac{1}{2\pi} \int_0^t \frac{h(u)}{\sqrt{t-u}}\, du.$$

8.14.6 Solve each of the following equations for y:

(.7) $$y * \langle \cos t \rangle = 1.$$

Answer: $y = \delta(t) + t$.

(.8) $$y * f = f.$$

Answer: $y = \delta(t)$.

(.9) $$y * \langle e^{-t} \rangle = 1 + te^{-t}.$$

Answer: $y = \delta(t) + 1 + e^{-t}$.

8.14.10 Let m be an integer ≥ 1; find an operator y such that

$$y * \langle \sin t \rangle = t^m;$$

for which values of m is y a function-operator? When does y have a non-zero impulse part (6.44)?

Answers: $m \geq 2$, $m = 1$.

8.14.11 Given a number λ, find an operator y such that

$$\langle \sin t \rangle * y = \lambda t + \sin t;$$

for which value of λ is y a function-operator?

Answer: $y = (\lambda + 1)\,\delta(t) + \lambda t$.

8.14.12 Given two numbers λ and c, find an operator y such that

$$\langle \sin t \rangle * y = \lambda t + ct^2;$$

for which values of λ is y a function-operator? When does y have a non-zero impulse part?

Answer: $y = ct^2 + \lambda t + 2c + \lambda \delta(t)$.

8.14.13 Let $f(\ )$ be a $\mathcal{K}$-function, and let y be the operator such that

$$\langle \cos t \rangle * y = f.$$

Given $f(\ )$, find the operator y; what is the impulse part (6.44) of y when $f(\ )$ has no jumps on $[0, \infty)$ and $f'(\ )$ is a $\mathcal{K}$-function?

Hint: use 3.8.4.

Answers: $y = Df + D^{-1}f$; the impulse part $= f(0-)\, D$.

8.14.14 If $g()$ is an entering function, its translate is the function $g_\alpha()$ defined by

$$g_\alpha(\tau) = g(\tau - \alpha) \qquad (-\infty < \tau < \infty).$$

Use 6.53 to prove that

$$[\![\delta(t - \alpha)]\!] * [\![g]\!] = [\![g_\alpha]\!] \qquad (-\infty < \alpha < \infty).$$

8.14.15 Suppose that $p \geq 1$. Use 8.7 and [8.11] to find a particular solution of the equation

$$y''(\tau) - 6y'(\tau) + 9y(\tau) = \tau^{-p} \qquad (\text{all } \tau > 0).$$

Hint: solve the equation

$$(D^2 - 6D + 9)\, y = \mathrm{t}^{-p}.$$

Answer:

$$y = \mathrm{FP} \int_0^t \frac{u\, e^{3u}}{(t-u)^p}\, du.$$

8.14.16 Prove that

$$\mathrm{FP}\int_0^t \frac{u\,\mathrm{e}^{3u}}{(t-u)^{5/2}}\,\mathrm{d}u = \frac{4}{3}\,t^{-1/2} + 4\int_0^t \frac{(3u+2)\,\mathrm{e}^{3u}}{\sqrt{t-u}}\,\mathrm{d}u.$$

Hint: use 3.22 and [8.2].

Indexed Functions

Application-oriented readers may skip the following §§ 8.15—28; this material deals with the notion of *value of an operator at a point.* The basic idea is due to A. ERDÉLYI [E 1].

8.15 **Lemma.** *Let k be an integer ≥ 1; the function $Y_k(\,)$ is defined by*

$$(8.16)\qquad Y_k(t) = \mathsf{T}_0(t)\,\frac{t^{k-1}}{(k-1)!} \qquad (-\infty < t < \infty).$$

If $g(\,)$ is an entering function, we have

$$(8.17)\qquad [\![Y_k * g]\!] = D^{-k}\,[\![g]\!]$$

and the equation

$$(8.18)\qquad \left(\frac{\mathrm{d}}{\mathrm{d}\tau}\right)^k (Y_k * g)(\tau) = g(\tau)$$

holds at each point τ where $g(\,)$ is continuous.

Proof:

$$[\![Y_k * g]\!] = [\![Y_k]\!]\,D^{-1}\,[\![g]\!] \qquad \text{by 1.27}$$

$$= \left[\!\!\left[\left\{\frac{t^{k-1}}{(k-1)!}\right\}\right]\!\!\right] D^{-1}\,[\![g]\!] \qquad \text{by [8.16]}$$

$$= \left\langle\frac{t^{k-1}}{(k-1)!}\right\rangle D^{-1}\,[\![g]\!] \qquad \text{by 2.1}$$

$$= D^{-k+1} D^{-1}\,[\![g]\!] \qquad \text{by 3.20:}$$

this concludes the proof of 8.17. If $k = 1$ then $Y_1(\,) = \mathsf{T}_0(\,)$, so that

$$\frac{\mathrm{d}}{\mathrm{d}\tau}(Y_1 * g)(\tau) = \frac{\mathrm{d}}{\mathrm{d}\tau}(\mathsf{T}_0 * g)(\tau) = \frac{\mathrm{d}}{\mathrm{d}\tau}\int_{-\infty}^{\tau} g(u)\,\mathrm{d}u = g(\tau): \tag{1}$$

the middle equation is from 0.26. Thus, we have proved 8.18 in case $k = 1$. To prove 8.18 for any integer k, suppose that 8.18 holds for some integer $k = m \geq 2$; from 0.37 (with $\psi = Y_{m+1}$) we have

$$\frac{\mathrm{d}}{\mathrm{d}\tau}(Y_{m+1} * g)(\tau) = g * Y'_{m+1}(\tau) = g * Y_m(\tau): \tag{2}$$

note that $\psi'(t) = Y_m(t) = t^{m-1}/(m-1)!$, so that $\psi'(\,)$ is a continuous function (since $m \geq 2$). Note that

$$\left(\frac{\mathrm{d}}{\mathrm{d}\tau}\right)^{m+1}(Y_{m+1} * g)(\tau) = \left(\frac{\mathrm{d}}{\mathrm{d}\tau}\right)^{m}\frac{\mathrm{d}}{\mathrm{d}\tau}(Y_{m+1} * g)(\tau)$$

$$= \left(\frac{\mathrm{d}}{\mathrm{d}\tau}\right)^{m}(Y_m * g)(\tau) = g(\tau):$$

the last two equations are from (2) and our induction hypothesis (8.18 with $k = m$).

8.19 Definitions. An *indexed function* is a pair (n, f) consisting of an integer $n \geq 1$ and a function-operator f. An indexed function (n, f) *attaches an operator* V *to a point* τ if

$$V = D^n[\![f]\!] \quad \text{and} \quad |f^{(n)}(\tau)| < \infty.$$

The number $f^{(n)}(\tau)$ is called *the value at* τ of the indexed function (n, f).

8.20 Theorem. *All indexed functions attaching an operator V to a point τ have the same value at τ.*

Proof. Let (m, f) and (n, h) be any two indexed functions attaching the operator V to the point τ: this means that

$$|f^{(m)}(\tau)| < \infty, \quad |h^{(n)}(\tau)| < \infty,$$

and $D^m[\![f]\!] = V = D^n[\![h]\!]$. Consequently,

$$D^{-n}[\![f]\!] = D^{-m}[\![h]\!]; \tag{3}$$

the equations

$$\llbracket Y_n * f \rrbracket = D^{-n} \llbracket f \rrbracket = D^{-m} \llbracket h \rrbracket = \llbracket Y_m * h \rrbracket \tag{4}$$

are from 8.17, (3), and 8.17. From (4) and 1.24 it now follows that

$$Y_n * f(\tau) = Y_m * h(\tau) \qquad (-\infty < \tau < \infty): \tag{5}$$

the equality *everywhere* is caused by the fact that both functions $Y_n * f(\,)$ and $Y_m * h(\,)$ are continuous (by 2.61). To conclude, note that

$$\left(\frac{\mathrm{d}}{\mathrm{d}\tau}\right)^m f(\tau) = \left(\frac{\mathrm{d}}{\mathrm{d}\tau}\right)^m \left[\left(\frac{\mathrm{d}}{\mathrm{d}\tau}\right)^n (Y_n * f)(\tau)\right] \qquad \text{by 8.18} \tag{6}$$

$$= \left(\frac{\mathrm{d}}{\mathrm{d}\tau}\right)^{n+m} (Y_m * h)(\tau) \qquad \text{by (5)} \tag{7}$$

$$= \left(\frac{\mathrm{d}}{\mathrm{d}\tau}\right)^n \left[\left(\frac{\mathrm{d}}{\mathrm{d}\tau}\right)^m (Y_m * h)(\tau)\right] \tag{8}$$

$$= \left(\frac{\mathrm{d}}{\mathrm{d}\tau}\right)^n h(\tau) \qquad \text{by 8.18.} \tag{9}$$

Consequently, $f^{(m)}(\tau) = h^{(n)}(\tau)$: this means that both indexed functions (m, f) and (n, h) have the same value at τ; the proof is complete.

8.21 **Definitions.** Consider an operator V and a real number τ: if there is an indexed function attaching V to the point τ, *we denote by* $\{V\}(\tau)$ *the value at* τ *of any one of the indexed functions attaching* V *to the point* τ (they all have the same value at τ, by 8.20).

We set $\{V\}(\tau) = \infty$ if (and only if) *there is no indexed function attaching* V *to the point* τ.

The number $\{V\}(\tau)$ will be called **the value at τ of the operator V.**

8.22 Thus, if $\{V\}(\tau) \neq \infty$ there is an indexed function attaching V to the point τ; by the above definition, its value at τ is precisely $\{V\}(\tau)$.

8.23 If there is an indexed function (n, f) such that $D^n \llbracket f \rrbracket = V$ and $|f^{(n)}(\tau)| < \infty$, then (n, f) attaches V to the point τ; from [8.21] it follows that $\{V\}(\tau) = f^{(n)}(\tau)$.

8.24 *Remark.* If the equation $V = D^n \llbracket f \rrbracket$ does not hold for some integer n and some entering function $f(\,)$, then there exist no indexed functions attaching V to any point τ: in consequence, $\{V\}(\tau) = \infty$ for every real value of τ. For example, this is the case when V is the operator that was discussed in 6.11.

8.25 **Theorem.** *If the equality* $V = D^n[\![f]\!]$ *holds for some integer* $n \geq 1$ *and for some entering function* $f(\,)$, *then* $\{V\}(\tau) = f^{(n)}(\tau)$ *at each point* τ *such that* $|f^{(n)}(\tau)| < \infty$.

Proof. Consider the pair (n, f): from our hypotheses and [8.19] it follows that it is an indexed function attaching V to the point τ; consequently, [8.21] gives

$$\{V\}(\tau) = \text{the value at } \tau \text{ of the indexed function } (n, f)$$

$$= f^{(n)}(\tau) \qquad \text{(by [8.19])}.$$

8.26 **Theorem.** *If* V_1 *and* V_2 *are operators, if* a *and* b *are numbers, then the equation*

$$\{a V_1 + b V_2\}(\tau) = a\,\{V_1\}(\tau) + b\,\{V_2\}(\tau) \tag{8.27}$$

holds whenever $\{V_1\}(\tau) \neq \infty$ *and* $\{V_2\}(\tau) \neq \infty$.

Proof. From our hypothesis $\{V_k\}(\tau) \neq \infty$ $(k = 1, 2)$ and 8.22 it follows the existence of two indexed functions (m, f) and (n, h) such that (m, f) attaches V_1 to the point τ and (n, h) attaches V_2 to the point τ; in view of [8.19], this implies that

$$|f^{(m)}(\tau)| < \infty, \qquad |h^{(n)}(\tau)| < \infty,$$

$$V_1 = D^m[\![f]\!], \quad \text{and} \quad V_2 = D^n[\![h]\!];$$

consequently, 8.23 gives

$$\{V_1\}(\tau) = f^{(m)}(\tau) \quad \text{and} \quad \{V_2\}(\tau) = h^{(n)}(\tau). \tag{1}$$

Therefore,

$$aV_1 + bV_2 = D^m[\![af]\!] + D^n[\![bh]\!] \tag{2}$$

$$= D^{m+n}(D^{-n}[\![af]\!] + D^{-m}[\![bh]\!]) \tag{3}$$

$$= D^{m+n}([\![Y_n * af]\!] + [\![Y_m * bh]\!]) \quad \text{by 8.17}. \tag{4}$$

Let $g(\,)$ be the function defined by

$$g(\tau) = Y_n * af(\tau) + Y_m * bh(\tau) \qquad (-\infty < \tau < \infty). \tag{5}$$

From (2)—(4) we see that

$$aV_1 + bV_2 = D^{m+n}\,[\![g]\!]. \tag{6}$$

Equation (5) implies that

$$\left(\frac{\mathrm{d}}{\mathrm{d}\tau}\right)^{m+n} g(\tau) = \left(\frac{\mathrm{d}}{\mathrm{d}\tau}\right)^{m} \left(\frac{\mathrm{d}}{\mathrm{d}\tau}\right)^{n} (Y_n * af)(\tau) + \left(\frac{\mathrm{d}}{\mathrm{d}\tau}\right)^{n} \left(\frac{\mathrm{d}}{\mathrm{d}\tau}\right)^{m} (Y_m * bh)(\tau),$$

so that 8.18 gives

$$g^{(m+n)}(\tau) = \left(\frac{\mathrm{d}}{\mathrm{d}\tau}\right)^{m} af(\tau) + \left(\frac{\mathrm{d}}{\mathrm{d}\tau}\right)^{n} bh(\tau). \tag{7}$$

From (6) and (7) we see that the indexed function $(m + n, g)$ attaches the operator $aV_1 + bV_2$ to the point τ; from [8.21] we therefore have

$$\begin{aligned} \{aV_1 + bV_2\}(\tau) &= g^{(m+n)}(\tau) \\ &= af^{(m)}(\tau) + bh^{(n)}(\tau) && \text{by (7)} \\ &= a\,\{V_1\}(\tau) + b\,\{V_2\}(\tau) && \text{by (1).} \end{aligned}$$

Exercise

8.28 Suppose that $\alpha \geq 0$ and let W be an operator; prove that

$$\{\mathsf{T}_\alpha W\}(t) = \{W\}(t - \alpha)$$

at all points t such that $\{W\}(t - \alpha) \neq \infty$.

Derivatives on the Open Interval (0, ∞)

The rest of this book deals with starting-value problems; initial conditions (that is, conditions at $t = 0-$) will not appear any more.

8.29 **Definitions.** Suppose that y is an operator. If k is an integer ≥ 0, we set

$$y^{(k)}(0+) \overset{\text{def}}{=} \lim_{\tau \to 0+} \{D^k y\}(\tau); \tag{8.30}$$

as usual, $\tau \to 0+$ means that τ approaches zero through positive values; the number $\{D^k y\}(\tau)$ is the value at τ of the operator $D^k y$ (see [8.21]). Further, we set

$$\frac{\mathrm{d}^m}{\mathrm{d}t^m} y \overset{\text{def}}{=} D^m y - \sum_{k=0}^{m-1} y^{(k)}(0+)\, D^{m-k}. \tag{8.31}$$

8.32 *Remarks.* In particular, if Y is an operator, 8.31 gives

$$\boxed{\frac{\mathrm{d}}{\mathrm{d}t} Y = DY - Y(0+)\, D}, \tag{8.33}$$

where

$$Y(0+) \overset{\text{def}}{=} Y^{(0)}(0+) = \lim_{\tau \to 0+} \{Y\}(\tau).$$

Definition 8.30 assigns to any operator y the number $y^{(k)}(0+)$; there is no function $y(\ )$ assigned to an arbitrary operator y.

Until further notice, suppose that $\alpha \geq 0$. If m is any integer ≥ 1, we have

$$\{D^m \mathsf{T}_\alpha\}(\tau) = 0 \qquad (\text{all } \tau \neq \alpha\text{: see 6.20}). \tag{8.34}$$

In particular,

$$\{D^m\}(\tau) = 0 \qquad (\text{all } \tau > 0). \tag{8.35}$$

From [8.31] and 8.26 we see that

$$\left\{\frac{\mathrm{d}^m}{\mathrm{d}t^m} y\right\}(\tau) = \{D^m y\}(\tau) - \sum_{k=0}^{m-1} y^{(k)}(0+)\, \{D^{m-k}\}(\tau) = \{D^m y\}(\tau):$$

the last equation is from 8.35. Consequently,

$$\boxed{\left\{\frac{\mathrm{d}^m}{\mathrm{d}t^m} y\right\}(\tau) = \{D^m y\}(\tau)} \qquad (\text{all } \tau > 0) \tag{8.36}$$

and

$$\text{(1)} \qquad \boxed{\lim_{\tau\to 0+}\left\{\frac{d^m}{dt^m}\,y\right\}(\tau) = y^{(m)}(0+)} \qquad \text{(by 8.36 and [8.30]).}$$

8.36.1 **Theorem.** *If* $n = 0, 1, 2, \ldots$, *then*

$$\text{(8.37)} \qquad \boxed{\frac{d}{dt}\left(\frac{d^n}{dt^n}\,y\right) = \frac{d^{n+1}}{dt^{n+1}}\,y}\,.$$

Proof. Note that

$$\frac{d}{dt}\left(\frac{d^n}{dt^n}\,y\right) = D\,\frac{d^n}{dt^n}\,y - y^{(n)}(0+)D \qquad \text{from 8.33 and (1)}$$

$$= D^{n+1} - \sum_{k=0}^{n-1} y^{(k)}(0+)\,D^{n-k}D - y^{(n)}(0+)D:$$

the last equation is from [8.31]. In consequence,

$$\frac{d}{dt}\left(\frac{d^n}{dt^n}\,y\right) = D^{n+1}\,y - \sum_{k=0}^{n} y^{(k)}(0+)\,D^{n+1-k} = \frac{d^{n+1}}{dt^{n+1}}\,y.$$

8.38 **Particular cases.** If $y = D^n\mathsf{T}_\alpha$ and $n \geq 1$ then

$$\text{(8.39)} \qquad y^{(k)}(0+) = \lim_{\tau\to 0+}\{D^n\mathsf{T}_\alpha\}(\tau):$$

this is by [8.30] and 8.34. Consequently,

$$\text{(8.40)} \qquad \frac{d^m}{dt^m}\,D^n\mathsf{T}_\alpha = D^{n+m}\mathsf{T}_\alpha \qquad (m = 0, 1, 2, \ldots).$$

Since

$$\text{(8.41)} \qquad \boxed{\delta(t-\alpha) = D\mathsf{T}_\alpha} \qquad \text{(see 6.54),}$$

it follows from 8.39—40 that

$$\delta(t-\alpha)^{(k)}(0+) = 0$$

and

$$\frac{\mathrm{d}^m}{\mathrm{d}t^m}\,\delta(t-\alpha) = D^m\,\delta(t-\alpha): \tag{8.42}$$

both m and k are integers ≥ 0.

8.43 **Derivatives of function-operators.** Until further notice, let y be the canonical operator $\langle y(t)\rangle$ of a $\mathcal{K}$-function $y(\,)$ whose m^{th} order derivative $y^{(m)}(\,)$ is also a $\mathcal{K}$-function. From 8.36 and 6.14.2 we see that the equations

$$\left\{\frac{\mathrm{d}^k}{\mathrm{d}t^k}\,y\right\}(\tau) = \{D^k y\}(\tau) = y^{(k)}(\tau) \qquad (\text{all } \tau > 0) \tag{8.44}$$

hold whenever the right-hand side has a meaning. Since $y^{(m)}(\,)$ is a $\mathcal{K}$-function, it is continuous in some interval of the form $(0, \lambda)$; consequently, 8.44 implies that

$$\lim_{\tau\to 0+}\{D^k y\}(\tau) = y^{(k)}(0+). \tag{2}$$

Definition 8.30 assigns to any operator the number $y^{(k)}(0+)$; from (2) we see that *the number* $y^{(k)}(0+)$ *is the starting value of the* k^{th} *order derivative* $y^{(k)}(\,)$ *of the function* $y(\,)$.

Thus, $\mathrm{d}^m y/\mathrm{d}t^m$ is the result of replacing in the definition (3.11) of $\partial^m y/\partial t^m$ the initial values $y^{(k)}(0-)$ by the starting values $y^{(k)}(0+)$.

8.44.1 **Recall:** a function $f()$ is said to be *continuous in the interval* $(0, \infty)$ if

$$f(\tau-) = f(\tau) = f(\tau+) \qquad (\text{for each } \tau > 0).$$

8.45 **Theorem.** *Suppose that* y *is the canonical operator* $\langle y(t)\rangle$ *of a function* $y(\,)$ *which is continuous in the interval* $(0,\infty)$. *If* $y'(\,)$ *is a* $\mathcal{K}$*-function, then* $|y(0+)| < \infty$ *and*

$$\frac{\mathrm{d}}{\mathrm{d}t}\,y = \langle y'(t)\rangle = y' = \left\langle\frac{\mathrm{d}}{\mathrm{d}t}\,y(t)\right\rangle. \tag{8.46}$$

Proof. The last two equations are merely alternative notations for the operator $\langle y'(t)\rangle$ (see [2.1]). The Fundamental Theorem of Calculus gives

$$\int_\lambda^1 y'(u)\,\mathrm{d}u = y(1) - y(\lambda) \qquad (0 < \lambda < 1);$$

letting $\lambda \to 0$, we obtain the conclusion $|y(0+)| < \infty$. To prove the first equation in 8.46, consider the function $F(\ )$ defined by

$$F(t) = \begin{cases} y(0+) & (t \leq 0) \\ y(t) & (t > 0): \end{cases} \tag{3}$$

it has no jumps on $[0, \infty)$ (see 2.26); consequently, we can combine [8.33] with the Derivation Property (2.24) to obtain

$$\frac{\mathrm{d}}{\mathrm{dt}} y = DF - F(0-)\, D = \langle F'(t) \rangle = \langle y'(t) \rangle:$$

the last equation is immediate from (3). This concludes the proof.

8.47 Theorem. *Suppose that y is the operator $\langle y(t) \rangle$ of a function $y(\ )$ such that $y^{(m-1)}(\ )$ is continuous in the interval $(0, \infty)$. If $y^{(m)}(\ )$ is a $\mathcal{K}$-function, then*

$$\frac{\mathrm{d}^k}{\mathrm{dt}^k} y = y^{(k)} = \left\langle \frac{\mathrm{d}^k}{\mathrm{d}t^k} y(t) \right\rangle \qquad (0 \leq k \leq m). \tag{8.48}$$

Proof. We proceed by induction: suppose that 8.48 holds for some integer $k = n \leq m - 1$;

$$\frac{\mathrm{d}^n}{\mathrm{dt}^n} y = y^{(n)}. \tag{4}$$

The equations

$$\frac{\mathrm{d}^{n+1}}{\mathrm{dt}^{n+1}} y = \frac{\mathrm{d}}{\mathrm{dt}} \left(\frac{\mathrm{d}^n}{\mathrm{dt}^n} y \right) = \frac{\mathrm{d}}{\mathrm{dt}} y^{(n)} = \left\langle \frac{\mathrm{d}}{\mathrm{d}t} y^{(n)}(t) \right\rangle = \langle y^{(n+1)}(t) \rangle$$

are from 8.37, (4), and 8.45; note that $y^{(n)}(\ )$ is continuous on $(0, \infty)$ (since $n \leq m - 1$ and since $y^{(m-1)}(\ )$ is continuous on $(0, \infty)$). We have shown that 8.48 holds for $k = n + 1$ whenever it holds for $k = n$: the proof is concluded by glancing at 8.45.

8.49 The general case. Let a_k $(k = 0, 1, 2, \ldots, n)$ and c_k $(k = 0, 1, 2, \ldots, \ldots, n - 1)$ be sequences of numbers such that $n \geq 1$; further, let h be a given operator. If y is a perfect operator such that

$$y^{(k)}(0+) = c_k \qquad (0 \leq k < n) \tag{8.50}$$

and

$$\left[a_n \frac{\mathrm{d}^n}{\mathrm{dt}^n} + a_{n-1} \frac{\mathrm{d}^{n-1}}{\mathrm{dt}^{n-1}} + \cdots + a_1 \frac{\mathrm{d}}{\mathrm{dt}} + a_0 \right] y = h, \tag{8.51}$$

we can proceed exactly as in 7.24 to obtain the equation

$$y = \frac{h + (\sum a_k D^k \,|||\, \sum c_k/D^k)}{a_n D^n + a_{n-1} D^{n-1} + \cdots + a_1 D + a_0}. \tag{8.52}$$

Consequently, there is at most one solution of the problem 8.50—51: it is the operator determined by the equation 8.52. Recall that

$$(\sum a_k D^k \,|||\, \sum c_k/D^k)$$

is the polynomial $p(D)$ obtained by carrying out the multiplication

$$(a_n D^n + a_{n-1} D^{n-1} + \cdots + a_0)\left(c_0 + \frac{c_1}{D} + \cdots + \frac{c_{n-1}}{D^{n-1}}\right)$$

and discarding all function-operators in the resulting sum: see 7.22.

8.53 **Theorem.** *If h is an operator such that the equation*

$$0 = \lim_{\tau \to 0+} \left\{\frac{D^k h}{a_n D^n + \cdots + a_1 D + a_0}\right\}(\tau) \tag{8.54}$$

holds for $k = 0, 1, 2, \ldots, n-1$, then the equation 8.52 *determines a perfect operator y satisfying* 8.50—51.

Proof. Set

$$G \overset{\text{def}}{=} a_n D^n + \cdots + a_1 D + a_0. \tag{8.55}$$

It can be shown that there exists an infinitely differentiable function $A_c()$ such that

$$A_c = \frac{(G \,|||\, \sum c_k/D^k)}{G}. \tag{8.56}$$

Consequently, 8.52 can be written

$$y = \frac{h}{G} + A_c. \tag{8.57}$$

It is easy to verify that 8.57 (*alias* 8.52) implies 8.51 — **provided the starting conditions** (8.50) **are satisfied.** Let y be the perfect operator determined by 8.52: since our aim is to prove that y satisfies 8.50—51, it only remains to prove the starting conditions (8.50). To that effect, we set

$$F \overset{\text{def}}{=} \frac{h}{G}, \tag{8.58}$$

so that our hypothesis 8.52 (*alias* 8.57) becomes

$$y = F + A_c,$$

whence

$$y^{(k)}(0+) = F^{(k)}(0+) + A_c^{(k)}(0+) = 0 + c_k \qquad (k = 0, 1, 2, \ldots, n-1):$$

the last equation is from [8.58], [8.30], 8.54, and 7.32. Consequently, the operator y satisfies the starting conditions 8.50; therefore, y satisfies 8.50—51.

8.59 **Some non-standard cases.** The equations 8.50 — 51 have no solutions in some very simple and important cases. For example, there is no perfect operator y such that

$$y(0+) = 1 \quad \text{and} \quad \frac{\mathrm{d}}{\mathrm{d}t}\, y = D; \tag{5}$$

to verify that there is no such operator y, observe that (5) implies $Dy - D = D$, whence $y = \langle 2 \rangle$, which gives the contradiction

$$2 = \langle 2 \rangle(0+) = y(0+) = 1:$$

see 6.18. Note that 8.53 is not applicable to (5).

8.60 **The standard case.** Until further notice, let h be the canonical operator of a $\mathcal{K}$-function $h()$. It can be verified that the equation 8.52 determines a $\mathcal{K}$-function $y()$ satisfying 8.50—51.

Let $g()$ be the Green's function of the problem:

$$g = \frac{D}{G} \qquad \text{(see 7.33.2)}, \tag{8.61}$$

where G is as in 8.55. Note that

$$(8.62)\qquad \frac{h}{G}=\frac{D}{G}D^{-1}h=\frac{D}{G}*h=g*h:$$

the last two equations are from [8.8] and 8.61. From 8.62 and 8.10.2 it now follows that

$$\frac{h}{G}=\left\langle\int_0^t g(t-u)\,h(u)\,\mathrm{d}u\right\rangle;$$

consequently, the solution 8.52 of the problem 8.50—51 can be written

$$(8.63)\qquad y=\int_0^t g(t-u)\,h(u)\,\mathrm{d}u+\frac{(G\;|||\;\sum c_k/D^k)}{G},$$

where $g()$ is the Green's function of the problem.

8.64 A standard problem. To solve the equation $y^{(4)}+2y''+y=0$ subject to the starting conditions

$$(6)\qquad y(0+)=1,\quad y'(0+)=0,\quad y^{(2)}(0+)=-1,\quad\text{and}\quad y^{(3)}(0+)=0.$$

Setting $h=0$ in 8.63, we see that

$$y=\frac{(D^4+2D^2+1\;|||\;1+0/D-1/D^2+0/D^3)}{D^4+2D^2+1}=\frac{D^4-D^2+2D^2}{(D^2+1)^2};$$

whence

$$(7)\qquad y=\frac{D^2(D^2+1)}{(D^2+1)^2}$$

determines a solution of the equation

$$(8)\qquad \left[\frac{\mathrm{d}^4}{\mathrm{d}t^4}+2\frac{\mathrm{d}^2}{\mathrm{d}t^2}+1\right]y=0$$

which satisfies the starting conditions (6). Cancelling the common factor in (7), we obtain

$$y=\langle\cos t\rangle\qquad\text{(by 3.28)}:$$

since $y()$ is infinitely differentiable, we may use 8.47 to conclude that (8) is equivalent to the equation

$$y^{(4)}+2y^{(2)}+y=0.$$

8.65 **Classical solutions.** A function $y(\,)$ is called a *classical solution* of the equation

$$(8.66)\qquad a_n y^{(n)}(t) + a_{n-1} y^{(n-1)}(t) + \cdots + a_1 y'(t) + a_0 y(t) = h(t)$$

if $y^{(n-1)}(\,)$ is continuous in the interval $(0, \infty)$ and $y(\,)$ satisfies the equation at each point where $y^{(n)}(\,)$ and $h(\,)$ are continuous. If $y(\,)$ is a classical solution of 8.66, it follows from 2.65 that

$$(9)\qquad a_n y^{(n)} + a_{n-1} y^{(n-1)} + \cdots + a_1 y' + a_0 y = h.$$

On the other hand, 8.66 implies that $y^{(n)}(\,)$ is a linear combination of $h(\,)$ and $y^{(k)}(\,)$ $(k = 0, 1, 2, \ldots, n-1)$; consequently, $y^{(n)}(\,)$ is a $\mathcal{K}$-function, so we may use 8.47 to infer from (9) that

$$(10)\qquad \left[a_n \frac{\mathrm{d}^n}{\mathrm{d}t^n} + \cdots + a_1 \frac{\mathrm{d}}{\mathrm{d}t} + a_0\right] y = h;$$

the continuity hypothesis on $y^{(n-1)}(\,)$ implies that each one of the functions $y^{(k)}(\,)$ $(0 \leq k \leq n-1)$ is continuous on the interval $(0, \infty)$.

8.67 **In conclusion:** if y is the canonical operator $\langle y(t) \rangle$ of a classical solution of 8.66, it satisfies the operator-equation (10). Thus, *each classical solution is found among the solutions of the operator-equation*; conversely, it can be shown that each solution of the operator-equation (10) is a solution of 8.66: **the two equations** (10) **and** 8.66 **are equivalent** (in case h is the canonical operator of a $\mathcal{K}$-function!).

Duhamel's Formula

8.67.1 Suppose that $h()$ is a function having at most a finite number of discontinuities in an interval $(0, \lambda)$; moreover, suppose that

$$\int_0^x |h(u)|\,\mathrm{d}u < \infty \qquad \text{whenever} \qquad 0 < x < \lambda.$$

8.67.2 **Theorem.** *The equation*

$$(8.67.3)\qquad a_n y^{(n)}(\tau) + \cdots + a_1 y'(\tau) + a_0 y(\tau) = h(\tau) \qquad (0 < \tau < \lambda)$$

always has a solution of the form

$$y(t) = \int_0^t g(t-u)\, h(u)\, du \qquad (0 < t < \lambda), \tag{8.67.4}$$

where $g()$ *is the Green's function.*

8.67.5 *Remarks.* The above theorem applies to the case where $h()$ is not a $\mathcal{K}$-function; the resulting procedure is even more general than the one discussed in 8.7. For example, consider the equation

$$y'(t) - y(t) = e^t (\lambda - t)^{-3} \qquad (0 < t < \lambda): \tag{11}$$

recall that the Green's function $g()$ is the function whose operator is given by

$$g = D[a_n D^n + \cdots + a_1 D + a_0]^{-1} \qquad \text{(see 7.33.2)}.$$

In the problem (11) we see that $g = D/(D-1)$, so that 8.67.4 becomes

$$y(t) = \int_0^t e^{t-u} e^u (\lambda - u)^{-3}\, du = \left(\frac{2^{-1}}{(\lambda - t)^2} - \frac{2^{-1}}{\lambda^2}\right) e^t.$$

8.67.6 **Theorem.** *If* $y()$ *is any solution of the equation* 8.67.3, *there exists a sequence of numbers* c_k $(k = 0, 1, 2, \ldots, n-1)$ *such that*

$$y(t) = A_c(t) + \int_0^t g(t-u)\, h(u)\, du \qquad (0 < t < \lambda); \tag{8.67.7}$$

as before, $A_c()$ *is the function determined by* 8.55—56.

8.67.8 **Definition.** The expression 8.67.7 is often called the *general solution* of the equation 8.67.3.

Exercises

8.67.9 Use 8.67.6 to find the general solution of the equation

$$y''(t) + 2y'(t) + y(t) = e^{-t}/t \qquad (0 < t < \infty).$$

Answer: $c_0 \exp(-t) + c_1 t \exp(-t) + t \exp(-t) \log t$.

8.67.10 Use 8.67.2 to find a particular solution of the equation

$$y''(t) + y(t) = \tan t \qquad (0 < t < \pi/2).$$

Answer: $-(\cos t)(\log|\sec t \pm \tan t|)$.

8.67.11 Use 8.67.6 to find the general solution of the equation

$$y''(t) + y(t) = \sec t \qquad (0 < t < \pi/2).$$

Answer: $(c_0 + \log|\cos t|)\cos t + (c_1 + t)\sin t$.

8.67.12 Use 8.67.6 to find the general solution of the equation

$$y'(t) - y(t) = e^t/(7 - t)^3 \qquad (0 < t < 7).$$

Answer: $[c_0 + (1/2)(7 - t)^{-2} - (1/98)]\, e^t$.

Chapter 4

This chapter begins by discussing derivatives with respect to a real parameter x. For a pre-view of what is to come, consider the canonical operator T_x of the step-function $\mathsf{T}_x()$ defined by

$$\mathsf{T}_x(\tau) = \begin{cases} 0 & (\tau \leq x) \\ 1 & (\tau > x); \end{cases}$$

we suppose $x \geq 0$. As we shall see, its derivative with respect to x is the negative of the Dirac Delta:

$$\frac{\mathrm{d}}{\mathrm{d}x} \mathsf{T}_x = - D\mathsf{T}_x = -\delta(t - x).$$

§ 9. Partial Differential Equations

Given an interval I, consider a rule assigning to each x in I a perfect operator y_x. If λ is in the interval I, we have

$$\boxed{\frac{\mathrm{d}}{\mathrm{d}t} y_\lambda = Dy_\lambda - y_\lambda(0+)\, D} \qquad \text{(by [8.33])} \tag{9.0.0}$$

and

$$\boxed{\frac{\mathrm{d}^2}{\mathrm{d}t^2} y_\lambda = D^2 y_\lambda - y_\lambda(0+)\, D^2 - y_\lambda'(0+)\, D} \qquad \text{(by [8.31])}, \tag{9.0.1}$$

where

$$y_\lambda(0+) = \lim_{\tau \to 0+} \{y_\lambda\}(\tau) \tag{9.0.2}$$

and

$$y_\lambda'(0+) = \lim_{\tau\to 0+} \{Dy_\lambda\}(\tau) \qquad \text{(by [8.30])}. \tag{9.0.3}$$

In case y_λ is the operator $\langle y_\lambda(t)\rangle$ of a $\mathcal{K}$-function $y_\lambda()$, let the function $y_\lambda^{(k)}()$ be defined by

$$y_\lambda^{(k)}(t) = \frac{\partial^k}{\partial t^k} y_\lambda(t) \qquad (-\infty < t < \infty);$$

if $y_\lambda^{(1)}()$ is a $\mathcal{K}$-function, it follows from 8.45 that

$$\frac{\mathrm{d}}{\mathrm{d}t} y_\lambda = \langle y_\lambda^{(1)}(t)\rangle = \left\langle \frac{\partial}{\partial t} y_\lambda(t)\right\rangle \tag{9.1}$$

— provided $y_\lambda()$ is continuous in the interval $(0, \infty)$; further, if $y_\lambda^{(1)}()$ is continuous in the interval $(0, \infty)$, then the equation

$$\frac{\mathrm{d}^2}{\mathrm{d}t^2} y_\lambda = \left\langle \frac{\partial^2}{\partial t^2} y_\lambda(t)\right\rangle \tag{9.2}$$

holds whenever $y_\lambda^{(2)}()$ is a $\mathcal{K}$-function: see 8.47.

Differentiation with Respect to x

9.3 If λ is in an interval I, we set

$$y_\lambda^{[1]} \overset{\text{def}}{=} \lim_{x\to\lambda} \frac{1}{x-\lambda}[y_x - y_\lambda]. \tag{1}$$

In view of [6.6], Equation (1) means that

$$y_\lambda^{[1]} \cdot \varphi() = \lim_{x\to\lambda} \frac{1}{x-\lambda}[(y_x\cdot\varphi - y_\lambda\cdot\varphi)()] = \left[\frac{\partial}{\partial x}(y_x\cdot\varphi)()\right]_{x=\lambda}$$

for every test-function $\varphi()$; consequently, $y_\lambda^{[1]}$ is the mapping that assigns to each test-function $\varphi()$ the function $y_\lambda^{[1]}\cdot\varphi()$ defined by the equation

$$y_\lambda^{[1]}\cdot\varphi(\tau) = \left[\frac{\partial}{\partial x}(y_x\cdot\varphi)(\tau)\right]_{x=\lambda} \qquad (-\infty < \tau < \infty). \tag{9.4}$$

9.5 **Definition.** If a and b are real numbers, we set

$$\boxed{\frac{\mathrm{d}}{\mathrm{d}x}\, y_{ax+b} \overset{\text{def}}{=} a\, y^{[1]}_{ax+b}}\,. \tag{9.6}$$

9.7 If W is any operator, then

$$\frac{\mathrm{d}}{\mathrm{d}x}\,[y_{ax+b} W] = \left[\frac{\mathrm{d}}{\mathrm{d}x}\, y_{ax+b}\right] W; \tag{9.8}$$

this is an immediate consequence of the definitions (the proof is entirely similar to the proof of 6.9). As usual, we define

$$\frac{\mathrm{d}^2}{\mathrm{d}x^2}\, y_{ax+b} = \frac{\mathrm{d}}{\mathrm{d}x}\, B_x, \quad \text{where} \quad B_x = \frac{\mathrm{d}}{\mathrm{d}x}\, y_{ax+b}.$$

It is easily seen that

$$\frac{\mathrm{d}^2}{\mathrm{d}x^2}\,[\psi(x)\, W] = \psi''(x)\, W. \tag{9.9}$$

Setting $a = 1$ and $b = 0$ in [9.6]:

$$\frac{\mathrm{d}}{\mathrm{d}x}\, y_x = y^{[1]}_x \tag{9.10}$$

— that is,

$$y^{[1]}_\lambda = \frac{\mathrm{d}}{\mathrm{d}x}\, y_x\Big|_{x=\lambda}. \tag{9.11}$$

From [9.4] it follows immediately that

$$\left[\frac{\mathrm{d}}{\mathrm{d}x}\, y_x\right] \cdot \varphi() = \frac{\partial}{\partial x}\,[y_x \cdot \varphi]\,() \qquad \text{for every test-function } \varphi\,().$$

Z Until further notice, we shall consider a $\mathcal{K}$-function $y_x()$ depending on the parameter $x > 0$; its canonical operator $\langle y_x(t)\rangle$ will often be written y_x:

$$y_x \overset{\text{def}}{=} \langle y_x(t)\rangle \qquad \text{(as in [2.1]).}$$

We shall begin by describing some conditions which imply

$$\frac{\mathrm{d}}{\mathrm{d}x}\, y_x = \left\langle \frac{\partial}{\partial x}\, y_x(t)\right\rangle.$$

9.12 Theorem. *If there exists a $\mathcal{K}$-function $Y_\lambda(\,)$ depending on the parameter $\lambda > 0$ such that*

$$\left|\frac{\partial}{\partial x}\, y_x(t)\right| \le Y_\lambda(t) \quad \text{whenever} \quad \begin{cases}\lambda/2 < x < 3\lambda/2\\ 0 < t < \infty,\end{cases} \tag{9.13}$$

then

$$\frac{d}{d\lambda}\langle y_\lambda(t)\rangle = \left\langle \frac{\partial}{\partial\lambda}\, y_\lambda(t)\right\rangle \qquad (\text{all } \lambda > 0)$$

Proof: see 15.24.

9.14 Example. Suppose that $\psi(\,)$ is a continuous function in the open interval $(0, \infty)$: in consequence, there exists a positive number N_λ depending on the parameter $\lambda > 0$ such that

$$|\psi(x)| \le N_\lambda \qquad \text{whenever} \qquad \lambda/2 < x < 3\lambda/2. \tag{1}$$

As we shall see, diffusion problems involve the following type of equations:

$$\frac{\partial}{\partial x}\, y_\lambda(t) = [\psi(x) f_0(t)] \exp\left(f_1(t) - \frac{x^2}{4t}\right) \qquad (\text{all } x > 0), \tag{9.15}$$

where $f_k(\,)$ $(k = 0, 1)$ are $\mathcal{K}$-functions and $t > 0$. To verify 9.13, take $\lambda > 0$ and observe that the relation $\lambda/2 < x < 3\lambda/2$ implies

$$\left|\frac{\partial}{\partial x}\, y_x(t)\right| \le Y_\lambda(t) \overset{\text{def}}{=} [N_\lambda f_0(t)] \exp\left(f_1(t) - \frac{\lambda^2/4}{4t}\right)$$

(all $t > 0$): see (1). Since $Y_\lambda(\,)$ is a $\mathcal{K}$-function, we find from 9.12 that

$$\frac{d}{dx}\langle y_x(t)\rangle = \left\langle \frac{\partial}{\partial x}\, y_x(t)\right\rangle \qquad (\text{all } x > 0).$$

9.15.1 **Another example.** Suppose that the function

$$(x, t) \mapsto \frac{\partial}{\partial x}\, y_x(t)$$

is continuous in the two-dimensional region

$$\{(x, t) : x > 0, \;\; t > 0\};$$

consequently, there exists a number N_λ depending on the parameter $\lambda > 0$ such that

$$\left|\frac{\partial}{\partial x} y_x(t)\right| \leq N_\lambda \quad \text{whenever} \quad \begin{cases} \lambda/2 < x < 3\lambda/2 \\ 0 < t < \infty : \end{cases}$$

this implies 9.13 (with $Y_\lambda() = N_\lambda$), whence

$$\frac{\mathrm{d}}{\mathrm{d}x} y_x = \left\langle \frac{\partial}{\partial x} y_x(t) \right\rangle \qquad (\text{all } x > 0).$$

9.16 **Theorem.** *If p is a number and*

(1) $$p_\lambda \overset{\text{def}}{=} \mathrm{e}^{-p\lambda} \mathsf{T}_\lambda \qquad (0 \leq \lambda < \infty),$$

then

(2) $$p_\lambda^{[1]} = -(D + p)\, p_\lambda.$$

Proof. Equation (1) means that p_λ is the operator

$$p_\lambda = \langle \exp(-p\lambda) \rangle\, \mathsf{T}_\lambda.$$

Let $\varphi()$ be any test-function; from (1) it follows that $p_\lambda \cdot \varphi()$ is the function defined by

(3) $$p_\lambda \cdot \varphi(\tau) = \mathrm{e}^{-p\lambda} [\mathsf{T}_\lambda \cdot \varphi](\tau) = \mathrm{e}^{-p\lambda} \varphi(\tau - \lambda) \qquad (-\infty < \tau < \infty):$$

see 5.1. From [9.4] and (3) we see that

$$p_\lambda^{[1]} \cdot \varphi(\tau) = \left[\frac{\partial}{\partial x} \mathrm{e}^{-px} \varphi(\tau - x)\right]_{x=\lambda}$$

$$= -p\mathrm{e}^{-p\lambda} \varphi(\tau - \lambda) - \mathrm{e}^{-p\lambda} \varphi'(\tau - \lambda),$$

so that (3) now gives

$$p_\lambda^{[1]} \cdot \varphi() = -p(p_\lambda \cdot \varphi)() - p_\lambda \cdot (D \cdot \varphi)(),$$

whence $p_\lambda^{[1]} = -p p_\lambda - p_\lambda D$ (from 0.10), and Conclusion (2) is now at hand.

9.17 **Definition.** *If p is a number and $s > 0$, we set*

$$(9.18)\qquad p_s^u \overset{\text{def}}{=} e^{-spu}\, \mathsf{T}_{su} \qquad (0 \le u < \infty).$$

9.18.1 **Theorem.** *If $0 \le b < \infty$, then*

$$(9.19)\qquad \boxed{\frac{d}{dx}\, p_s^{\pm x+b} = \mp\, s\,(D+p)\, p_s^{\pm x+b}} \qquad (0 \le x < b).$$

Proof. Note that

$$\begin{aligned}
\frac{d}{dx}\, p_s^{\pm x+b} &= \frac{d}{dx}\, p_{\pm sx+sb} && \text{since } p_s^u = p_{su} \text{ (see (1))}\\
&= \pm\, s p^{[1]}_{\pm sx+sb} && \text{by [9.6]}\\
&= \mp\, s\,(D+p)\, p_{\pm sx+sb} && \text{by (2)}\\
&= \mp\, s\,(D+p)\, p_s^{\pm x+b} && \text{since } p_s^u = p_{su} \text{ (see (1))}.
\end{aligned}$$

9.19.1 **Note:**

$$\boxed{\text{if} \quad p = 0, \quad \text{then} \quad p_s^u = \mathsf{T}_{su}}\,.$$

Vibrating String

From 9.19 we see that

$$\left[s^2(D+p)^2 - \frac{d^2}{dx^2}\right] p_s^{\pm x+b} = 0;$$

consequently, if X and Y are any two perfect operators, it follows immediately from 9.8 that

$$(9.20)\qquad \left[s^2(D+p)^2 - \frac{d^2}{dx^2}\right](X p_s^x + Y p_s^{l-x}) = 0 \qquad (0 < x < l),$$

where l is any positive number. Until further notice, suppose that $p = 0$: consequently, $p_s^x = \mathsf{T}_{sx}$ and 9.20 becomes

$$(9.21)\qquad \boxed{\left[s^2(D+p)^2 - \frac{d^2}{dx^2}\right](X\,\mathsf{T}_{sx} + Y\,\mathsf{T}_{sl-sx}) = 0} \qquad (0 < x < l).$$

Setting $p = 0 = b$ in 9.19, we obtain

$$\frac{d}{dx} \mathsf{T}_{sx} = -sD\mathsf{T}_{sx} = -\, s\delta(t - sx) \qquad (\text{for } sx \geq 0).$$

the last equation is from 8.41.

9.22 **Displacements in a string.** Consider a string with end-points at $x = 0$ and $x = l$. If $s = \sqrt{d/T}$ (where T is the tension, and d is the mass per unit length), the equation

$$s^2 \frac{d^2}{dt^2} y_x = \frac{d^2}{dx^2} y_x \qquad (0 < x < l) \tag{4}$$

governs the vertical displacement y_x. More precisely, if $y(x, t)$ is the second coordinate of a point of the string whose first coordinate is x at a time $t > 0$, then the equation

$$y_x(t) = y(x, t) \qquad (\text{all } t > 0)$$

defines an operator $y_x = \langle y_x(t) \rangle$ satisfying (4).

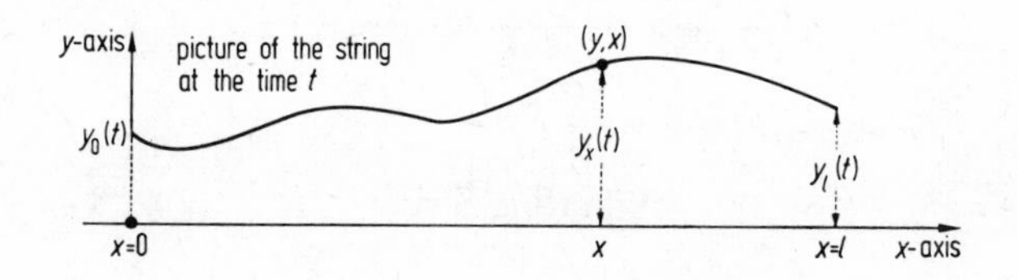

In view of [9.0.1], Equation (4) implies that

$$\left[s^2 D^2 - \frac{d^2}{dx^2}\right] y_x = s^2 [y_x(0+)D^2 + y_x'(0+)D] \qquad (0 < x < l). \tag{5}$$

9.23 **Finite string.** Suppose that $l < \infty$, and let $(f(\,), F(\,))$ be a pair of $\mathcal{K}$-functions describing the motion of the end-points of the string:

$$y_0(t) = f(t) \quad \text{and} \quad y_l(t) = F(t) \qquad (\text{all } t > 0). \tag{6}$$

It will be convenient to suppose that

$$f(t) = F(t) = 0 \quad \text{for } t \leq 0. \tag{7}$$

Finally, let the string be at rest on the segment $I = [0, l]$ when $t = 0+$:

$$y_x(0+) = y'_x(0+) = 0 \qquad (0 < x < l); \tag{8}$$

thus, Equation (5) becomes

$$\left[s^2 D^2 - \frac{d^2}{dx^2}\right] y_x = 0 \qquad (0 < x < l). \tag{9}$$

From 9.21, we find that the equation

$$y_x = X\mathsf{T}_{sx} + Y\mathsf{T}_{sl-sx} \tag{10}$$

implies (9); let us determine the pair (X, Y) of perfect operators such that Equation (10) implies the boundary conditions (6). Setting $x = 0$ and $x = l$ in (10), we can use 5.4 and (6) to obtain

$$f = X + Y\mathsf{T}_{sl} \quad \text{and} \quad F = X\mathsf{T}_{sl} + Y;$$

solving for X and Y:

$$X = \frac{f - \mathsf{T}_{sl} F}{1 - \mathsf{T}_{2sl}} \quad \text{and} \quad Y = \frac{F - \mathsf{T}_{sl} f}{1 - \mathsf{T}_{2sl}}; \tag{11}$$

substituting into (10):

$$y_x = \frac{[\mathsf{T}_{sx} - \mathsf{T}_{2sl-sx}] f + [\mathsf{T}_{sl-sx} - \mathsf{T}_{sl+sx}] F}{1 - \mathsf{T}_{2sl}}.$$

This is our answer: retracing our steps, we see that the boundary conditions (6) and the operator-equation (9) are satisfied; if the starting conditions (8) are satisfied (and they are, as we shall proceed to verify in the case $f = 0$) it then follows directly from (5) that Equation (4) is satisfied. Let us write our answer more explicitely in the case $f = 0$:

$$y_x = \frac{\mathsf{T}_{sl-sx} F}{1 - \mathsf{T}_{2sl}} - \frac{\mathsf{T}_{sl+sx} F}{1 - \mathsf{T}_{2sl}}; \tag{12}$$

from 5.27 it follows that

$$y_x = \sum_{k=0}^{\infty} F(t - 2ksl - [sl - sx]) - \sum_{k=0}^{\infty} F(t - 2ksl - [sl + sx]);$$

that is,

$$y_x(t) = \sum_{k=0}^{\infty} \left[F\left(t - \frac{2kl + l - x}{1/s}\right) - F\left(t - \frac{2kl + l + x}{1/s}\right)\right] \qquad (\text{all } t > 0).$$

Let us now verify the starting conditions (8). From (7) and 5.24.6 we see that

$$y_x(t) = F\left(t - \frac{l - x}{1/s}\right) - F\left(t - \frac{l + x}{1/s}\right) \qquad (\text{if } 0 < t < sl).$$

Recall that $0 < x < l$; if α is a positive number less than $s(l \pm x)$, it again follows from 5.24.6 that $y_x(t) = 0$ for all $t < \alpha$: therefore, the starting conditions (8) are satisfied.

A Diffusion Problem

9.23.1 **Theorem.** *Let I be an interval $[0, l]$, and consider a rule assigning to any x in I a perfect operator B_x such that the equation*

$$\frac{\mathrm{d}^2}{\mathrm{d}x^2} B_x = d^2 B_x \qquad (0 < x < l) \tag{9.24}$$

holds for some number d^2. If V is a perfect operator, then

$$\boxed{\left[V - s^{-2} \frac{\mathrm{d}^2}{\mathrm{d}x^2}\right] \frac{B_x}{V - d^2/s^2} = B_x} \qquad (0 < x < l). \tag{9.25}$$

Proof. Of course, we suppose that the operator $V - d^2/s^2$ is invertible; we set

$$W = \left(V - \frac{d^2}{s^2}\right)^{-1} \tag{1}$$

and note that

$$\left[V - s^{-2} \frac{\mathrm{d}^2}{\mathrm{d}x^2}\right] W B_x = V W B_x - s^{-2} \frac{\mathrm{d}^2}{\mathrm{d}x^2} W B_x$$

$$= W \left[V B_x - s^{-2} \frac{\mathrm{d}^2}{\mathrm{d}x^2} B_x\right] \qquad \text{by 9.8}$$

$$= W [V B_x - s^{-2} d^2 B_x] \qquad \text{by 9.24}$$

$$= W \left[V - \frac{d^2}{s^2}\right] B_x = B_x:$$

the last equation is from (1).

9.26 **Remark.** Suppose that the operator B_x is invertible; the required number d^2 is determined by the equation

$$d^2 = B_x^{-1}\left[\frac{\mathrm{d}^2}{\mathrm{d}x^2} B_x\right].$$

9.27 **Temperature in a rod.** Let s be the reciprocal of the square root of the thermal diffusivity of a rod of length l whose left end-point is located at $x = 0$; we suppose that the heat can only flow through the ends of the bar.

Let $\psi_0(x)$ be the starting temperature at a point x; the equations

$$y_x(0+) = \psi_0(x) \quad \text{with} \quad \frac{\mathrm{d}}{\mathrm{d}t} y_x = s^{-2} \frac{\mathrm{d}^2}{\mathrm{d}x^2} y_x \qquad (0 < x < l) \tag{2}$$

govern the temperature y_x at a point x. In view of [9.0.0], the equations (2) imply that

$$\left[D - s^{-2}\frac{\mathrm{d}^2}{\mathrm{d}x^2}\right] y_x = \psi_0(x)\, D \qquad (0 < x < l). \tag{3}$$

From 9.25 we see that

$$\left[D - s^{-2}\frac{\mathrm{d}^2}{\mathrm{d}x^2}\right] \frac{\psi_0(x)\, D}{D - d^2/s^2} = \psi_0(x)\, D, \tag{4}$$

where d^2 is a number such that

$$\frac{\mathrm{d}^2}{\mathrm{d}x^2}\psi_0(x)\, D = d^2 \psi_0(x)\, D \qquad (0 < x < l). \tag{5}$$

Suppose that the starting temperature $\psi_0(x)$ is given by the equation

$$\psi_0(x) = \alpha \sin \frac{m\pi x}{l} \qquad (0 \leq x \leq l). \tag{6}$$

From (6) and 9.26 it follows immediately that the equation (5) holds for the number $d^2 = -(m\pi)^2/l^2$; Equation (4) becomes

$$\left[D - s^{-2}\frac{\mathrm{d}^2}{\mathrm{d}x^2}\right] \frac{\psi_0(x)\, D}{D + (m\pi/l s)^2} = \psi_0(x)\, D \qquad (0 < x < l). \tag{7}$$

Equation (7) means that the equation

(8) $$y_x = \frac{\varphi_0(x)\,D}{D + (m\pi/l s)^2} \qquad (0 < x < l)$$

determines a solution of (3). This solution (8) can also be written as follows:

(9) $$y_x = \varphi_0(x)\,\frac{D}{D + (m\pi/l s)^2} = \varphi_0(x)\left\langle \exp\left(-\left(\frac{m\pi}{l s}\right)^2 t\right)\right\rangle:$$

see 3.21. Observe that y_x (as defined by (9)) satisfies the starting condition $y_x(0+) = \varphi_0(x)$ for $0 < x < l$: it now follows from [9.0.0] and (3) that (9) determines a solution of (2). Indeed, (9) determines a solution of (3); but (3) implies (2) whenever the starting condition is satisfied by y_x, since (3) can be written

$$D y_x - \varphi_0(x)\,D = s^{-2}\frac{d^2}{dx^2}\,y_x,$$

and the left-hand side of this equation equals dy_x/dt (by [9.0.0]) whenever $y_x(0+) = \varphi_0(x)$.

Wave Problems

The two preceding examples illustrate the method and ideas that will be used to solve more complicated problems.

9.28 **The auxiliary solution.** Let a_k ($k = 0, 1, 2, 3$) be given numbers; to solve the problem

(1) $$y_x(0+) = \varphi_0(x), \quad y_x'(0+) = \varphi_1(x)$$

with

(2) $$\left[a_2\frac{d^2}{dt^2} + a_1\frac{d}{dt} + a_0 - s^{-2}\frac{d^2}{dx^2}\right] y_x = h_x \qquad (0 < x < l)$$

(the $\varphi_k(x)$ ($k = 0, 1$) are given numbers, and h_x is a given operator), we proceed as follows:

Step (i): observe that (1)—(2) imply

(3) $$\left[a_2 D^2 + a_1 D + a_0 - s^{-2}\frac{d^2}{dx^2}\right] y_x = B_x,$$

where $B_x = h_x + a_2\psi_0(x)\, D^2 + a_2\,\psi_1(x)\, D + a_1\,\psi_0(x)\, D$ (this fact comes directly from the definitions 9.0.0 and 9.0.1).

Step (ii): look for a number d^2 such that $\frac{d^2}{dx^2}\, B_x = d^2 B_x$, where B_x is the right-hand side of Equation (3).

Step (iii): replace $\frac{d^2}{dx^2}$ by d^2 in (3) and solve the resulting equation

$$[a_2 D^2 + a_1 D + a_0 - s^{-2} d^2]\, y_x = B_x;$$

this gives

$$y_x = \frac{B_x}{(a_2\, D^2 + a_1 D + a_0) - d^2/s^2}\,. \tag{4}$$

Setting $V = (a_2 D^2 + a_1 D + a_0)$ in 9.25, it follows immediately that (4) implies (3). The right-hand side of (4) will be called the **auxiliary solution** of the problem (1)—(2).

9.29 *Remark.* If $h_x(\,)$ is a $\mathcal{K}$-function, it is not hard to verify that the equation (4) actually implies (1)—(2): it determines a solution of our problem.

Thus, *the auxiliary solution is a solution of* (1)—(2) when h_x is the canonical operator of a $\mathcal{K}$-function $h_x()$; however, boundary-value problems require the additional information given by the following

Z 9.30 **Theorem.** *Consider the starting-value problem*

$$y_x(0+) = \psi_0(x), \quad y_x'(0+) = \psi_1(x) \tag{9.31}$$

with

$$\left[\left(p + \frac{d}{dt}\right)^2 - s^{-2}\,\frac{d^2}{dx^2}\right] y_x = h_x \qquad (0 < x < l): \tag{9.32}$$

the $\psi_k(x)$ $(k = 0, 1)$ are given numbers, and the $h_x(\,)$ are $\mathcal{K}$-functions. Let S_x $(0 < x < l)$ be the auxiliary solution of this problem; if $X(\,)$ and $Y(\,)$ are any two $\mathcal{K}$-functions, the equation

$$y_x = S_x + X p_s^x + Y p_s^{l-x} \qquad (0 < x < l) \tag{9.33}$$

determines a solution of this problem (9.31—32).

Remarks. In other words, 9.33 implies 9.31—32; it could also be said that 9.33 determines a two-parameter family (the parameters are X and Y) of solutions of 9.31—32. We shall not prove this theorem.

9.34 **Concerning the number** d^2: all the problems in this book can be solved by using

$$(9.35.0) \qquad d^2 = \begin{cases} 0 & (\text{if } B_x = 0 \text{ whenever } 0 < x < l) \\ B_x^{-1}\left[\frac{d^2}{dx^2} B_x\right] & (\text{otherwise}). \end{cases}$$

For example, if $B_x = A + Bx$, then 9.35.0 and 9.8—9 give $d^2 = 0$.

9.35.1 **Important remark.** Theorem 9.30 could be stated in a much stronger form. In case the h_x are not function-operators, it is best to verify that 9.33 implies the starting conditions (9.31): it can easily be shown that this is sufficient to guarantee that 9.33 determines a solution of 9.31—32. In other words, the two-parameter family 9.33 is a family of solutions whenever it satisfies the starting conditions — and this is true when the h_x are *any* perfect operators.

The Infinite Interval

Z 9.36 In case $l = \infty$ we set $p_s^\infty = 0$ and observe that $p_s^{l-x} = 0$ for any $x < \infty$: this amounts to discarding the last term in 9.33. Let S_x $(0 \leq x < \infty)$ be the auxiliary solution of the problem

$$(9.37) \qquad y_x(0+) = \varphi_0(x), \qquad y_x'(0+) = \varphi_1(x)$$

with

$$(9.38) \qquad \left[\left(p + \frac{d}{dt}\right)^2 - s^{-2}\frac{d^2}{dx^2}\right] y_x = h_x \qquad (0 < x < \infty).$$

If $X(\,)$ and $Y(\,)$ are any $\mathcal{K}$-functions, Theorem 9.30 guarantees that the equation

$$(9.39) \qquad y_x = S_x + X p_s^x \qquad (0 \leq x < \infty)$$

determines a solution of 9.37—38.

9.40.0 **Problem I.** Given a real number p, given a number c and a $\mathcal{K}$-function $f()$, to solve the problem

$$(1) \qquad y_x(0+) = y'_x(0+) = 0$$

with

$$(2) \qquad \left(p + \frac{\mathrm{d}}{\mathrm{d}t}\right)^2 y_x = s^{-2} \frac{\mathrm{d}^2}{\mathrm{d}x^2} y_x - c \qquad (0 < x < \infty),$$

$$(3) \qquad y_0(t) = f(t) \qquad (\text{all } t > 0).$$

We begin by finding the auxiliary solution (as in 9.28). **Step (i):** observe that (1)—(2) imply

$$(4) \qquad (p + D)^2 y_x - s^{-2} \frac{\mathrm{d}^2}{\mathrm{d}x^2} y_x = -c.$$

Step (ii): we look for a number d^2 such that $\frac{\mathrm{d}^2}{\mathrm{d}x^2} B_x = d^2 B_x$ with $B_x = -c$ (from (4)); clearly, such a number is $d^2 = 0$. **Step (iii):** we replace $\frac{\mathrm{d}^2}{\mathrm{d}x^2}$ by d^2 in (4) and solve the resulting equation

$$(p + D)^2 y_x - s^{-2} 0 = -c.$$

In consequence, the auxiliary solution is $-c/(p + D)^2$: from 9.39 we see that the equation

$$(5) \qquad y_x = \frac{-c}{(p + D)^2} + X p_s^x \qquad (0 \leq x < \infty)$$

determines a solution of (1)—(2). Let us determine X so that (5) satisfies our boundary condition (3). Setting $x = 0$ in (5):

$$f = y_0 = \frac{-c}{(p + D)^2} + X \qquad (\text{by (3) and [9.18]});$$

consequently, $X = f + c/(p + D)^2$. Substituting this value of X into (5), we obtain

$$y_x(t) = \mathrm{e}^{-sxp} [f(t - sx) + g(t - sx)] - g(t) \qquad (\text{all } t > 0),$$

where

$$g(t) = -c \frac{\mathrm{e}^{-pt} - 1}{p^2} - \left(\frac{ct}{p}\right) \mathrm{e}^{-pt} \qquad (\text{see } 4.32.1).$$

9.40.1 **Problem II.** Given a number $s > 0$, to solve the problem

(1) $$y_x(0+) = e^{-x},\ y_x'(0+) = 0 \quad \text{with} \quad \frac{d^2}{dt^2}\, y_x = s^{-2}\, \frac{d^2}{dx^2}\, y_x$$

$(0 < x < \infty)$ and with the boundary condition

(2) $$\frac{d}{dx}\, y_x\Big|_{x=0} = 0.$$

We begin by finding the auxiliary solution of (1). **Step (i):** observe that Equation (1) implies

(3) $$\left[D^2 - s^{-2}\,\frac{d^2}{dx^2}\right] y_x = e^{-x} D^2.$$

Step (ii): we look for a number d^2 such that $\frac{d^2}{dx^2}\, B_x = d^2 B_x$ with $B_x = e^{-x} D^2$; clearly, such a number is $d^2 = 1$ (see 9.35.0 and 9.9). **Step (iii):** we replace $\frac{d^2}{dx^2}$ by d^2 in (3) and solve the resulting equation

$$[D^2 - s^{-2} d^{-2}]\, y_x = e^{-x} D^2;$$

this gives our auxiliary solution $e^{-x} D^2/(D^2 - s^{-2})$: recall that $d^2 = 1$. From 9.39 we now see that the equation

(4) $$y_x = \frac{e^{-x} D^2}{D^2 - s^{-2}} + X p_s^x \qquad (0 \le x < \infty)$$

determines a solution of (1). Let us determine X so that (4) satisfies our boundary condition (2): applying d/dx to both sides of (4), we obtain

$$\frac{d}{dx}\, y_x = \frac{d}{dx}\left[e^{-x}\, \frac{D^2}{D^2 - s^{-2}}\right] + X\, \frac{d}{dx}\, p_s^x$$

$$= -\, e^{-x}\, \frac{D^2}{D^2 - s^{-2}} - s X D p_s^x:$$

the last equation is from 9.9, 9.8, and 9.19 (with $p = 0$); setting $x = 0$, Condition (2) and 3.25 give

$$X = \left\langle -\sinh \frac{t}{s} \right\rangle:$$

substituting into (4), we can use 3.28 and 5.6 to write

$$y_x(t) = \mathrm{e}^{-x} \cosh \frac{t}{s} - \mathsf{T}_0(t - sx) \sinh\left(\frac{t}{s} - x\right) \quad (\text{all } t > 0):$$

to see this, note that $p = 0$, so that 9.19.1 shows that $p_s^x = \mathsf{T}_{sx}$.

Classical Solutions

9.40.2 **Definition.** A solution y of the operator-problem 9.37—38 will be called *classical* if the operators y_x are the canonical operators of $\mathcal{K}$-functions $y_x()$ *and* if the functions $Y_k()$ $(k = 1, 2)$ defined by the equations

$$(9.40.3) \qquad Y_1(x, t) = \frac{\partial^2}{\partial x^2} y_x(t) \quad \text{and} \quad Y_2(x, t) = \frac{\partial^2}{\partial t^2} y_x(t)$$

are both continuous in the two-dimensional region

$$\{(x, t) : x > 0, \quad t > 0\}.$$

9.40.4 **Theorem.** *Suppose that* $-\infty < p < \infty$ *and let* $h_x()$ *be a* $\mathcal{K}$*-function depending on the parameter* $x > 0$; *if* y *is a classical solution of the operator-problem*

$$(9.37) \qquad y_x(0+) = \varphi_0(x), \quad y'_x(0+) = \varphi_1(x),$$

with

$$(9.38) \qquad \left[\left(p + \frac{\mathrm{d}}{\mathrm{d}t}\right)^2 - s^{-2} \frac{\mathrm{d}^2}{\mathrm{d}x^2}\right] y_x = h_x \qquad (0 < x < \infty),$$

it then follows from 9.1—2 *and* 9.15.1 *that*

$$\frac{\mathrm{d}^k}{\mathrm{d}t^k} y_x = \left\langle \frac{\partial^k}{\partial t^k} y_x(t) \right\rangle \qquad (k = 1, 2)$$

and

$$\frac{\mathrm{d}^2}{\mathrm{d}x^2} y_x = \left\langle \frac{\partial^2}{\partial x^2} y_x(t) \right\rangle \qquad (\text{all } x > 0).$$

9.40.5 **Consequently,** *the operator-problem* 9.37—38 *is equivalent to the function problem*

$$y_x(0+) = \varphi_0(x), \quad y_x'(0+) = \varphi_1(x)$$

with

$$\left[\left(p + \frac{\partial}{\partial t}\right)^2 - s^{-2}\frac{\partial^2}{\partial x^2}\right] y_x(t) = h_x(t) \qquad (\text{all } t > 0, \text{ all } x > 0).$$

9.40.6 **Problem III.** Given an operator f, to solve the problem

$$(1) \qquad y_x(0+) = y_x'(0+) = 0 \quad \text{with} \quad \left[\left(p + \frac{\mathrm{d}}{\mathrm{d}t}\right)^2 - s^{-2}\frac{\mathrm{d}^2}{\mathrm{d}x^2}\right] y_x = 0$$

$(0 < x < \infty)$, and $y_0 = f$.

We begin by finding the auxiliary solution (as in 9.28). Step (i): observe that (1) implies

$$(2) \qquad \left[(p + D)^2 - s^{-2}\frac{\mathrm{d}^2}{\mathrm{d}x^2}\right] y_x = 0 \qquad (0 < x < \infty).$$

Step (ii): we look for a number d^2 such that $\frac{\mathrm{d}^2}{\mathrm{d}x^2} B_x = d^2 B_x$, where B_x is the right-hand side of the equation (2); clearly, we can take $d^2 = 0$. Step (iii): we replace $\mathrm{d}^2/\mathrm{d}x^2$ by d^2 in (2) and solve the resulting equation:

$$[(p + D)^2 - 0]\, y_x = 0,$$

which gives $y_x = 0$: *the auxiliary solution is the operator* 0. From 9.39 we see that the equation

$$(3) \qquad y_x = 0 + X p_s^x \qquad (0 < x < \infty)$$

determines a solution of (1). Let us determine the operator X so that (3) satisfies the boundary condition ($y_0 = f$):

$$f = y_0 = X p_s^0 = X 1 = X \qquad (\text{by } [9.18]);$$

substituting into (3):

$$(4) \qquad \boxed{y_x = f p_s^x = \mathrm{e}^{-psx} \mathsf{T}_{sx} f} \qquad (\text{by } [9.18]).$$

9.40.7 Until further notice, we take $p = 0$ and let $f(\ )$ be a $\mathcal{K}$-function such that $f(t) = 0$ for $t < 0$: from (4) and 5.25 it follows that the equation

$$y_x(t) = f(t - sx) \qquad \text{(all } t > 0, \text{ all } x > 0) \tag{5}$$

determines a solution of (1) such that $y_0(t) = f(t)$ for all $t > 0$.

In case the function $f''(\ )$ is continuous, it is easily seen that y **is a classical solution**; indeed, (5) implies

$$\frac{\partial^2}{\partial t^2} y_x(t) = f''(t - sx) \quad \text{and} \quad \frac{\partial^2}{\partial x^2} y_x(t) = s^2 f''(t - sx); \tag{6}$$

our hypothesis (that $f''()$ be continuous on $(-\infty, \infty)$) therefore guarantees that the equations 9.40.3 satisfy the condition required in Definition 9.40.2.

9.40.8 To consider an example where $f''(\ )$ is not continuous, let $f(\ )$ be the function defined by

$$f(t) = \begin{cases} m^2 t/2 & (0 < t < 2/m) \\ 0 & \text{(otherwise)}, \end{cases} \tag{7}$$

where m is a given number. To simplify matters, suppose that $s = 1$. When $x = 0$ the graph of the solution

$$y_x(t) = f(t - x) \tag{8}$$

given by (5) for the operator-problem (1) has the following shape:

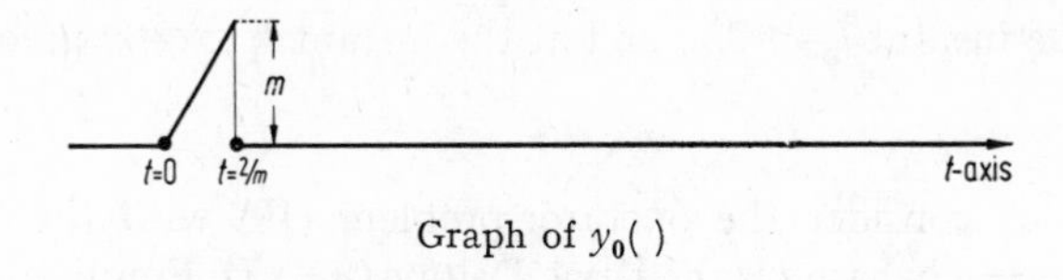

Graph of $y_0(\ )$

When $x = \lambda > 0$, the graph of (8) is obtained by a right-translation of the preceding graph:

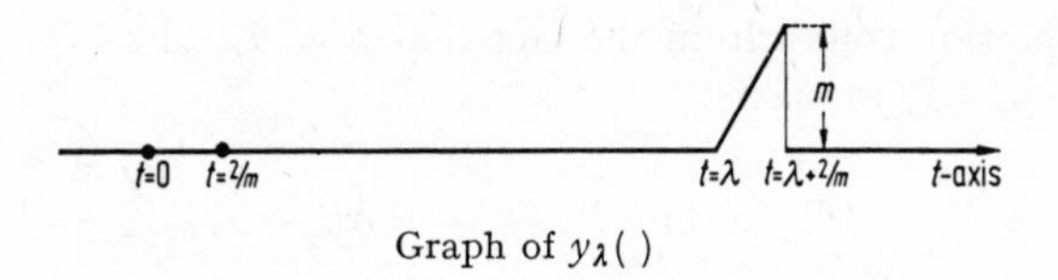

Graph of $y_\lambda(\ )$

indeed, at the time $t = \lambda + \tau$ we have $y_\lambda(t) = f(\tau) = y_0(\tau)$. The partial derivatives (6) do not exist when $t = x$ and $t = x + 2/m$; thus, the equation

$$\frac{\partial^2}{\partial t^2} y_x(t) = \frac{\partial^2}{\partial x^2} y_x(t) \qquad (0 < x < \infty) \tag{9}$$

fails at these values of t; nevertheless, the operator-equations

$$y_x(0+) = y_x'(0+) = 0, \quad \text{and} \quad \frac{\mathrm{d}^2}{\mathrm{d}t^2} y_x = \frac{\mathrm{d}^2}{\mathrm{d}x^2} y_x \quad (0 < x < \infty) \tag{10}$$

are satisfied.

The operator-equations (10) govern the displacement y_x of a string with end-points at $x = 0$ and at $x = \infty$; our boundary condition $y_0 = f$ (which is satisfied by (8)) means that the left end-point is raised from the origin with velocity $m^2/2$ (in view of the nature of $f(\,)$: see (7)), whereupon it snaps instantaneously back to the origin. We can combine (7) with (8) to obtain

$$y_x(t) = 2^{-1} m^2 (t - x) \qquad \left(\text{when } t - \frac{2}{m} < x < t\right),$$

and $y_x(t) = 0$ otherwise.

The graph of the string is the graph of $y_x(t)$ as a function of x:

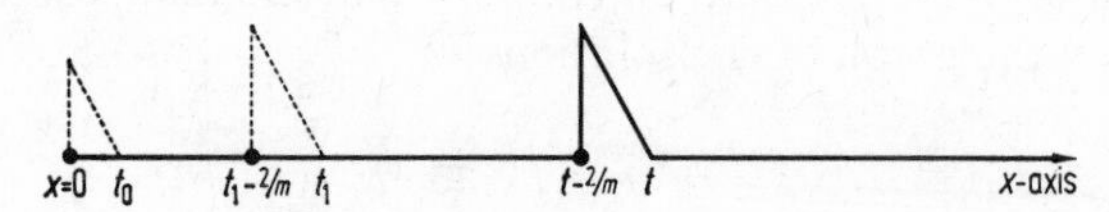

Shape of the string at the instant $t > 2/m$

The two dotted triangles in the above figure indicate the shape of the string at the instant $t_0 < 2/m$ and at the instant $t_1 > 2/m$ preceding the instant t.

9.40.9 Let us consider the operator-problem (10) with the boundary condition $y_0 = f$ in case f is the Dirac Delta $\delta(t) = D$. From (4) we obtain

$$y_x = \mathrm{e}^{-psx} \mathsf{T}_{sx} D = D\mathsf{T}_x \qquad (0 < x < \infty): \tag{11}$$

the last equation comes from the fact that $p = 0$ and $s = 1$. In view of 6.45—46 and 7.1, we have

$$D\mathsf{T}_x = \lim_{m \to \infty} f(t - x) = \delta(t - x),$$

where $f()$ is the function defined by (7). Clearly, (11) is not a classical solution, the equation (9) is entirely devoid of meaning, but (11) is a *physically meaningful* solution of the operator-problem (10): in fact, $D\mathsf{T}_x$ is an impulse of magnitude $= 1$ applied at the time $t = x$ (see 6.42); from (11) and 6.43 we see that

$$\{y_x\}(\tau) = 0 \quad \text{when} \quad \tau \neq x. \tag{12}$$

Since we chose to consider the operator-problem (10) as representing the displacement of a semi-infinite string, we can make the following observations: at the instant τ the displacement is zero everywhere except at the point $x = \tau$ (because of (12)), while the point x is subjected to an impulse of magnitude $= 1$; we could also say that the situation represents the motion of an *impulse of unit magnitude travelling at unit velocity*.

The Finite Interval

For the reader's future convenience, we gather some results here, before working out some more problems: we assume $-\infty < p < \infty$ (as before) and $s > 0$ throughout. Recall that

$$\boxed{p_s^u \overset{\text{def}}{=} \langle e^{-sup} \rangle \mathsf{T}_{su}} \qquad (0 \leq u < \infty); \tag{9.41}$$

further, if $0 \leq b < \infty$, then

$$\boxed{\frac{d}{dx}\, p_s^{\pm x+b} = \mp\, s(D + p)\, p_s^{\pm x+b}} \qquad (0 \leq x \leq b): \tag{9.42}$$

see 9.19. From 5.4 we see that

$$\boxed{x = 0 \ \text{ implies } \ p_s^x = p_s^0 = 1}; \tag{9.43}$$

moreover,

$$\boxed{p_s^x p_s^\lambda = p_s^{x+\lambda}} \qquad (\text{for } x \geq 0 \text{ and } \lambda \geq 0\text{: see 5.2}). \tag{9.44}$$

From [9.41] and 2.19 it follows that

$$(9.45)\qquad \frac{1}{p_s^x} = e^{sxp} [\![\mathsf{T}_{-sx}]\!] \qquad (0 \leq x < \infty).$$

If V is an operator, it follows from 8.28 and [9.41] that the equation

$$(9.46)\qquad \{p_s^x V\}(t) = e^{-sxp} \{V\}(t - sx)$$

holds at all points t such that the right-hand side is $\neq \infty$. If $g()$ is a $\mathcal{K}$-function such that $g(t) = 0$ for $t \leq 0$, then

$$(9.47)\qquad p_s^x \langle g(t) \rangle = e^{-sxp} \langle g(t - sx) \rangle \qquad \text{(see 5.25)}.$$

Also, note that

$$(9.48)\qquad p = 0 \quad \text{implies} \quad p_s^x = \mathsf{T}_{sx}.$$

Until further notice, we suppose that $p = 0$. Let $g()$ be a $\mathcal{K}$-function such that $g(t) = 0$ for $t \leq 0$; it follows from 5.29 and 9.48 that the equation

$$(9.49)\qquad \boxed{\frac{p_s^\lambda g}{1 - b p_s^m} = \left\langle \sum_{k=0}^{\infty} b^k g\left(t - \frac{km + \lambda}{1/s}\right) \right\rangle}$$

holds for any numbers b, $\lambda \geq 0$, and $m > 0$.

9.50 **Theorem.** *Let $h()$ be a function which is continuous in the interval $(0, \infty)$. If*

$$(9.51)\qquad (n-1)\, m + \lambda < \frac{t}{s} < nm + \lambda \qquad (n = 1, 2, 3, \ldots),$$

then the equation

$$(9.52)\qquad \left\{ \frac{p_s^\lambda h}{1 - b p_s^m} \right\}(\tau) = \sum_{k=0}^{n-1} b^k h\left(\tau - \frac{km + \lambda}{1/s}\right)$$

holds for all $\tau > 0$.

Proof. See 9.49, 6.17, and 5.30.1 (with $c = b$, $\lambda = ms$, and $x = \lambda s$).

9.53 Vibrations of a finite bar. Consider a uniform bar whose end-points are located at $x = 0$ and at $x = l < \infty$. The equation

$$(1)\qquad \frac{\mathrm{d}^2}{\mathrm{d}t^2}\, y_x = s^{-2}\, \frac{\mathrm{d}^2}{\mathrm{d}x^2}\, y_x \qquad (0 < x < l)$$

governs the longitudinal displacement y_x of the cross-section of the bar at a point x. Suppose that the bar is fixed at $x = 0$ and is subjected to a longitudinal force F at the end-point $x = l$:

$$(2)\qquad y_0 = 0 \qquad \text{and} \qquad F = \frac{\mathrm{d}}{\mathrm{d}x}\, y_x \big|_{x=l}.$$

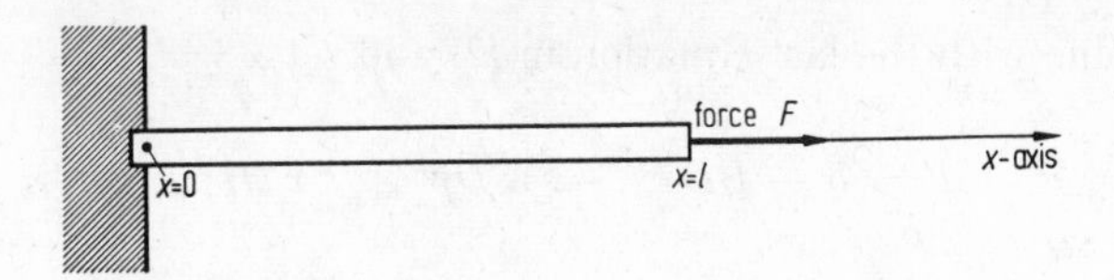

Let us impose the starting conditions

$$(3)\qquad y_x(0+) = a + bx \quad \text{and} \quad y_x'(0+) = A + Bx \qquad (0 < x < l).$$

We begin by finding the auxiliary solution (as in 9.28). **Step (i):** observe that the equations (1) and (3) imply

$$(4)\qquad \left[D^2 - s^{-2}\frac{\mathrm{d}^2}{\mathrm{d}x^2}\right] y_x = (a + bx)\, D^2 + (A + Bx)\, D$$

(by [9.0.0] and [9.0.1]).

Step (ii): look for a number d^2 such that $\frac{\mathrm{d}^2}{\mathrm{d}x^2} B_x = d^2 B_x$, where B_x is the right-hand side of (4): clearly, we can take $d^2 = 0$ (see 9.26). **Step (iii):** we replace $\mathrm{d}^2/\mathrm{d}x^2$ by d^2 in (4) and solve the resulting equation:

$$[D^2 - s^{-2}0]\, y_x = (a + bx)\, D^2 + (A + Bx)\, D \qquad (0 < x < l):$$

the auxiliary solution is therefore the operator

$$(a + bx) + (A + Bx)\, D^{-1}.$$

Let $X(\,)$ and $Y(\,)$ be any two $\mathcal{K}$-functions; from 9.30 it follows that the equation

$$(5)\quad y_x = (a + bx) + (A + Bx)\, D^{-1} + X p_s^x + Y p_s^{l-x} \qquad (0 < x < l)$$

determines a solution of our problem (1): note that $p = 0$. Let us determine X and Y so that (5) satisfies the boundary conditions (2). Setting $x = 0$ in (5), the first condition in (2) gives

$$0 = a + AD^{-1} + X + Yp_s^l \qquad \text{(by 9.43)}. \tag{6}$$

Next, we apply $\frac{\mathrm{d}}{\mathrm{d}x}$ to both sides of (5) and use 9.42 to obtain

$$\frac{\mathrm{d}}{\mathrm{d}x} y_x = b + BD^{-1} - sXDp_s^x + sYDp_s^{-x+l};$$

now combine with the last equation in (2) and set $x = l$:

$$F = b + BD^{-1} - sXDp_s^l + sYDp_s^0,$$

which gives

$$sD[-Xp_s^l + Y] = F - b - BD^{-1}. \tag{7}$$

The problem is completed by solving the system (6)—(7) for X and Y and substituting the values obtained into (5). Now follows a sketch of the calculations in the case $a = 0 = A$. Equation (6) now implies

$$X = -Yp_s^l \qquad \text{(since } a = 0 = A). \tag{8}$$

Substituting (8) into (7), we can use 9.44 to obtain

$$Y(p_s^{2l} + 1) = s^{-1}D^{-1}(F - b - BD^{-1}), \tag{9}$$

which gives

$$Y = \frac{G}{1 + p_s^{2l}}, \tag{10}$$

where

$$G = (sD)^{-1}(F - b - BD^{-1}). \tag{11}$$

From (8) and (10) we see that

$$X = \frac{-p_s^l G}{1 + p_s^{2l}}. \tag{12}$$

Next, we substitute (12) and (10) into (5):

$$y_x = bx + BxD^{-1} + \frac{-p_s^l G}{1 + p_s^{2l}} p_s^x + \frac{G p_s^{l-x}}{1 + p_s^{2l}} \qquad (0 < x < l);$$

recall our assumption $a = 0 = A$. Another application of 9.44 gives

$$(13)\qquad y_x = bx + BxD^{-1} - \frac{p_s^{l+x} G}{1 + p_s^{2l}} + \frac{p_s^{l-x} G}{1 + p_s^{2l}} \qquad (0 < x < l).$$

In view of 9.49, Equation (13) implies

$$(14)\qquad y_x = bx + Bxt - \sum_{k=0}^{\infty} (-1)^k G\Big(t - \frac{k\,2l + l + x}{1/s}\Big)$$

$$+ \sum_{k=0}^{\infty} (-1)^k G\Big(t - \frac{k\,2l + l - x}{1/s}\Big):$$

recall that $p = 0$. From (11), 2.65, 3.19, and 3.15 we also have

$$G(\tau) = s^{-1} \int_0^{\tau} F(u)\, du - \frac{b\tau}{s} - B\,\frac{\tau^2}{2s} \qquad (\text{all } \tau > 0).$$

9.54 **Particular case.** Let the bar be at rest when $t = 0+$, and let the applied force F be a number m: we may therefore use (13) with $b = 0$ and $B = 0$; Equation (13) becomes

$$(15)\qquad y_x = \frac{p_s^{l-x} - p_s^{l+x}}{1 + p_s^{2l}}\, G = m\,(sD)^{-1}\, \frac{p_s^{l-x} - p_s^{l+x}}{1 + p_s^{2l}} \qquad (0 < x < l):$$

the last equation is from (11) and the fact that $F = m$. Recall that $p = 0$, so that $p_s^x = \mathsf{T}_{sx}$ (see 9.48).

The above form of the answer is more informative than (14); *it brings out facts that are not easily recognizable in the Fourier series solution of this problem.* Multiplying numerator and denominator of (15) by $(1 - p_s^{2l})$, we may use 9.44 to write

$$(16)\qquad y_x = \frac{m\, s^{-1} h_x}{1 - \mathsf{T}_{4sl}},$$

where

$$(17)\qquad h_x = D^{-1}\,[\mathsf{T}_{sl-sx} - \mathsf{T}_{sl+sx} - \mathsf{T}_{3sl-sx} + \mathsf{T}_{3sl+sx}].$$

From 5.14—16 it now follows that the graph of $h_x(\,)$ has the following shape:

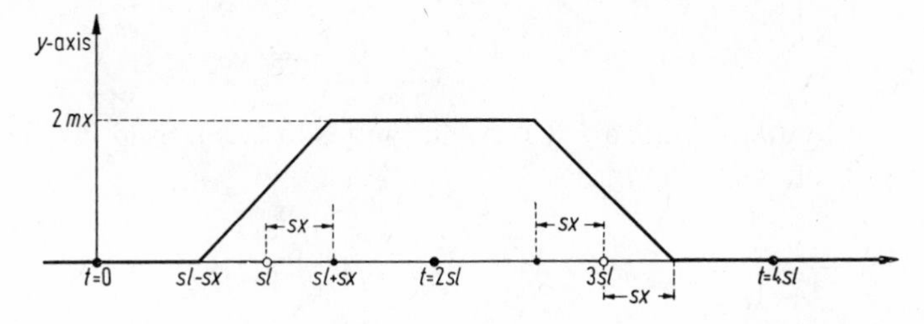

In view of (15) and 5.35 (with $\lambda = 4sl$), we can infer from (16) that the function $y_x(\,)$ has period $= 4sl$:

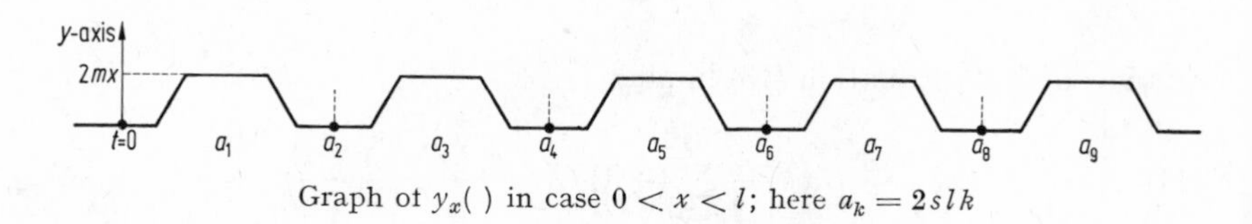

Graph of $y_x(\,)$ in case $0 < x < l$; here $a_k = 2slk$

At the point $x = l$ we see that the displacement $y_l(\,)$ varies with time according to the following graph:

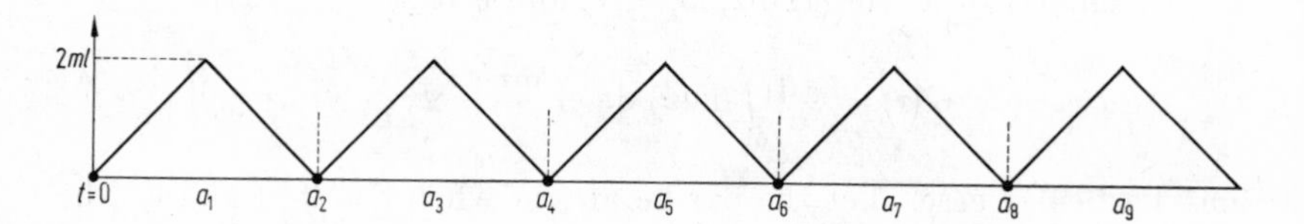

9.55 **Vibrating string.** Let the starting displacement and velocity of a finite string be given by the equations

$$(1) \qquad \varphi_0(x) = \alpha \sin \frac{m\pi x}{l} \quad \text{and} \quad \varphi_1(x) = A \sin \frac{m\pi x}{l} \qquad (0 < x < l).$$

Consequently, the corresponding starting-value problem is

$$(2) \qquad y_x(0+) = \varphi_0(x), \quad y_x'(0+) = \varphi_1(x) \quad \text{with} \quad \frac{d^2}{dt^2}\, y_x = s^{-2} \frac{d^2}{dx^2}\, y_x.$$

Further, let the string be fixed at $x = 0$ and $x = l$:

$$(3) \qquad y_0 = y_l = 0.$$

Shape of the string at the time $t = 0+$

We begin by finding the auxiliary solution (as in 9.28). Step (i): observe that (2) implies

$$(4) \qquad \left[D^2 - s^{-2} \frac{d^2}{dx^2}\right] y_x = \varphi_0(x)\, D^2 + \varphi_1(x)\, D.$$

Step (ii): we look for a number d^2 such that $\frac{\mathrm{d}^2}{\mathrm{d}x^2} B_x = d^2 B_x$, where B_x is the right-hand side of (4); since

$$B_x = \psi_0(x) D^2 + \psi_1(x) D$$

is invertible, we have

$$d^2 = B_x^{-1}\left[\frac{\mathrm{d}^2}{\mathrm{d}x^2} B_x\right] = B_x^{-1}\left[-\left(\frac{m\pi}{l}\right)^2 B_x\right] = -\left(\frac{m\pi}{l}\right)^2 \qquad \text{(by 9.26)}.$$

Step (iii): we replace $\mathrm{d}^2/\mathrm{d}x^2$ by d^2 in (4) and solve the resulting equation:

$$[D^2 - s^{-2} d^2]\, y_x = \psi_0(x)\, D^2 + \psi_1(x)\, D.$$

The auxiliary solution is therefore the operator

$$\frac{\psi_0(x)\, D^2 + \psi_1(x)\, D}{D^2 + (m\pi/sl)^2}.$$

Let $X(\ )$ and $Y(\ )$ be any two $\mathcal{K}$-functions; from 9.30 it follows that the equation

$$(5) \qquad y_x = \frac{\psi_0(x)\, D^2 + \psi_1(x)\, D}{D^2 + (m\pi/sl)^2} + X p_s^x + Y p_s^{l-x} \qquad (0 \leq x \leq l)$$

determines a solution of our problem (2). Let us determine X and Y so that (5) satisfies the boundary conditions (3). Setting $x = 0$ and $x = l$ in (5), we obtain

$$(6) \qquad 0 = X + Y p_s^l \qquad \text{and} \qquad 0 = X p_s^l + Y$$

(since $\psi_k(0) = \psi_k(l) = 0$ for $k = 0, 1$). Solving the system (6) for X and Y:

$$(7) \qquad X(1 - p_s^{2l}) = 0 = Y(1 - p_s^{2l}).$$

Since the operator $(1 - p_s^{2l})$ is invertible (see 5.31), we can multiply by $(1 - p_s^{2l})^{-1}$ both sides of the equation (7) to obtain $X = Y = 0$; substituting into (5):

$$y_x = \psi_0(x) \frac{D^2}{D^2 + a^2} + \psi_1(x) \frac{D}{D^2 + a^2} \qquad (\text{if } a = m\pi/sl)$$

$$= \alpha \sin\left(\frac{m\pi x}{l}\right) \cos(at) + \frac{A}{a} \sin\left(\frac{m\pi x}{l}\right) \sin(at):$$

the last equation is from 3.25, 3.28, and (1).

Exercises

In the following exercises, y'_x is an abbreviation of $\mathrm{d}y_x/\mathrm{d}t$ (as defined in [9.0.0]) and

$$y''_x \text{ is an abbreviation of } \frac{\mathrm{d}^2}{\mathrm{d}t^2}\, y_x$$

as defined in 9.0.1. Recall that

$$y_\lambda^{[1]} = \frac{\mathrm{d}}{\mathrm{d}x}\, y_x \big|_{x=\lambda} \qquad \text{(see 9.11).}$$

9.56.0 Solve the problem

$$y_x(0+) = \mathrm{e}^{ax}, \quad y'_x(0+) = 0 \text{ with } y''_x = c^2 \frac{\mathrm{d}^2}{\mathrm{d}x^2}\, y_x \qquad (0 < x < \infty)$$

subject to the condition $y_0^{[1]} = 0$.

Answer: $\mathrm{e}^{ax} \cosh(act) + \mathsf{T}_0(t - x/c) \sinh(act - ax)$.

9.56.1 Consider an electrical transmission line of length l; if the line constants are R, L and C, the equation

$$LC \frac{\mathrm{d}^2}{\mathrm{d}t^2}\, y_x + (RC + GL) \frac{\mathrm{d}}{\mathrm{d}t}\, y_x + RGy_x = \frac{\mathrm{d}^2}{\mathrm{d}x^2}\, y_x \qquad (0 < x < l)$$

governs the potential y_x at a point x. Assuming $G = RC/L$, $l = \infty$, and zero starting conditions ($y_x(0+) = y'_x(0+) = 0$), find y in case $y_0(\,)$ is a given $\mathcal{K}$-function $f(\,)$.

Answer: $y_x = \left[\exp\left(-xR\sqrt{\frac{C}{L}}\right)\right] f\left(t - x\sqrt{LC}\right) \mathsf{T}_0\left(t - x\sqrt{LC}\right)$.

9.56.2 Solve the problem

$$y_x(0+) = y'_x(0+) = 0 \text{ with } y''_x = c^2 \frac{\mathrm{d}^2}{\mathrm{d}x^2}\, y_x \qquad (0 < x < l < \infty)$$

subject to the boundary conditions $y_0 = f$ and $y_l = 0$, where $f(\,)$ is a $\mathcal{K}$-function.

Answer: $\sum_{k=0}^{\infty} \left\{ f\left(t - \frac{2kl + x}{c}\right) - f\left(t - \frac{2kl - x + 2l}{c}\right) \right\}$.

9.56.3 Solve:

$$y_x(0+) = ax, \quad y'_x(0+) = 0, \quad y''_x = c^2 \frac{d^2}{dx^2} y_x \qquad (0 < x < l < \infty)$$

subject to the condition $y_0 = y_l = 0$.

Answer: $ax + al \sum_{k=0}^{\infty} \left\{ \mathsf{T}_0\left(t - \frac{2kl + x + l}{c}\right) - \mathsf{T}_0\left(t - \frac{2kl - x + l}{c}\right) \right\}.$

9.56.4 Solve:

$$y_x(0+) = 0, \quad y'_x(0+) = 2, \quad y''_x = (1/9) \frac{d^2}{dx^2} y_x \qquad (0 < x < \infty)$$

subject to the condition $y_0(t) = \sin t$ (all $t > 0$).

Answer: $2t + [\sin(t - 3x) - 2(t - 3x)]\, \mathsf{T}_0(t - 3x)$.

9.56.5 Solve:

$$y_x(0+) = 0, \quad y'_x(0+) = -2, \quad 16 y''_x = \frac{d^2}{dx^2} y_x \qquad (0 < x < \infty)$$

subject to the condition $y_0(t) = t$ (all $t > 0$).

Answer: $-2t + 3(t - 4x)\, \mathsf{T}_0(t - 4x)$.

9.56.6 Solve:

$$y_x(0+) = ax, \quad y'_x(0+) = -1, \quad 16 y''_x = \frac{d^2}{dx^2} y_x \qquad (0 < x < \infty)$$

subject to the condition $y_0(t) = t^2$ (all $t > 0$).

Answer: $ax - t + \mathsf{T}_0(t - 4x)\, [(t - 4x)^2 + (t - 4x) - ax]$.

9.56.7 Solve:

$$y_x(0+) = y'_x(0+) = 0, \quad s^2 [y''_x + 2m] = \frac{d^2}{dx^2} y_x \qquad (0 < x < \infty)$$

subject to the condition $y_0 = 0$.

Answer: $y_x(t) = \begin{cases} -mt^2 & (t \le sx) \\ -ms(2tx - sx^2) & (t > sx). \end{cases}$

9.56.8 Solve:

$$y_x(0+) = ax, \quad y_x'(0+) = 0, \quad y_x'' = \frac{\mathrm{d}^2}{\mathrm{d}x^2} y_x \qquad (0 < x < l < \infty)$$

subject to the condition $y_0 = 0 = y_l^{[1]}$.

Answer: $ax + a \sum_{k=0}^{\infty} (-1)^k \{h(t - 2kl - l - x) - h(t - 2kl - l + x)\}$,

where $h(t) = t\mathsf{T}_0(t)$.

9.56.9 Solve:

$$y_x(0+) = y_x'(0+) = 0, \quad y_x'' = \frac{\mathrm{d}^2}{\mathrm{d}x^2} y_x \qquad (0 < x < l < \infty)$$

subject to the conditions

$$y_0 = 0 \quad \text{and} \quad \frac{\mathrm{d}}{\mathrm{d}x} y_x\Big|_{x=l} = \begin{cases} a & (t \leq \lambda) \\ a + b & (t > \lambda). \end{cases}$$

Answer: $a \sum_{k=0}^{\infty} (-1)^k \{h(t - 2kl - l + x) - h(t - 2kl - l - x)\}$

$+ b \sum_{k=0}^{\infty} (-1)^k \{h(t - 2kl - l - \lambda + x) - h(t - 2kl - l - \lambda - x)\}$,

where $h(t) = t\mathsf{T}_0(t)$.

9.56.10 Solve:

$$y_x(0+) = \sin\left(\frac{3\pi x}{l}\right), \quad y_x'(0+) = 0, \quad s^2 y_x'' = \frac{\mathrm{d}^2}{\mathrm{d}x^2} y_x \qquad (0 < x < l),$$

where $l < \infty$, subject to the condition $y_0 = y_l = 0$.

Answer: $\cos\left(\frac{3\pi t}{sl}\right) \sin\left(\frac{3\pi x}{l}\right)$.

9.56.11 Solve:

$$y_x(0+) = 0, \quad y'_x(0+) = \sin\left(\frac{5\pi x}{l}\right), \quad s^2 y''_x = \frac{d^2}{dx^2}\, y_x \qquad (0 < x < l)$$

where $l < \infty$, subject to the condition $y_0 = y_l = 0$.

Answer: $\frac{sl}{5\pi} \sin\left(\frac{5\pi x}{l}\right) \sin\left(\frac{5\pi t}{sl}\right)$.

9.56.12 Given two numbers c and p, solve the problem

$$y_x(0+) = y'_x(0+) = 0,$$

with

$$\left[\frac{d^2}{dt^2} + 2p\frac{d}{dt} + p^2 - \frac{d^2}{dx^2}\right] y_x = -c \qquad (0 < x < \infty)$$

and

$$y_0 = 0.$$

Hint: use 3.30.1.

Answer:

$$y_x = \begin{cases} g(t) & (0 < t < x) \\ g(t) - e^{-pt}\, g(t - x) & (t > x), \end{cases}$$

where

$$g(t) = \frac{c}{p^2}\,[e^{-pt} + pt e^{-pt} - 1].$$

9.56.13 Given a $\mathcal{K}$-function $G(\,)$, solve the problem:

$$y_x(0+) = y'_x(0+) = 0, \qquad y''_x + G = c^2 \frac{d^2}{dx^2}\, y_x \qquad (0 < x < l)$$

where $l < \infty$, subject to the condition $y_0 = y_l = 0$.

Answer: $-h(t) + \sum_{k=0}^{\infty} \left\{h\left(t - \frac{kl + x}{c}\right) + h\left(t - \frac{kl + l + x}{c}\right)\right\},$

where

$$h(t) = \mathsf{T}_0(t) \int_0^t (t-u)\, G(u)\, \mathrm{d}u \qquad \text{(see 3.17)}.$$

9.56.14 Again given a $\mathcal{K}$-function $G(\ )$, solve the problem:

$$y_x(0+) = y_x'(0+) = 0,$$

$$y_x'' + 2p y_x' + p^2 y_x = -G + c^2 \frac{\mathrm{d}^2}{\mathrm{d}x^2} y_x \qquad (0 < x < \infty)$$

subject to the condition $y_0 = 0$.

Answer:

$$y_x = -h(t) + \mathrm{e}^{-px/c}\, h(t - x/c),$$

where

$$h(t) = \mathsf{T}_0(t)\, \mathrm{e}^{-pt} \int_0^t (t-u)\, \mathrm{e}^{pu}\, G(u)\, \mathrm{d}u.$$

9.56.15 Given a function $F(\ )$, solve the problem:

$$y_x(0+) = y_x'(0+) = 0, \quad s^2 y_x'' = \frac{\mathrm{d}^2}{\mathrm{d}x^2} y_x \qquad (0 < x < \infty)$$

subject to the condition

$$\frac{\mathrm{d}}{\mathrm{d}x} y_x\Big|_{x=0} = F.$$

Answer:

$$y_x(t) = -\frac{1}{s}\, \mathsf{T}_0(t - sx) \int_0^{t-sx} F(u)\, \mathrm{d}u.$$

9.56.16 Consider a rule assigning to any x in an open interval I an operator y_x such that $y^{[1]} = 0$ for all x in I. Prove that there exists an operator V such that $y_x = V$ for all x in I.

§ 10. Diffusion Problems

This section consists mostly of worked examples of boundary-value problems involving equations of the form

$$\left[\left(p + \frac{\mathrm{d}}{\mathrm{d}t}\right) - s^{-2}\frac{\mathrm{d}^2}{\mathrm{d}x^2}\right] y_x = h_x \qquad (0 < x < l \leq \infty);$$

such problems will be solved by procedures entirely analogous to the ones that we applied to wave problems. To begin with, let cerf be the function defined by

(10.0) $$\operatorname{cerf}\lambda = \frac{2}{\sqrt{\pi}}\int_{\lambda}^{\infty}\exp(-u^2)\,\mathrm{d}u \qquad (0 \leq \lambda \leq \infty):$$

note that

$$\operatorname{cerf} 0 = 1 \quad \text{and} \quad \operatorname{cerf}\infty = 0.$$

Given a number p, let $\hat{p}_x(\,)$ be the function defined by

(10.1) $$\hat{p}_x(\,) \overset{\text{def}}{=} \left\{\mathrm{e}^{-pt}\operatorname{cerf}\frac{x}{2\sqrt{t}}\right\}(\,) \qquad (0 \leq x \leq \infty):$$

see [2.0]. Consequently, $\hat{p}_x(t) = 0$ for $t \leq 0$ and

$$\hat{p}_x(\tau) = \int_{x/2\sqrt{\tau}}^{\infty}\exp(-p\tau - u^2)\,\mathrm{d}u \qquad (\text{for } \tau > 0).$$

From [10.1] and [2.1] we see that

$$\boxed{\hat{p}_\lambda = \left\langle \mathrm{e}^{-pt}\operatorname{cerf}\frac{\lambda}{2\sqrt{t}}\right\rangle} \qquad (0 \leq \lambda \leq \infty).$$

From [10.0] and 3.21 we have

(10.2) $$\boxed{\hat{p}_0 = \langle \mathrm{e}^{-pt}\rangle = \frac{D}{D+p}}$$

and

(10.2.1) $$\hat{p}_\infty = 0.$$

The equation

$$\hat{p}_x(0+) = 0 \qquad \text{(when } 0 < x < \infty) \tag{10.3}$$

is also an immediate consequence of the definitions. The equation

$$\frac{\mathrm{d}}{\mathrm{d}t} \hat{p}_x = D\hat{p}_x \qquad (0 < x < \infty) \tag{10.4}$$

comes directly from [9.0.0] and 10.3. It is easily verified that

$$\frac{\partial}{\partial x} \hat{p}_x(t) = \frac{-1}{\sqrt{\pi t}} \exp\left(-pt - \frac{x^2}{4t}\right) \qquad \text{when} \begin{cases} 0 < x < \infty \\ 0 < t < \infty. \end{cases} \tag{1}$$

Since (1) has precisely the form 9.15, it follows from 9.14 that

$$\left\langle \frac{\partial}{\partial x} \hat{p}_x(t) \right\rangle = \frac{\mathrm{d}}{\mathrm{d}x} \hat{p}_x \qquad (0 < x < \infty), \tag{2}$$

so that (1) gives

$$\frac{\mathrm{d}}{\mathrm{d}x} \hat{p}_x = \left\langle \frac{-1}{\sqrt{\pi t}} \exp\left(-pt - \frac{x^2}{4t}\right) \right\rangle. \tag{10.5}$$

Setting $y_x(t) = \partial \hat{p}_x(t)/\partial x$, we obviously have

$$\frac{\partial}{\partial x} y_x(t) = \frac{\partial^2}{\partial x^2} \hat{p}_x(t) = \frac{x t^{-3/2}}{2\sqrt{\pi}} \exp\left(-pt - \frac{x^2}{4t}\right): \tag{3}$$

the last equation is immediate from (1). Since (3) has the form 9.15, it follows from 9.14 that

$$\left\langle \frac{\partial}{\partial x} \left[\frac{\partial}{\partial x} \hat{p}_x(t) \right] \right\rangle = \frac{\mathrm{d}}{\mathrm{d}x} \left\langle \frac{\partial}{\partial x} \hat{p}_x(t) \right\rangle = \frac{\mathrm{d}}{\mathrm{d}x} \left[\frac{\mathrm{d}}{\mathrm{d}x} \hat{p}_x \right]: \tag{4}$$

the last equation is from (2). Consequently, the equations

$$\frac{\mathrm{d}^2}{\mathrm{d}x^2} \hat{p}_x = \left\langle \frac{\partial^2}{\partial x^2} \hat{p}_x(t) \right\rangle = \left\langle \frac{x t^{-3/2}}{2\sqrt{\pi}} \exp\left(-pt - \frac{x^2}{4t}\right) \right\rangle \tag{10.6}$$

are from (4) and (3). Next, note that

(5) $$\frac{\partial}{\partial t}\hat{p}_x(t) + p\hat{p}_x(t) = \frac{x t^{-3/2}}{2\sqrt{\pi}} \exp\left(-pt - \frac{x^2}{4t}\right) \qquad (\text{all } t > 0);$$

from 10.6 it therefore follows that

(6) $$\left\langle\frac{\partial}{\partial t}\hat{p}_x(t)\right\rangle + p\hat{p}_x = \frac{\mathrm{d}^2}{\mathrm{d}x^2}\hat{p}_x.$$

On the other hand, from (5) we see that the function $\hat{p}_x^{(1)}(\,)$ defined by the equation

$$\hat{p}_x^{(1)}(t) = \frac{\partial}{\partial t}\hat{p}_x(t)$$

is continuous in $(0, \infty)$: we may therefore use 9.1 to infer from (6) that

(10.7) $$\frac{\mathrm{d}}{\mathrm{dt}}\hat{p}_x + p\hat{p}_x = \frac{\mathrm{d}^2}{\mathrm{d}x^2}\hat{p}_x \qquad (0 < x < \infty).$$

10.8 **Theorem.** *Suppose that* $s > 0$, *and let* S_x $(0 < x < l)$ *be the auxiliary solution of the problem*

(10.9) $$y_x(0+) = \psi(x)$$

with

(10.10) $$\left[\left(p + \frac{\mathrm{d}}{\mathrm{dt}}\right) - s^{-2}\frac{\mathrm{d}^2}{\mathrm{d}x^2}\right] y_x = h_x \qquad (0 < x < l):$$

the $\psi(x)$ *are given numbers and the* h_x *are the operators of given* $\mathcal{K}$*-functions* $h_x(\,)$. *If* $X(\,)$ *and* $Y(\,)$ *are any two* $\mathcal{K}$*-functions, the equation*

(10.11) $$y_x = S_x + X\hat{p}_{sx} + Y\hat{p}_{sl-sx} \qquad (0 \le x \le l)$$

implies 10.9—10; *in other words, it determines a solution of the problem in question.*

10.12.0 *Remarks.* Recall that p is an arbitrary number. To find the auxiliary solution of 10.9—10, we proceed as in 9.28. Step (i): we use [9.0.0] to observe that 10.10 implies

(1) $$\left[(p + D) - s^{-2}\frac{\mathrm{d}^2}{\mathrm{d}x^2}\right] y_x = h_x + \psi(x)\,D \qquad (0 < x < l).$$

Step (ii): we look for a number d^2 such that $\frac{\mathrm{d}^2}{\mathrm{d}x^2} B_x = d^2 B_x$, where B_x is the right-hand side of (1). Step (iii): replace $\mathrm{d}^2/\mathrm{d}x^2$ by d^2 in (1) and solve the resulting equation:

$$[(p + D) - s^{-2} d^2]\, y_x = h_x + \psi(x)\, D.$$

Consequently, the auxiliary solution is the operator

$$S_x = \frac{h_x + \psi(x) D}{(p + D) - d^2/s^2} \qquad (0 < x < l). \tag{10.12.1}$$

Since $\frac{\mathrm{d}^2}{\mathrm{d}x^2} B_x = d^2 B_x$ with $B_x = h_x + \psi(x)\, D$, we can apply 9.25 to obtain

$$\left[(p + D) - s^{-2} \frac{\mathrm{d}^2}{\mathrm{d}x^2}\right] S_x = h_x + \psi(x)\, D \qquad (0 < x < l). \tag{10.12.2}$$

10.12.3 The following fact will be needed in the proof of 10.8. Let $F(\ )$ and $G(\ )$ be $\mathcal{K}$-functions; *if one of these functions is regulated, then*

$$F * G(0+) = 0. \tag{2}$$

To verify (2), we begin by observing that

$$F * G(0+) = \lim_{\tau \to 0+} \{F * G\}(\tau) \qquad \text{by [9.0.2]} \tag{3}$$

$$= \lim_{\tau \to 0+} \{F D^{-1} G\}(\tau) \qquad \text{by [8.8]}. \tag{4}$$

On the other hand, 2.9 gives

$$F D^{-1} G = [\![\{F(t)\} * \{G(t)\}]\!], \tag{5}$$

and 2.61 guarantees that the function $\{F(t)\} * \{G(t)\}(\ )$ is continuous; consequently, (5) implies

$$\{F D^{-1} G\}(\tau) = \{F(t)\} * \{G(t)\}(\tau) \qquad \text{(by 6.15)}. \tag{6}$$

Combining (3)—(4) with (6):

$$F * G(0+) = \{F(t)\} * \{G(t)\}(0+) = 0;$$

the last equation is from 2.63.

10.12.4 *Proof of* 10.8. From 10.11 and 10.12.1 we see that

$$y_x = \psi(x) \frac{D}{D - (d^2/s^2 - p)} + H_x,$$

where

$$H_x = h_x D^{-1} \frac{D}{D - (d^2/s^2 - p)} + X D^{-1}(D\hat{p}_{sx}) + Y D^{-1}(D\hat{p}_{sl-sx}).$$

Setting $a = d^2/s^2 - p$, we can use 3.21 to write

$$y_x = \langle \psi(x) e^{at} \rangle + H_x \tag{7}$$

and

$$H_x = h_x * \langle e^{at} \rangle + X * (D\hat{p}_{sx}) + Y * (D\hat{p}_{sl-sx}) \qquad \text{(by [8.8])}. \tag{8}$$

From 10.4, 10.7, and 10.6 we see that $D\hat{p}_\lambda$ is the operator of a regulated $\mathcal{K}$-function when $0 \leq \lambda \leq \infty$: we may therefore conclude from 10.12.3 and (8) that

$$H_x(0+) = 0 \qquad (0 < x < l). \tag{9}$$

On the other hand, (7) gives

$$y_x(t) = \psi(x)\, e^{at} + H_x(t);$$

letting $t \to 0+$, it follows from (9) that

$$y_x(0+) = \psi(x) \qquad (0 < x < l). \tag{10}$$

Thus, we have verified the starting condition 10.9; it remains to prove 10.10. Let b be a real number such that $0 < \pm\, sx + b < \infty$: the equations

$$\frac{d^2}{dx^2} \hat{p}_{\pm sx+b} = (\pm s)^2 \frac{d^2}{d\lambda^2} \hat{p}_\lambda|_{\lambda = \pm sx+b} = s^2 [D\hat{p}_{\pm sx+b} + p\hat{p}_{\pm sx+b}]$$

are from [9.6], 10.7, and 10.4: consequently,

$$\left[(p + D) - s^{-2} \frac{d^2}{dx^2}\right] \hat{p}_{\pm sx+b} = 0 \qquad (0 < \pm\, sx + b < \infty).$$

Choosing the positive sign and $b = 0$, we get

$$\left[(p + D) - s^{-2} \frac{d^2}{dx^2}\right] \hat{p}_{sx} = 0 \qquad (0 < x < l);$$

choosing the negative sign and $b = sl$:

$$\left[(p + D) - s^{-2} \frac{d^2}{dx^2}\right] \hat{p}_{-sx+sl} = 0 \qquad (0 < x < l).$$

In view of 9.8, the preceding two equalities can be combined as follows (when X and Y are any two operators):

$$\text{(11)} \qquad \left[(p + D) - s^{-2} \frac{d^2}{dx^2}\right] (X\hat{p}_{sx} + Y\hat{p}_{sl-sx}) = 0 \qquad (0 < x < l),$$

so that 10.12.2 gives

$$\left[(p + D) - s^{-2} \frac{d^2}{dx^2}\right] (S_x + X\hat{p}_{sx} + Y\hat{p}_{sl-sx}) = h_x + \psi(x)\, D + 0;$$

in other words, 10.11 implies

$$\left[(p + D) - s^{-2} \frac{d^2}{dx^2}\right] y_x = h_x + \psi(x)\, D \qquad (0 < x < l);$$

re-arranging:

$$h_x = Dy_x - \psi(x)\, D + py_x - s^{-2} \frac{d^2}{dx^2}\, y_x$$

$$= [Dy_x - y_x(0+)D] + py_x - s^{-2} \frac{d^2}{dx^2}\, y_x \qquad \text{(by (10))}$$

$$= \frac{d}{dt}\, y_x + py_x - s^{-2} \frac{d^2}{dx^2}\, y_x \qquad \text{(by [9.0.0])}$$

$$= \left[\left(p + \frac{d}{dt}\right) - s^{-2} \frac{d^2}{dx^2}\right] y_x \qquad (0 < x < l).$$

This concludes the proof of 10.10; in view of (10), this concludes the proof of 10.8.

10.13.0 **Concerning the auxiliary solution.** In the problems to follow, the numbers $\psi(x)$ and the functions $h_x(\,)$ will always be such that either

$$h_x + \psi(x) D = 0 \qquad (0 < x < l)$$

(in which case $d^2 = 0$), or else the operator $h_\lambda + \psi(\lambda) D$ is invertible for some λ with $0 < \lambda < l$, in which case

(10.13.1)
$$d^2 = \frac{1}{h_\lambda + \psi(\lambda) D} \left(\frac{d^2}{d\lambda^2}\right) [h_\lambda + \psi(\lambda)] .$$

For example, if both $h_x()$ and $\psi(x)$ are independent of x, then 10.13.1 gives $d^2 = 0$.

The Infinite Interval

10.13.2 In case $l = \infty$, we have $sl - sx = \infty - sx = \infty$ when $x < \infty$ (since $0 < s < \infty$): in view of the fact that $\hat{p}_\infty = 0$ (see 10.2.0), we therefore have $\hat{p}_{sl-sx} = 0$, whence Equation 10.11 becomes

(10.13.3)
$$y_x = S_x + X\hat{p}_{sx} \qquad (0 \leq x < \infty).$$

Let S_x $(0 \leq x < \infty)$ be the auxiliary solution (9.28) of the problem

(10.13.4)
$$y_x(0+) = \psi(x)$$

with

(10.13.5)
$$\left[p + \frac{d}{dt} - s^{-2} \frac{d^2}{dx^2}\right] y_x = h_x \qquad (0 < x < l);$$

if $X()$ is any $\mathcal{K}$-function, Theorem 10.8 guarantees that Equation 10.13.3 determines a solution of this problem.

10.13.6 **For example,** given a $\mathcal{K}$-function $f()$ and a number α, let us solve the starting-value problem

(1) $$y_x(0+) = \alpha \quad \text{with} \quad \left[p + \frac{d}{dt} - s^{-2} \frac{d^2}{dx^2}\right] y_x = 0 \qquad (0 < x < \infty)$$

subject to the boundary condition

(2)
$$y_0(t) = f(t) \qquad (\text{all } t > 0).$$

We begin by finding the auxiliary solution (as in 9.28). Step (i): observe that (1) implies

(3)
$$\left[(p + D) - s^{-2} \frac{d^2}{dx^2}\right] y_x = \alpha D .$$

Step (ii): we determine the number d^2 such that $\frac{\mathrm{d}^2}{\mathrm{d}x^2} B_x = d^2 B_x$, where B_x is the right-hand side αD of the equation (3); we have

$$d^2 = (\alpha D)^{-1} \frac{\mathrm{d}^2}{\mathrm{d}x^2} (\alpha D) = 0 \qquad \text{(see 10.13.1)}.$$

Step (iii): we replace $\mathrm{d}^2/\mathrm{d}x^2$ by d^2 in the equation (3) and solve the resulting equation $[(p + D) - s^{-2} 0]\, y_x = \alpha D$; the auxiliary solution is therefore the operator

$$S_x = \frac{\alpha D}{D + p}.$$

Let $X(\,)$ be any $\mathcal{K}$-function; from 10.13.2 it follows that the equation

$$y_x = \frac{\alpha D}{D + p} + X \hat{p}_{sx} \qquad (0 \leq x < \infty) \tag{4}$$

determines a solution of our problem (1). Let us determine X so that (4) satisfies the boundary condition (2). Setting $x = 0$ in (4), Condition (2) gives

$$f = \frac{\alpha D}{D + p} + X \hat{p}_0 = \frac{D}{D + p} (\alpha + X): \tag{5}$$

the second equation is from 10.2. Solving (5) for X and substituting into (4):

$$y_x = \alpha \frac{D}{D + p} - \alpha \hat{p}_{sx} + f D^{-1} (D + p)\, \hat{p}_{sx} \qquad (0 \leq x < \infty);$$

by means of 3.21 and [10.1] this can be written

$$y_x = \alpha e^{-pt} \left(1 - \operatorname{cerf} \frac{s x}{2\sqrt{t}}\right) + f * (D + p)\, \hat{p}_{sx} \qquad \text{(see [8.8])}. \tag{6}$$

The second equation in (1) gives the concentration y_x of moisture at a point x of a thin plate with end-points at $x = 0$ and at $x = \infty$.

10.14 **Particular case.** From 10.7 and 10.4 we see that

$$D \hat{p}_\lambda + p \hat{p}_\lambda = \frac{\mathrm{d}}{\mathrm{d}\lambda} \hat{p}_\lambda = \frac{\lambda t^{-3/2}}{2\sqrt{\pi}} \exp\left(-pt - \frac{\lambda^2}{4t}\right) \qquad (0 < \lambda < \infty): \tag{7}$$

the last equation is from 10.6; consequently,

$$(D + p)\, \hat{p}_{sx} = \frac{s x t^{-3/2}}{2\sqrt{\pi}} \exp\left(-pt - \frac{s^2 x^2}{4t}\right) \qquad (0 < x < \infty). \tag{10.15}$$

Accordingly, the last term in (6) can be written

$$(8) \qquad \boxed{f * (D + p)\, \hat{p}_{sx} = \int_0^t f(t - u)\, \frac{s x u^{-3/2}}{2\sqrt{\pi}} \exp\left(-pu - \frac{s^2 x^2}{4t}\right) du:}$$

we have used 8.10.2 and omitted the angular brackets on the right-hand sides. We note from (6) that (8) is the answer to our problem (1)—(2) in the case $\alpha = 0$.

10.16 **A useful formula.** *If λ and p are finite numbers > 0, then*

$$(10.17) \qquad \boxed{\frac{D + p}{D}\, \hat{p}_\lambda = \frac{1}{2} \sum_{z = \pm \sqrt{p}} e^{\lambda z} \operatorname{cerf}\left(\frac{\lambda}{2\sqrt{t}} + z\sqrt{t}\right)}\,;$$

we have omitted the angular brackets on the right-hand side. To prove 10.17, set

$$(10.18) \qquad p_\lambda(t) \stackrel{\text{def}}{=} \frac{1}{2} \sum_{z = \pm\sqrt{p}} e^{\lambda z} \operatorname{cerf}\left(\frac{\lambda}{2\sqrt{t}} + z\sqrt{t}\right);$$

the equations

$$(10.19) \qquad \frac{\partial}{\partial t}\, p_\lambda(t) = \frac{\lambda t^{-3/2}}{2\sqrt{\pi}} \exp\left(-pt - \frac{\lambda^2}{4t}\right)$$

$$(10.19.1) \qquad = \frac{\partial}{\partial t}\, \hat{p}_\lambda(t) + p\hat{p}_\lambda(t)$$

are easily verified (see 10.40.8 and Equation (5) below 10.6).

From 10.19 and 10.15 (with $sx = \lambda$) we see that

$$(10.19.2) \qquad (D + p)\, \hat{p}_\lambda = \left\langle \frac{\partial}{\partial t}\, \hat{p}_\lambda(t) \right\rangle;$$

further, 9.1 gives

$$(10.19.3) \qquad (D + p)\, \hat{p}_\lambda = \frac{\mathrm{d}}{\mathrm{d}t}\, p_\lambda.$$

On the other hand, it is easily verified that

$$(10.20) \qquad p_\lambda(0+) = 0. \qquad (0 < \lambda < \infty);$$

we may therefore combine 10.19.3 with [9.0.0] to conclude that

(10.20.1) $$(D + p)\, \hat{p}_\lambda = D p_\lambda.$$

Conclusion 10.17 is immediate from 10.20.1 and [10.18].

10.21 **Temperature in a rod.** The equation

(9) $$\left[p + \frac{\mathrm{d}}{\mathrm{d}t} - s^{-2} \frac{\mathrm{d}^2}{\mathrm{d}x^2}\right] y_x = 0 \qquad (0 < x < \infty)$$

governs the temperature y_x at a point x of a rod with end-points at $x = 0$ and at $x = \infty$; the number p is different from zero in the case where each point of the rod loses heat at a rate proportional to its temperature. Suppose that the starting temperature of the rod is zero; from (4) we see that the equation

(10) $$y_x = X \hat{p}_{sx} \qquad (0 < x < l)$$

determines (for any $\mathcal{K}$-function $X(\,)$) a solution of (9) having starting value zero. Consider the case where the temperature of the left end-point varies according to the relation

(11) $$y_0(t) = \begin{cases} a & (t \leq \lambda) \\ a + b & (t > \lambda \geq 0). \end{cases}$$

Let us determine X in (10) so that (10) satisfies the boundary condition (11). Setting $x = 0$ in (10) and combining with (11):

$$a + b\mathsf{T}_\lambda = X \hat{p}_0 = X \frac{D}{D + p} \qquad \text{(by 10.2)};$$

we now solve this for X and substitute into (10):

(12) $$y_x = a\left(\frac{D + p}{D}\, \hat{p}_{sx}\right) + b\mathsf{T}_\lambda \left(\frac{D + p}{D}\, \hat{p}_{sx}\right);$$

we may now use 10.17 to express (12) in terms of the function cerf; in particular, if $b = 0$ then

$$y_x = \frac{a}{2}\left[\mathrm{e}^{sx\sqrt{p}} \operatorname{cerf}\left(\frac{sx}{2\sqrt{t}} + \sqrt{pt}\right) + \mathrm{e}^{-sx\sqrt{p}} \operatorname{cerf}\left(\frac{sx}{2\sqrt{t}} - \sqrt{pt}\right)\right].$$

Insulated Rod

10.22 If the sides of the rod (described in 10.21) are insulated, the equations

$$(10.23)\quad y_x(0+) = \alpha \quad \text{with} \quad \frac{\mathrm{d}}{\mathrm{d}t}\, y_\iota = s^{-2}\, \frac{\mathrm{d}^2}{\mathrm{d}x^2}\, y_x \qquad (0 < x < \infty)$$

govern the heat y_x when the starting temperature has the value α. Comparing 10.23 with our problem (1) in 10.13.6, we see from (5) that the equation

$$(10.24)\quad y_x = \alpha + X\hat{p}_{sx} \qquad (p = 0, \quad 0 \leq x < \infty)$$

defines a solution of 10.23 for any $\mathcal{K}$-function $X(\,)$. The following abbreviations will be used:

$$(10.25)\quad \sqrt{D} \overset{\text{def}}{=} \frac{1}{\sqrt{\pi}}\, \langle t^{-1/2} \rangle$$

and

$$(10.26)\quad \mathrm{q}^\lambda(\,) \overset{\text{def}}{=} \left\{ \operatorname{cerf} \frac{\lambda}{2\sqrt{t}} \right\}(\,). \qquad (0 \leq \lambda < \infty).$$

From 3.16 we see that

$$(10.27)\quad \left(\sqrt{D}\right)^2 = D.$$

Henceforth, $p = 0$, the equations:

$$(10.28)\quad \boxed{\mathrm{q}^\lambda = \hat{p}_\lambda \quad \text{and} \quad \mathrm{q}^0 = 1}$$

are immediate from [10.26], [10.1], and 10.2.

10.29 **Theorem:**

$$(10.30)\quad \boxed{\langle t^{-1/2} \rangle\, \mathrm{q}^\lambda = \frac{1}{\sqrt{t}} \exp\left(-\frac{\lambda^2}{4t}\right)} \qquad (0 < \lambda < \infty),$$

$$(10.31)\quad \boxed{\frac{\mathrm{d}}{\mathrm{d}x}\, \mathrm{q}^{sx} = -s\sqrt{D}\, \mathrm{q}^{sx}} \qquad (0 < x < \infty),$$

and

$$D^{-1}\sqrt{D}\,\mathrm{q}^{\lambda} = 2\sqrt{\frac{t}{\pi}}\exp\left(\frac{-\lambda^2}{4t}\right) - \lambda\,\mathrm{cerf}\,\frac{\lambda}{2\sqrt{t}}. \tag{10.32}$$

Proof. To begin with 10.30, note that 10.15 and 10.28 give

$$D\mathrm{q}^{\lambda} = \left\langle \frac{\lambda t^{-3/2}}{2\sqrt{\pi}}\exp\left(\frac{-\lambda^2}{4t}\right)\right\rangle \qquad (\text{since } p = 0);$$

we may therefore use 3.15 to obtain

$$\langle t^{-1/2}\rangle\,\mathrm{q}^{\lambda} = \langle t^{-1/2}\rangle * D\mathrm{q}^{\lambda} = \frac{\lambda}{2\sqrt{\pi}}\int_0^t \frac{u^{-3/2}\,\mathrm{e}^{-\lambda^2/4u}}{\sqrt{t-u}}\,\mathrm{d}u;$$

the change of variable $v = (\lambda^2/4)[u^{-1} - t^{-1}]$ yields

$$\frac{\lambda^2}{4u} = v + \frac{\lambda^2}{4t}, \qquad \sqrt{t-u} = 2\lambda^{-1}\sqrt{vtu},$$

and $\mathrm{d}u = (-4u^2/\lambda^2)\,\mathrm{d}v$; therefore,

$$\langle t^{-1/2}\rangle\,\mathrm{q}^{\lambda} = \frac{\lambda}{2\sqrt{\pi}}\int_0^\infty \frac{u^{-3/2}\,\mathrm{e}^{-v}\,\mathrm{e}^{-\lambda^2/4t}}{2\lambda^{-1}\sqrt{vtu}}\left(\frac{-4u^2}{\lambda^2}\right)\mathrm{d}v$$

$$= \frac{\mathrm{e}^{-\lambda^2/4t}}{\sqrt{\pi t}}\int_0^\infty \mathrm{e}^{-v}\,\frac{\mathrm{d}v}{\sqrt{v}} = \frac{2\mathrm{e}^{-\lambda^2/4t}}{\sqrt{\pi t}}\int_0^\infty \mathrm{e}^{-w^2}\,\mathrm{d}w;$$

Conclusion 10.30 is now at hand. The equations

$$\frac{\mathrm{d}}{\mathrm{d}x}\,\mathrm{q}^{sx} = \frac{\mathrm{d}}{\mathrm{d}x}\,\hat{p}^{sx} = s\hat{p}^{[1]}_{sx} = \frac{-s}{\sqrt{\pi t}}\exp\left(-\frac{s^2x^2}{4t}\right) \tag{10.32.1}$$

are from 10.28, [9.6], and 10.5 (with x replaced by sx). From 10.32.1 and 10.30 it follows that

$$\frac{\mathrm{d}}{\mathrm{d}x}\,\mathrm{q}^{sx} = \frac{-s}{\sqrt{\pi}}\,\langle t^{-1/2}\rangle\,\mathrm{q}^{sx}: \tag{1}$$

Conclusion 10.31 is now immediate from 10.25. It only remains to verify 10.32. From [10.25], 3.14 and 10.30 we see that

$$D^{-1}\sqrt{D}\,q^{\lambda}=D^{-1}\frac{\langle t^{-1/2}\rangle}{\sqrt{\pi}}\,q^{\lambda}=\frac{1}{\sqrt{\pi}}\int_0^t e^{-\lambda^2/4u}\frac{du}{\sqrt{u}};$$

next, the change of variable $w=\lambda/2\sqrt{u}$ gives $2\sqrt{u}=\lambda/w$, $du/\sqrt{u}=d(\lambda w^{-1})$, and

$$D^{-1}\sqrt{D}q^{\lambda}=-\int_a^{\infty} e^{-w^2}\,d\left(\frac{\lambda w^{-1}}{\sqrt{\pi}}\right)\qquad\left(\text{with } a=\frac{\lambda}{2\sqrt{t}}\right). \tag{2}$$

Integrating by parts:

$$\int_a^{\infty} e^{-w^2}\,d\left(\frac{\lambda w^{-1}}{\sqrt{\pi}}\right)=\left[\frac{\lambda w^{-1}}{\sqrt{\pi}}\,e^{-w^2}\right]_a^{\infty}-\int_a^{\infty}\frac{\lambda w^{-1}}{\sqrt{\pi}}\,d(e^{-w^2}). \tag{3}$$

From (2)—(3) we now see that

$$D^{-1}\sqrt{D}q^{\lambda}=\frac{\lambda a^{-1}}{\sqrt{\pi}}\,e^{-a^2}-\frac{2\lambda}{\sqrt{\pi}}\int_a^{\infty} e^{-w^2}\,dw;$$

substituting $a=\lambda/2\sqrt{t}$, we obtain 10.32 as an immediate consequence of [10.0].

10.33 **Return to the insulated rod.** If the left end of the rod (in 10.22) is subject to a constant heat influx, there is a number m such that

$$-m=\frac{d}{dx}\,y_x\big|_{x=0}. \tag{4}$$

Let us determine X in 10.24 to satisfy (4). To that effect, apply d/dx to both sides of 10.24:

$$\frac{d}{dx}\,y_x=X\frac{d}{dx}\,\hat{p}_{sx}=X\frac{d}{dx}\,q^{sx}=-Xs\sqrt{D}\,q^{sx}: \tag{5}$$

the two last equations are from 10.28 and 10.31. Equations (4)—(5) and 10.28 give

$$-m=-s\sqrt{D}\,X; \tag{6}$$

using 10.27, we can solve (6) for X and substitute into 10.24:

$$y_x=\alpha+\frac{m}{s}\,D^{-1}\sqrt{D}\,q^{sx},$$

and 10.32 now yields

$$y_x = \alpha + \frac{2m\sqrt{t}}{s\sqrt{\pi}} \exp\left(\frac{-s^2x^2}{4t}\right) - mx \operatorname{cerf} \frac{sx}{2\sqrt{t}} .$$

10.34 **If the left end is imperfectly cooled,** there exists a number a such that

$$say_0 = \frac{d}{dx} y_x|_{x=0}. \tag{7}$$

From 10.24 we know that the equation

$$y_x = \alpha + X q^{sx} \qquad (0 \le x < \infty) \tag{8}$$

defines a solution of the corresponding initial-value problem; let us determine X to satisfy (7). Setting $x = 0$ in (8), we get

$$sa[\alpha + Xq^0] = \frac{d}{dx} y_x|_{x=0} = -sX\sqrt{D} \qquad \text{(by (7) and 10.31)}.$$

Re-arranging:

$$X\left[a + \sqrt{D}\right] = -a\alpha;$$

multiplying both sides by $(a - \sqrt{D})$, we obtain

$$X = -a\alpha \frac{a - \sqrt{D}}{a^2 - D};$$

substituting into (8):

$$y_x = \alpha - \alpha \frac{\sqrt{D} - a}{D - a^2} a q^{sx} \qquad (0 \le x < \infty).$$

The explicit answer

$$y_x = \alpha - \alpha\left[q^{sx}(t) - \exp(a^2t + sxa) \operatorname{cerf}\left(\frac{sx}{2\sqrt{t}} + a\sqrt{t}\right)\right]$$

is immediate from 10.38.

10.35 Three formulas. If $p > 0$ and $-\infty < a < \infty$, then

$$\text{(10.36)} \qquad \boxed{\frac{D}{D-p}\,\mathrm{q}^{\lambda} = \frac{1}{2}\,\mathrm{e}^{pt} \sum_{z=\pm\sqrt{p}} \mathrm{e}^{\lambda z}\,\mathrm{cerf}\left(\frac{\lambda}{2\sqrt{t}} + z\sqrt{t}\right)},$$

$$\text{(10.37)} \qquad \boxed{\frac{a^2}{D-a^2}\,\mathrm{q}^{sx} = -\mathrm{q}^{sx}(t) + \frac{1}{2}\exp(a^2 t) \sum_{z=\pm a} \mathrm{e}^{sxz}\,\mathrm{cerf}\left(\frac{sx}{2\sqrt{t}} + z\sqrt{t}\right)},$$

$$\text{(10.38)} \qquad \boxed{\frac{\sqrt{D}-a}{D-a^2}\,a\mathrm{q}^{sx} = \mathrm{q}^{sx}(t) - \exp(a^2\,t + sxa)\,\mathrm{cerf}\left(\frac{sx}{2\sqrt{t}} + a\sqrt{t}\right)}.$$

Initial remark. Let y_λ be the operator $\langle y_\lambda(t)\rangle$ of a $\mathcal{K}$-function $y_\lambda(\,)$ such that the function $y'_\lambda(\,)$ defined by

$$\text{(0)} \qquad y'_\lambda(t) = \frac{\partial}{\partial t}\,y_\lambda(t) \qquad \text{(all } t > 0)$$

is continuous in the interval $(0, \infty)$; if $y_\lambda(0+) = 0$ it follows from 9.1 and [9.0.0] that

$$\left\langle \frac{\partial}{\partial t}\,y_\lambda(t)\right\rangle = \frac{\mathrm{d}}{\mathrm{d}t}\,y_\lambda = Dy_\lambda;$$

thus, Dy_λ is the canonical operator of the function $y'_\lambda()$ that is defined by (0):

$$\text{(1)} \qquad Dy_\lambda = \left\langle \frac{\partial}{\partial t}\,y_\lambda(t)\right\rangle.$$

Proof of 10.36. From [10.26] and [10.1] it obviously follows that

$$\text{(2)} \qquad \mathrm{q}^{\lambda}(t) = \mathrm{e}^{pt}\,\hat{p}_\lambda(t) \qquad \text{(all } t > 0).$$

On the other hand, since $\hat{p}_\lambda(0+) = 0$ (by 10.3), we have $\mathrm{q}^{\lambda}(0+) = 0$; in consequence, we may apply our initial remark:

$$\text{(3)} \qquad D\mathrm{q}^{\lambda} = \left\langle \frac{\partial}{\partial t}\,\mathrm{q}^{\lambda}(t)\right\rangle.$$

Consequently,

$$\text{(4)} \qquad D\mathrm{q}^\lambda = \left\langle \frac{\partial}{\partial t}\, \mathrm{e}^{pt}\, \hat{p}_\lambda(t) \right\rangle \qquad \text{(by (3) and (2))}$$

$$= \left\langle \mathrm{e}^{pt} \left[\frac{\partial}{\partial t}\, \hat{p}_\lambda(t) + p\hat{p}_\lambda(t) \right] \right\rangle$$

$$\text{(5)} \qquad = \left\langle \mathrm{e}^{pt}\, \frac{\partial}{\partial t}\, p_\lambda(t) \right\rangle = \langle \mathrm{e}^{pt}\, p'_\lambda(t) \rangle:$$

the two last equations are from 10.19.1—10.19 and (0). We are now in a position to prove 10.36; observe that

$$\text{(6)} \qquad \frac{D}{D-p}\, \mathrm{q}^\lambda = \langle \mathrm{e}^{pt} \rangle\, D^{-1}\, (D\mathrm{q}^\lambda) \qquad \text{(from 3.21)}$$

$$= \langle \mathrm{e}^{pt} \rangle\, D^{-1}\, \langle \mathrm{e}^{pt} p'_\lambda(t) \rangle \qquad \text{(by (4)—(5))}$$

$$= \int_0^t \mathrm{e}^{pt-pu}\, \mathrm{e}^{pu}\, p'_\lambda(u)\, \mathrm{d}u \qquad \text{(by 3.15)}$$

$$\text{(7)} \qquad = \mathrm{e}^{pt}\, [p_\lambda(t) - p_\lambda(0+)] = \mathrm{e}^{pt} p_\lambda(t):$$

the last two equations are from the Fundamental Theorem of Calculus and from 10.20. Conclusion 10.36 is immediate from (6)—(7) and 10.18—10.17.

To obtain 10.37, note that

$$\frac{a^2}{D-a^2}\, \mathrm{q}^\lambda = \left(-1 + \frac{D}{D-a^2} \right) \mathrm{q}^\lambda = -\mathrm{q}^\lambda + \frac{D}{D-a^2}\, \mathrm{q}^\lambda,$$

and set $p = |a|^2$ in 10.36. It remains to prove 10.38. Observe that 10.31 implies

$$\frac{\sqrt{D}}{D-a^2}\, a\mathrm{q}^{sx} = \frac{-1}{sa}\, \frac{\mathrm{d}}{\mathrm{d}x}\, \frac{a^2 \mathrm{q}^{sx}}{D-a^2},$$

so that 10.37 gives

$$\text{(8)} \qquad \frac{\sqrt{D}}{D-a^2}\, a\mathrm{q}^{sx} = -\frac{1}{2}\, \exp(a^2 t) \left(\sum_{z=\pm a} \frac{sz}{sa}\, \mathrm{e}^{sxz} F_z \right) - W_x,$$

where

(9)
$$F_z \overset{\text{def}}{=} \operatorname{cerf}\left(\frac{sx}{2\sqrt{t}} + z\sqrt{t}\right)$$

and

(10)
$$W_x \overset{\text{def}}{=} \frac{1}{sa}\left[-\frac{d}{dx}\,q^{sx} + \frac{1}{2}\exp(a^2 t)\sum_{z=\pm a} e^{sxz}\frac{\partial}{\partial x}F_z\right].$$

Subtracting 10.37 from (8):

$$\frac{\sqrt{D}-a}{D-a^2}\,aq^{sx} = q^{sx} - \frac{1}{2}\exp(a^2 t)\left(\sum_{z=\pm a}\left(\frac{z}{a}+1\right)e^{sxz}F_z\right) - W_x.$$

Since $z = \pm a$, the middle summation becomes

$$\sum_{z=\pm a}\left(\frac{z}{a}+1\right)e^{sxz}F_z = 0 + 2e^{sxa}F_a,$$

whence

$$\frac{\sqrt{D}-a}{D-a^2}\,aq^{sx} = q^{sx} - \exp(a^2 t)\operatorname{cerf}\left(\frac{sx}{2\sqrt{t}} + a\sqrt{t}\right) - W_x.$$

Conclusion 10.38 will be obtained by proving that $W_x = 0$. To that effect, an easy calculation based on [10.0] gives

$$\frac{\partial}{\partial x}\operatorname{cerf}\left(\frac{sx}{2\sqrt{t}} + z\sqrt{t}\right) = \frac{-s}{\sqrt{\pi t}}\exp\left(-\left[\frac{sx}{2\sqrt{t}} + z\sqrt{t}\right]^2\right)$$

$$= e^{-sxz}\left[\frac{-s}{\sqrt{\pi t}}\exp\left(\frac{-s^2x^2}{4t}\right)\right]\exp(-z^2 t);$$

consequently, we may use (9) to write

$$\sum_{z=\pm a} e^{sxz}\frac{\partial}{\partial x}F_z = \left[\frac{-s}{\sqrt{\pi t}}\exp\left(\frac{-s^2x^2}{4t}\right)\right]\sum_{z=\pm a}\exp(-z^2 t),$$

whence

(11)
$$\frac{1}{2}\exp(a^2 t)\sum_{z=\pm a} e^{sxz}\frac{\partial}{\partial x}F_z = \frac{-s}{\sqrt{\pi t}}\exp\left(\frac{-s^2x^2}{4t}\right);$$

the conclusion $W_x = 0$ now comes directly from (10)—(11) and 10.32.1.

10.39 Exponential decay. Given a number a, to solve the problem

(1) $$y_x(0+) = 0 \text{ with } \frac{\mathrm{d}}{\mathrm{d}t} y_x - s^{-2} \frac{\mathrm{d}^2}{\mathrm{d}x^2} y_x = \mathrm{e}^{-sax} \qquad (0 < x < \infty)$$

subject to the boundary condition

(2) $$y_0 = 0.$$

We begin by finding the auxiliary solution (as in 9.28). Step (i): observe that (1) implies

(3) $$\left[D - s^{-2} \frac{\mathrm{d}^2}{\mathrm{d}x^2}\right] y_x = \mathrm{e}^{-sax} \qquad (0 < x < \infty).$$

Step (ii): we compute the number d^2 by means of the formula

$$d^2 = \mathrm{e}^{sax} \frac{\mathrm{d}^2}{\mathrm{d}x^2} [\mathrm{e}^{-sax}] = s^2 a^2 \qquad \text{(by 10.13.1 and 9.9)}.$$

Step (iii): we replace $\mathrm{d}^2/\mathrm{d}x^2$ by d^2 in the equation (3) and solve the resulting equation

$$[D - s^{-2}(s^2 a^2)]\, y_x = \mathrm{e}^{-sax};$$

the auxiliary solution is therefore the operator

$$S_x = \frac{\mathrm{e}^{-sax}}{D - a^2} \qquad (0 < x < \infty).$$

Let $X(\,)$ be any $\mathcal{K}$-function; from 10.13.2 it follows that the equation

(4) $$y_x = \frac{\mathrm{e}^{-sax}}{D - a^2} + X \mathrm{q}^{sx} \qquad (0 \leq x < \infty)$$

determines a solution of our problem (1) (since $p = 0$, we have used 10.28). Let us determine X so that (4) satisfies the boundary condition (2): comparing (4) and (2), we can use the fact that $\mathrm{q}^0 = 1$ (see 10.28) to obtain

$$X = -\frac{1}{D - a^2}$$

substituting into (4), Formula 3.24 yields

$$y_x = a^{-2}\left[\mathrm{e}^{-sax}(\mathrm{e}^{a^2 t} - 1) - \frac{a^2}{D - a^2}\mathrm{q}^{sx}\right],$$

so that 10.37 gives

$$y_x = \frac{e^{-sax}}{a^2}\,(e^{a^2 t} - 1) + \frac{q^{sx}(t)}{a^2} - \frac{e^{a^2 t}}{2a^2} \sum_{z=\pm a} e^{sxz} \operatorname{cerf}\left(\frac{sx}{2\sqrt{t}} + z\sqrt{t}\right).$$

Exercises

10.40.0 Consider an electrical transmission line with end-points at $x = 0$ and at $x = \infty$ having negligible inductance; if the line constants are R, C, and G, the equation

$$RC\,\frac{d}{dt}\,y_x + RGy_x = \frac{d^2}{dx^2}\,y_x \qquad (0 < x < \infty)$$

governs the potential y_x at a point x. Find y in case the starting potential is zero and the potential at $x = 0$ is

$$y_0(t) = a \qquad (\text{for any } t > 0),$$

where a is a given number. *Hints*: Imitate 10.21; the zero starting potential condition means that

$$y_x(0+) = 0 \qquad (\text{for any } x > 0).$$

Answer: $y_x = a h_x$, where

$$h_x(t) = \frac{1}{2}\,e^{x\sqrt{RG}} \operatorname{cerf}\left(\frac{x}{2}\sqrt{\frac{RC}{t}} + \sqrt{\frac{Gt}{C}}\right)$$
$$+ \frac{1}{2}\,e^{-x\sqrt{RG}} \operatorname{cerf}\left(\frac{x}{2}\sqrt{\frac{RC}{t}} - \sqrt{\frac{Gt}{C}}\right).$$

10.40.1 Same problem, except that

$$y_x(0+) = \alpha \;\; (\text{if } 0 < x < \infty) \;\text{ and }\; y_0 = 0.$$

Answer: $y_x = \alpha G_x$, where

$$G_x(t) = e^{-Gt/C}\left(1 - \operatorname{cerf}\frac{x}{2}\sqrt{\frac{RC}{t}}\right).$$

10.40.2 Same problem, except that

$$y_x(0+) = \alpha \quad (\text{if } 0 < x < \infty) \quad \text{and} \quad y_0(t) = \begin{cases} a & (0 < t < \lambda) \\ a + b & (\lambda < t < \infty). \end{cases}$$

Answer: $y_x(t) = \alpha G_x(t) + a h_x(t) + b \mathsf{T}_0(t - \lambda)\, h_x(t - \lambda)$, where G_x and h_x are the function-operators introduced in 10.40.1 and 10.40.0.

10.40.3 Given two numbers α and a, solve the problem

$$RC \frac{\mathrm{d}}{\mathrm{d}t} y_x = \frac{\mathrm{d}^2}{\mathrm{d}x^2} y_x \quad \text{with } y_x(0+) = \alpha \qquad (0 < x < \infty),$$

and subject to the boundary condition $y_0 = a$.

Answer: $y_x = \alpha + (a - \alpha) \operatorname{cerf} \frac{x}{2} \sqrt{\frac{RC}{t}}$.

10.40.4 Solve the problem

$$y_x(0+) = \alpha, \quad \frac{\mathrm{d}}{\mathrm{d}t} y_x = s^{-2} \frac{\mathrm{d}^2}{\mathrm{d}x^2} y_x \qquad (0 < x < \infty)$$

subject to the condition

$$y_0(t) = \begin{cases} a & (0 < t \le \lambda) \\ 0 & (t > \lambda). \end{cases}$$

Answer:

$$y_x(t) = \alpha \left(1 - \operatorname{cerf} \frac{s x}{2\sqrt{t}}\right) + a \left(\operatorname{cerf} \frac{s x}{2\sqrt{t}} - \mathsf{T}_\lambda(t) \operatorname{cerf} \frac{s x}{2\sqrt{t - \lambda}}\right).$$

10.40.5 Solve the same problem as 10.40.4, except that the starting condition now is $y_x(0+) = \alpha + Bx$.

Answer:

$$y_x(t) = Bx + \alpha \left(1 - \operatorname{cerf} \frac{s x}{2\sqrt{t}}\right) + a \left(\operatorname{cerf} \frac{s x}{2\sqrt{t}} - \mathsf{T}_\lambda(t) \operatorname{cerf} \frac{s x}{2\sqrt{t - \lambda}}\right).$$

10.40.6 Given a number a, solve the problem

$$s^2 \left[\frac{\mathrm{d}}{\mathrm{d}t} + p\right] y_x = \frac{\mathrm{d}^2}{\mathrm{d}x^2} y_x, \quad \text{with } y_x(0+) = \alpha \qquad (0 < x < \infty),$$

and with the boundary condition

$$y_0(t) = a \qquad (\text{all } t > 0).$$

Answer:

$$\alpha e^{-pt}\left(1 - \operatorname{cerf}\frac{sx}{2\sqrt{t}}\right) + \frac{a}{2} e^{sx\sqrt{p}} \operatorname{cerf}\left(\frac{sx}{2\sqrt{t}} + \sqrt{pt}\right)$$

$$+ \frac{a}{2} e^{-sx\sqrt{p}} \operatorname{cerf}\left(\frac{sx}{2\sqrt{t}} - \sqrt{pt}\right).$$

10.40.7 Verify that

$$\frac{\partial}{\partial t} \operatorname{cerf}\left(\frac{\lambda}{2\sqrt{t}} + z\sqrt{t}\right) = \left(\frac{\lambda t^{-3/2}}{2\sqrt{\pi}} - \frac{z t^{-1/2}}{\sqrt{\pi}}\right) \exp\left(-\frac{\lambda^2}{4t} - \lambda z - z^2 t\right).$$

10.40.8 Prove 10.19. *Hint*: use 10.40.7 and

$$\sum_{z=\pm\sqrt{p}} (A + zB) = 2A.$$

The Finite Interval

Given a $\mathcal{K}$-function $h(\,)$, a pair of positive (and finite) numbers (s, l), and a pair of numbers (α, b). This section is devoted to solving the problem

$$(10.41) \quad y_x(0+) = \alpha + bx \text{ with } \left[\frac{d}{dt} - s^{-2}\frac{d^2}{dx^2}\right] y_x = h \quad (0 < x < l)$$

subject to various boundary conditions. We begin by finding its auxiliary solution. Step (i): observe that 10.41 implies

$$(1) \qquad \left[D - s^{-2}\frac{d^2}{dx^2}\right] y_x = h + (\alpha + bx)\, D \qquad (\text{by } [9.0.0]).$$

Step (ii): we compute the number d^2 by means of the formula

$$d^2 = \frac{1}{h + (\alpha + bx)\, D}\left(\frac{d^2}{dx^2}\right)[h + (\alpha + bx)\, D] = 0:$$

see 10.13.1 and 9.28. Step (iii): we replace $\mathrm{d}^2/\mathrm{d}x^2$ by d^2 in the equation (1) and solve the resulting equation

$$[D - s^{-2} 0]\, y_x = h + (\alpha + b x)\, D;$$

the auxiliary solution is therefore the operator

$$S_x = D^{-1} h + (\alpha + b x).$$

Let $X(\,)$ and $Y(\,)$ be any two $\mathcal{K}$-functions: setting $p = 0$ in 10.8, we see that the equation

$$(10.42)\qquad y_x = D^{-1} h + (\alpha + b x) + X \mathrm{q}^{sx} + Y \mathrm{q}^{sl-sx} \qquad (0 \le x \le l)$$

implies 10.41: it determines a solution of our problem. Our first application will require the formula

$$(10.43)\qquad \boxed{\frac{\mathrm{d}}{\mathrm{d}x}\, \mathrm{q}^{sl-sx} = s \sqrt{D}\, \mathrm{q}^{sl-sx}} \qquad (0 < x < l).$$

Since $p = 0$, Formula 10.42 is expressed in terms of q^λ instead of $\hat{p}_\lambda$: recall that $\mathrm{q}^\lambda = \hat{p}_\lambda$ when $p = 0$: see 10.28. Setting $s = 1$ in 10.31, we find that

$$(2)\qquad \boxed{\frac{\mathrm{d}}{\mathrm{d}\lambda}\, \mathrm{q}^\lambda = -\sqrt{D}\, \mathrm{q}^\lambda} \qquad (0 < \lambda < l).$$

Let us verify Formula 10.43; from 10.28 we have

$$\frac{\mathrm{d}}{\mathrm{d}x}\, \mathrm{q}^{sl-sx} = \frac{\mathrm{d}}{\mathrm{d}x}\, \hat{p}^{-sx+sl}$$

$$= -s \frac{\mathrm{d}}{\mathrm{d}\lambda}\, \hat{p}_\lambda \big|_{\lambda = -sx+sl} \qquad \text{by [9.6] and 9.11}$$

$$= -s \frac{\mathrm{d}}{\mathrm{d}\lambda}\, \mathrm{q}^\lambda \big|_{\lambda = -sx+sl} = (-s)\left(-\sqrt{D}\right) \mathrm{q}^{-sx+sl}:$$

the two last equations are from 10.28 and (2): this completes the proof of 10.43.

10.44 **Insulated rod.** Given a $\mathcal{K}$-function $f(\,)$, consider the boundary-value problem

$$(1) \qquad y_x(0+) = \alpha \text{ with } \frac{\mathrm{d}}{\mathrm{d}t} y_x = s^{-2} \frac{\mathrm{d}^2}{\mathrm{d}x^2} y_x \qquad (0 < x < l),$$

$$(2) \qquad y_0 = f, \text{ and } 0 = \frac{\mathrm{d}}{\mathrm{d}x} y_x|_{x=l}.$$

The problem (1) governs the temperature y_x of an insulated rod of length l whose starting temperature is the number α; the first equation in (2) states that the temperature at the end-point $x = 0$ is given by the operator f; the last equation in (2) states that *no heat flows through the end* $x = l$. Since 10.42 (with $h = 0$ and $b = 0$) determines a solution of (1), it only remains to determine the parameters X and Y so that the equation

$$(3) \qquad y_x = \alpha + X\mathrm{q}^{sx} + Y\mathrm{q}^{sl-sx} \qquad (0 \leq x \leq l)$$

implies the boundary conditions (2). Applying $\mathrm{d}/\mathrm{d}x$ to both sides of (3), we may use 10.31 and 10.43 to obtain

$$\frac{\mathrm{d}}{\mathrm{d}x} y_x = -sX\sqrt{D}\,\mathrm{q}^{sx} + sY\sqrt{D}\,\mathrm{q}^{sl-sx};$$

combining with the last equation in (2):

$$(4) \qquad 0 = -X\mathrm{q}^{sl} + Y.$$

Setting $x = 0$ in (3), the first boundary condition gives

$$(5) \qquad f - \alpha = X + Y\mathrm{q}^{sl}.$$

Solving (4)—(5) for X and Y, we obtain

$$X = \frac{f - \alpha}{1 + \mathrm{q}^{sl}\mathrm{q}^{sl}} \qquad \text{and} \qquad Y = \frac{(f - \alpha)\,\mathrm{q}^{sl}}{1 + \mathrm{q}^{sl}\mathrm{q}^{sl}};$$

substituting into (3):

$$(6) \qquad y_x = \alpha + (f - \alpha)\left[\frac{\mathrm{q}^{sx}}{1 + \mathrm{q}^{sl}\mathrm{q}^{sl}} + \frac{\mathrm{q}^{sl}\mathrm{q}^{sl-sx}}{1 + \mathrm{q}^{sl}\mathrm{q}^{sl}}\right].$$

It is at this point that we need one last fact: if $a \geq 0$, $c \geq 0$, and $b \geq 0$, then

$$(10.45) \qquad \frac{\mathrm{q}^a \mathrm{q}^b}{1 \mp \mathrm{q}^c \mathrm{q}^c} = \left\langle \sum_{k=0}^{\infty} (\pm 1)^k \, \mathrm{q}^{a+b+2kc}(t) \right\rangle:$$

we shall not prove this. Note the analogy of 10.45 with the equation

$$\frac{q^a q^b}{1 \mp q^c q^c} = \sum_{k=0}^{\infty} (\pm 1)^k \, q^{a+b+2kc},$$

which holds whenever q is a number such that $0 < q < 1$. In view of [10.26], Formula 10.45 means that

$$(10.46) \qquad \boxed{\frac{\mathrm{q}^a \mathrm{q}^b}{1 \mp \mathrm{q}^c \mathrm{q}^c} = \sum_{k=0}^{\infty} (\pm 1)^k \operatorname{cerf} \frac{a+b+2kc}{2\sqrt{t}}};$$

we have omitted the angular brackets on the right-hand side.

10.47 **In case f is a number,** Equations (6) and 10.46 give

$$y_x = \alpha + (f - \alpha) \sum_{k=0}^{\infty} (-1)^k \left[\operatorname{cerf} \frac{x + 2kl}{(2\sqrt{t})/s} + \operatorname{cerf} \frac{2l - x + 2kl}{(2\sqrt{t})/s} \right].$$

This series converges rapidly for *small* values of t: that is, whenever the usual Fourier series expansion is impractical.

10.48 **Other conditions.** Let us replace the boundary condition (2) by

$$(7) \qquad y_0 = f \quad \text{and} \quad y_l = F:$$

both $f(\,)$ and $F(\,)$ are $\mathcal{K}$-functions. Setting $x = 0$ and $x = l$ in (3), we obtain

$$(8) \qquad f_0 \overset{\text{def}}{=} f - \alpha = X + Y\mathrm{q}^{sl}$$

and

$$F_0 \overset{\text{def}}{=} F - \alpha = X\mathfrak{q}^{sl} + Y. \tag{9}$$

Solving for X and Y:

$$X = \frac{f_0 - F_0\mathfrak{q}^{sl}}{1 - \mathfrak{q}^{sl}\mathfrak{q}^{sl}} \quad \text{and} \quad Y = \frac{F_0 - f_0\mathfrak{q}^{sl}}{1 - \mathfrak{q}^{sl}\mathfrak{q}^{sl}};$$

substituting into (3):

$$y_x = \alpha + f_0 \frac{\mathfrak{q}^{sx} - \mathfrak{q}^{sl}\mathfrak{q}^{sl-sx}}{1 - \mathfrak{q}^{sl}\mathfrak{q}^{sl}} + F_0 \frac{\mathfrak{q}^{sl-sx} - \mathfrak{q}^{sl}\mathfrak{q}^{sx}}{1 - \mathfrak{q}^{sl}\mathfrak{q}^{sl}}. \tag{10}$$

We can now use 10.46 to obtain the representation in terms of the function cerf. For example, if $f = \alpha$ and $F = 0$, then (10) becomes

$$y_x = \alpha - \alpha \left[\frac{\mathfrak{q}^{sl-sx}}{1 - \mathfrak{q}^{sl}\mathfrak{q}^{sl}} - \frac{\mathfrak{q}^{sl}\mathfrak{q}^{sx}}{1 - \mathfrak{q}^{sl}\mathfrak{q}^{sl}}\right];$$

a double application of 10.46 (with $c = sl$, $a = 0$, $b = sl - sx$ the first time, and then with $a = sl$, $b = sx$) gives

$$y_x = \alpha - \alpha \sum_{k=0}^{\infty} \left[\operatorname{cerf} \frac{l - x + 2kl}{(2\sqrt{t})/s} - \operatorname{cerf} \frac{l + x + 2kl}{(2\sqrt{t})/s}\right].$$

Comparison with the Laplace Transformation

In order to utilize Laplace transform techniques to solve the type of problems that have been discussed in the present chapter, it must be assumed at the beginning that the problem has a solution $y_x(t)$ satisfying three or four conditions of the following type:

$$\int_0^\infty e^{-st} \frac{\partial}{\partial x} y_x(t)\, dt = \frac{\partial}{\partial x} \int_0^\infty e^{-st} y_x(t)\, dt;$$

see, for example [S 2], p. 371.

Such assumptions are required in order to justify the calculations; they are much more restrictive than the assumptions required by our theorems 9.30 and 10.8. It sometimes happens (see p. 269 in [D 3]) that the solution obtained by Laplace transform techniques does *not* satisfy some of the required assumptions. Consequently, the result of the calculations suggested by the Laplace transform should be substituted back into the problem in order to verify that it is a solution. In short, Laplace transform techniques do *not* guarantee that the calculations will give a solution.

Exercises

In each of the following problems, y_x' is an abbreviation of

$$\frac{d}{dt} y_x \stackrel{\text{def}}{=} D y_x - y_x(0+)\, D \qquad \text{(see [9.0.0])}.$$

10.49.0 Given a number a; solve the problem

$$y_x' = \frac{d^2}{dx^2} y_x, \qquad y_x(0+) = 0 \; (0 < x < l = 1),\; y_0 = 0,\; y_1 = a.$$

Answer:

$$a \sum_{k=0}^{\infty} \left[\operatorname{cerf} \frac{2k+1-x}{2\sqrt{t}} - \operatorname{cerf} \frac{2k+1+x}{2\sqrt{t}} \right].$$

10.49.1 Solve the problem

$$y_x' = c^2 \frac{d^2}{dx^2} y_x, \qquad y_x(0+) = \alpha \;\; (0 < x < l < \infty),\; y_0 = 0, \text{ and } y_l = \alpha.$$

Answer:

$$\alpha + \alpha \sum_{k=0}^{\infty} \left[\operatorname{cerf} \frac{2kl - x + 2l}{2c\sqrt{t}} - \operatorname{cerf} \frac{2kl + x}{2c\sqrt{t}} \right].$$

10.49.2 Solve the problem

$$y_x' = c^2 \frac{d^2}{dx^2} y_x, \qquad y_x(0+) = 0 \;\; (0 < x < l < \infty),\; y_0 = a, \text{ and } y_l = 0.$$

Answer:

$$a \sum_{k=0}^{\infty} \operatorname{cerf} \frac{2kl + x}{2c\sqrt{t}} - a \sum_{k=1}^{\infty} \operatorname{cerf} \frac{2kl - x}{2c\sqrt{t}}.$$

10.49.3 Solve the problem

$$y'_x = c^2 \frac{d^2}{dx^2} y_x, \quad y_x(0+) = 1 \; (0 < x < l < \infty), \; y_0 = y_l = 0.$$

Answer:

$$1 + \sum_{k=0}^{\infty} \left[\operatorname{cerf} \frac{2(k+1)\,l - x}{2c\sqrt{t}} - \operatorname{cerf} \frac{2kl + x}{2c\sqrt{t}} \right] + \sum_{k=0}^{\infty} \left[\operatorname{cerf} \frac{2kl + l + x}{2c\sqrt{t}} - \operatorname{cerf} \frac{2kl + l - x}{2c\sqrt{t}} \right].$$

10.49.4 Solve the problem

$$y'_x = c^2 \frac{d^2}{dx^2} y_x, \quad y_x(0+) = 1 \; (0 < x < 1), \; y_0 = 0, \text{ and}$$

$$\frac{d}{dx} y_x \Big|_{x=1} = 0.$$

Answer:

$$1 - \sum_{k=0}^{\infty} (-1)^k \left[\operatorname{cerf} \frac{2k + x}{2c\sqrt{t}} + \operatorname{cerf} \frac{2k + 2 - x}{2c\sqrt{t}} \right].$$

10.49.5 Solve the problem

$$y'_x = \frac{d^2}{dx^2} y_x, \quad y_x(0+) = \alpha \; (0 < x < 1), \quad 0 = y_0, \text{ and}$$

$$\frac{d}{dx} y_x \Big|_{x=1} = 0.$$

Answer:

$$\alpha - \alpha \sum_{k=0}^{\infty} (-1)^k \left[\operatorname{cerf} \frac{2(k+1) - x}{2\sqrt{t}} + \operatorname{cerf} \frac{2k + x}{2\sqrt{t}} \right].$$

10.49.6 Solve the problem

$$y'_x = \frac{d^2}{dx^2} y_x, \quad y_x(0+) = \alpha \; (0 < x < 1), \; 0 = y_1, \text{ and}$$

$$\frac{d}{dx} y_x \Big|_{x=0} = 0.$$

Answer:

$$\alpha - \alpha \sum_{k=0}^{\infty} (-1)^k \left[\operatorname{cerf} \frac{2k+1-x}{2\sqrt{t}} + \operatorname{cerf} \frac{2k+1+x}{2\sqrt{t}}\right].$$

10.49.7 Given a $\mathcal{K}$-function $\varphi(\,)$; solve the problem

$$y'_x - \varphi = \frac{d^2}{dx^2} y_x, \quad y_x(0+) = 0 \ (0 < x < 1), \ 0 = y_1, \ \text{and}$$

$$\frac{d}{dx} y_x|_{x=0} = 0.$$

Answer: $y_x = D^{-1}\varphi - h_x * \varphi$, where

$$h_x = \sum_{k=0}^{\infty} (-1)^k \left[\operatorname{cerf} \frac{2k+1-x}{2\sqrt{t}} + \operatorname{cerf} \frac{2k+1+x}{2\sqrt{t}}\right].$$

10.49.8 Given a number a; solve the problem

$$y'_x = c^2 \frac{d^2}{dx^2} y_x, \quad y_x(0+) = \alpha \ (0 < x < l < \infty), \ y_l = a, \ \text{and}$$

$$\frac{d}{dx} y_x|_{x=0} = 0.$$

Answer:

$$\alpha + (a - \alpha) \sum_{k=0}^{\infty} (-1)^k \left[\operatorname{cerf} \frac{2kl + x + l}{2c\sqrt{t}} + \operatorname{cerf} \frac{2kl - x + l}{2c\sqrt{t}}\right].$$

10.49.9 Solve the problem

$$y_x = \frac{d^2}{dx^2} y_x, \quad y_x(0+) = 0 \quad (0 < x < l),$$

with $l < \infty$, and subject to the condition $y_0 = y_l = 1$.

Answer:

$$y_x(t) = \sum_{k=0}^{\infty} g_k(x, t) + h_k(x, t) \qquad (\text{all } t > 0),$$

where $g_k(x, t)$ and $h_k(x, t)$ are defined below.

10.49.10 Given a number a, solve the problem

$$y'_x = \frac{d^2}{dx^2} y_x, \quad y_x(0+) = \alpha \ (0 < x < l < \infty), \ y_0 = a, \ \text{and} \ y_l = 0.$$

Answer:

$$y_x = \alpha + (a - \alpha) \sum_{k=0}^{\infty} g_k(x, t) - \alpha \sum_{k=0}^{\infty} h_k(x, t),$$

where

$$g_k(x, t) = \operatorname{cerf} \frac{2kl + x}{2\sqrt{t}} - \operatorname{cerf} \frac{2kl - x + 2l}{2\sqrt{t}}$$

and

$$h_k(x, t) = \operatorname{cerf} \frac{2kl - x + l}{2\sqrt{t}} - \operatorname{cerf} \frac{2kl + x + l}{2\sqrt{t}}.$$

Chapter 5

This chapter is subdivided as follows: § 11 (Series of operators), § 12 (A functional calculus for D), § 13 (Non-linear equations), § 14 (Differential equations with polynomial coefficients).

A family of functions $w(\ )$ of the form

$$w(x) = \sum_{k=-\infty}^{\infty} w_k x^k \tag{1}$$

will be called a *functional calculus* for an operator X if the series

$$w(X) = \sum_{k=-\infty}^{\infty} w_k X^k \tag{2}$$

can be manipulated as if X were a number. Somewhat more precisely, we can say that a functional calculus (for X) is an algebra of functions $w(\ )$ of the form (1) such that the mapping $w(\ ) \mapsto w(X)$ is linear and transforms pointwise multiplication into operator multiplication:

$$v(\ )\, u(\ ) \mapsto v(X)\, u(X).$$

For example, let h be the canonical operator of a regulated $\mathcal{K}$-function; we shall prove that there is a functional calculus for the operator h/D; in particular, it turns out that

$$\frac{1}{1 - h/D} = \sum_{k=0}^{\infty} (h/D)^k. \tag{3}$$

As an immediate application of (3), let $g(\ )$ be a given regulated $\mathcal{K}$-function, and suppose that

$$y(t) - \int_0^t h(t-u)\, y(u)\, du = g(t) \qquad (\text{all } t > 0) \tag{4}$$

for some $\mathcal{K}$-function $y(\ )$; since (4) can be written in the form

$$(1 - h/D)\, y = g \qquad (\text{by } 3.15),$$

the conclusion

$$y = \sum_{k=0}^{\infty} (h/D)^k g$$

is immediate from (3). In the particular case $h = 1$, Equation (3) gives

(5) $$\frac{1}{1 - D^{-1}} = \sum_{k=0}^{\infty} D^{-k} = \sum_{k=0}^{\infty} \left\langle \frac{t^k}{k!} \right\rangle:$$

the last equation is from 3.20. If we can prove that

$$\sum_{k=0}^{\infty} \left\langle \frac{t^k}{k!} \right\rangle = \left\langle \sum_{k=0}^{\infty} \frac{t^k}{k!} \right\rangle,$$

then (5) obviously gives the familiar formula

$$\frac{D}{D-1} = \langle e^t \rangle \qquad \text{(compare with 3.21)}.$$

§ 11. Series of Operators

11.0 **Definitions.** A $\mathcal{K}_0$-function is a $\mathcal{K}$-function [1.36] that vanishes on the negative axis. Thus, a function $h(\,)$ is a $\mathcal{K}_0$-function if it has at most a finite number of discontinuities in every finite interval, if $h(t) = 0$ for $t \leq 0$ and if

$$\int_0^{\lambda} |h(u)|\, du < \infty \quad \text{whenever} \quad 0 < \lambda < \infty.$$

11.0.1 **Definition.** A function is said to be **regulated** if it is defined on $(-\infty, \infty)$ and has finite limits on both sides of every point (see [2.59]).

11.0.2 **Definition.** A sequence h_k $(k = 0, 1, 2, \ldots)$ is called **summable** if each h_k is the canonical operator $\langle h_k(t) \rangle$ of a regulated $\mathcal{K}_0$-function $h_k()$, and if the equation

(11.0.3) $$\sum_{k=0}^{\infty} h_k(\tau) = \lim_{n \to \infty} \sum_{k=0}^{n} h_k(\tau) \qquad (0 \leq \tau < \infty)$$

defines a regulated $\mathcal{K}$-function.

11.0.4 **Notation.** The function defined by 11.0.3 will be denoted by

$$\sum_{k=0}^{\infty} h_k();$$

as usual, its canonical operator is denoted $\left\langle \sum_{k=0}^{\infty} h_k(t) \right\rangle$.

11.0.5 **Definition.** A sequence h_k $(k = 0, 1, 2, \ldots)$ is called **$\mathcal{K}_0$-summable** if it is summable and if to every $x > 0$ there corresponds a regulated $\mathcal{K}$-function $H_x()$ such that

$$\left| \sum_{k=0}^{n} h_k(t) \right| \leq H_x(t) \quad \text{whenever} \begin{cases} 0 \leq t \leq x \\ n = 1, 2, 3, \ldots \end{cases} \tag{11.1}$$

11.2 **Theorem A.** *If f_k $(k = 0, 1, 2, \ldots)$ and g_k $(k = 0, 1, 2, \ldots)$ are $\mathcal{K}_0$-summable sequences, then*

$$\left\langle \sum_{k=0}^{\infty} f_k(t) \right\rangle \left\langle \sum_{k=0}^{\infty} g_k(t) \right\rangle = \lim_{n\to\infty} \left(\sum_{k=0}^{n} f_k \right) \left(\sum_{k=0}^{n} g_k \right). \tag{11.3}$$

Proof: see 15.35 in the Appendix.

11.3.1 **Definition.** If g_k $(k = 0, 1, 2, \ldots)$ is a sequence of operators, we set

$$\sum_{k=0}^{\infty} g_k \stackrel{\text{def}}{=} \lim_{n\to\infty} \sum_{k=0}^{n} g_k;$$

the notion of limit is defined in 6.4.0.

11.4 **Consequence.** *If a sequence g_k $(k = 0, 1, 2, \ldots)$ is $\mathcal{K}_0$-summable, then*

$$\left\langle \sum_{k=0}^{\infty} g_k(t) \right\rangle = \sum_{k=0}^{\infty} g_k = \sum_{k=0}^{n} \langle g_k(t) \rangle. \tag{11.5}$$

Proof. This comes directly from 11.2 and 2.17 by setting $f_0() = T_0()$ and $f_k() = 0$ for all $k > 0$.

11.5.1 *Remark.* In view of [11.0.5] and [11.0.2], the function

$$(6) \qquad \sum_{k=0}^{\infty} g_k()$$

is a $\mathcal{K}$-function: from 11.5, 2.1, and 1.15 it therefore follows that the operator

$$\sum_{k=0}^{\infty} g_k$$

is perfect: it is the canonical operator of the function (6).

11.6 If $h()$ is a regulated $\mathcal{K}_0$-function, to every $x > 0$ there corresponds a positive number $[h, x]$ such that

$$(11.7) \qquad |h(t)| \leq [h, x] \qquad \text{(all } t \leq x\text{): see 15.13.}$$

11.8 **Theorem.** *If $f()$ and $g()$ are regulated $\mathcal{K}_0$-functions, then $f * g()$ is a continuous $\mathcal{K}_0$-function.*

Proof. The function $f * g()$ is continuous (by 2.61); setting $\alpha = \beta = 0$ in 0.22, we see that it is also a $\mathcal{K}_0$-function.

Series of Impulses

Let s_k ($k = 0, 1, 2, \ldots$) be a sequence of numbers such that

$$(11.9) \qquad 0 = s_0 < s_k < s_{k+1} \qquad (\text{for } k = 1, 2, 3, \ldots),$$

and

$$(11.9.1) \qquad \lim_{k\to\infty} s_k = \infty.$$

11.10 **Definition.** For any real number $x \geq 0$, let $s(x)$ be the integer m such that

$$(11.10.1) \qquad s_m \leq x < s_{m+1}.$$

11.10.2 *Remarks.* From 11.10.1 and 11.9 it follows that $x < s_k$ when $k > m$, and $s_k \leq x$ whenever $k \leq m$; since $s(x)$ is the unique integer m determined by the inequalities 11.10.1, we have

$$k > s(x) \quad \text{implies} \quad x < s_k, \tag{11.10.3}$$

and

$$k \leq s(x) \quad \text{implies} \quad x \geq s_k. \tag{11.10.4}$$

11.11 **Theorem.** *If $f_k()$ $(k = 0, 1, 2, \ldots)$ is a sequence of regulated $\mathcal{K}$ functions, then the sequence*

$$\langle \mathsf{T}_0(t - s_k)\rangle f_k \qquad (k = 0, 1, 2, \ldots) \tag{1}$$

is $\mathcal{K}_0$-summable; moreover,

$$\sum_{k=0}^{\infty} \langle \mathsf{T}_0(t - s_k)\rangle f_k = \left\langle \sum_{k=0}^{s(t)} f_k(t - s_k)\right\rangle. \tag{11.12}$$

11.12.1 *Remark.* It is important to note that this theorem does not require that $f_k(t) = 0$ for $t \leq 0$.

11.12.2 *Proof of* 11.11. For $k = 0, 1, 2, \ldots$ let $h_k()$ be the function defined by

$$h_k(t) = \begin{cases} 0 & (t < s_k) \\ f_k(t - s_k) & (t \geq s_k). \end{cases} \tag{2}$$

From 3.12 it follows that

$$\langle \mathsf{T}_0(t - s_k)\rangle f_k = \langle h_k(t)\rangle \qquad (k = 0, 1, 2, \ldots). \tag{3}$$

Let x be any number > 0. If $n > s(x)$, then

$$\sum_{k=0}^{n} h_k(t) = \sum_{k=0}^{s(x)} h_k(t) + \sum_{k > s(x)} h_k(t). \tag{4}$$

We shall begin by verifying that

$$\sum_{k > s(x)} h_k(t) = 0 \qquad (0 \leq t \leq x). \tag{5}$$

To that effect, note that $x < s_k$ (by 11.10.3); the hypothesis $t \leq x$ therefore gives $t < s_k$, whence $h_k(t) = 0$ (by (2)). Having thus concluded the proof of (5), we can combine (5) and (4) to obtain

$$\sum_{k=0}^{n} h_k(t) = \sum_{k=0}^{s(x)} h_k(t) \quad \text{when} \begin{cases} 0 \leq t \leq x \\ n > s(x). \end{cases} \tag{6}$$

In the particular case $t = x$, this becomes

$$\sum_{k=0}^{n} h_k(t) = \sum_{k=0}^{s(t)} h_k(t) \quad \text{when} \begin{cases} 0 \leq t < \infty \\ n > s(t); \end{cases}$$

since $k \leq s(t)$, we can use 11.10.4 to conclude that $t \geq s_k$, whence $h_k(t) = f_k(t - s_k)$ (by (2)); therefore,

$$\sum_{k=0}^{n} h_k(t) = \sum_{k=0}^{s(x)} f_k(t - s_k) \quad \text{when} \begin{cases} 0 \leq t < \infty \\ n > s(t). \end{cases} \tag{7}$$

It is easily verified that 11.1 holds for

$$H_x(t) \overset{\text{def}}{=} \sum_{k=0}^{s(x)} |h_k(t)|:$$

indeed, 11.1 is obvious in case $n \leq s(x)$; in case $n > s(x)$ it follows from (6). Since $H_x()$ is a regulated $\mathcal{K}$-function, it results from 11.1 and (7) that (2) defines a $\mathcal{K}_0$-summable sequence: from 11.5 and (2) we see that

$$\sum_{k=0}^{\infty} \langle h_k(t) \rangle = \left\langle \sum_{k=0}^{\infty} h_k(t) \right\rangle: \tag{8}$$

Conclusion 11.12 is immediate from (8), (3), and (7).

11.13 **Particular cases.** If $G()$ is a $\mathcal{K}_0$-function, the equation

$$\delta(t - \alpha)\, G = D \mathsf{T}_\alpha G \tag{9}$$

is immediate from 6.54; multiplying both sides by D^{-1}, we can use [8.8] to write

$$\delta(t - \alpha) * G = \langle \mathsf{T}_0(t - \alpha) \rangle\, G. \tag{10}$$

In case $G() = \mathsf{T}_0()$, Equation (9) becomes

$$\delta(t - \alpha) = D\langle \mathsf{T}_0(t - \alpha)\rangle. \tag{11}$$

Z Until further notice, let s_k $(k = 0, 1, 2, \ldots)$ be a sequence of numbers satisfying 11.9.1—11.9. Let w_k $(k = 0, 1, 2, \ldots)$ be an *arbitrary* sequence of numbers; setting $f_k = w_k$ in 11.12, we obtain

$$\sum_{k=0}^{\infty} w_k \langle \mathsf{T}_0(t_0 - s_k)\rangle = \left\langle \sum_{k=0}^{s(t)} w_k \right\rangle. \tag{12}$$

Right-multiplying by D both sides of (12), we can use 6.9 and (11):

$$\boxed{\sum_{k=0}^{\infty} w_k \delta(t - s_k) = D\left\langle \sum_{k=0}^{s(t)} w_k \right\rangle}. \tag{11.14}$$

11.15 **An initial-value problem.** As in 8.49, consider the problem

$$\left[a_n \frac{\mathrm{d}^n}{\mathrm{d}t^n} + a_{n-1} \frac{\mathrm{d}^{n-1}}{\mathrm{d}t^{n-1}} + \cdots + a_1 \frac{\mathrm{d}}{\mathrm{d}t} + a_0\right] y = f, \tag{13}$$

$$y(0+) = y'(0+) = \cdots = y^{(n-1)}(0+) = 0. \tag{14}$$

From 8.52 we see that

$$y = fG^{-1}, \tag{15}$$

where $G = a_n D^n + a_{n-1} D^{n-1} + \cdots + a_1 D + a_0$. Recall that

$$\frac{D}{G} = g, \tag{11.15.1}$$

where $g()$ is **the Green's function** of the problem: see 8.61. Let us consider the particular case where

$$f = \sum_{k=0}^{\infty} w_k \delta(t - s_k): \tag{11.15.2}$$

obviously,

$$fG^{-1} = \sum_{k=0}^{\infty} w_k \delta(t - s_k) G^{-1} \qquad \text{by 6.9}$$

$$= \sum_{k=0}^{\infty} w_k \delta(t - s_k) * \frac{D}{G} \qquad \text{by [8.8]}$$

$$= \sum_{k=0}^{\infty} w_k \delta(t - s_k) * g \qquad \text{by 11.15.1.}$$

From 11.12 and (10) it now follows that the equation

$$y = \left\langle \sum_{k=0}^{s(t)} w_k g(t - s_k) \right\rangle \tag{16}$$

determines a solution of the problem (13)—(14) in case f is given by 11.15.2.

Let us consider the equation

$$\frac{d^n}{dt^n} y = \sum_{k=0}^{\infty} w_k \delta(t - s_k) \tag{17}$$

in case $y(0+) = y'(0+) = 0$. We can use (15) with $G = D^n$ to obtain

$$y = f * D^{n-1} = \left\langle \sum_{k=0}^{s(t)} w_k \frac{(t - s_k)^{n-1}}{(n-1)!} \right\rangle: \tag{18}$$

the last equation is from (16) and 3.20. When $n = 2$ the equation (17) governs the displacement y of a particle of unit mass subjected to hammer blows of magnitude $= w_k$ applied at $t = s_k$ $(k = 0, 1, 2, \ldots)$.

When $n = 1$ and $w_k = (-1)^k$ the equations (17)—(18) become

$$\frac{d}{dt} y = \sum_{k=0}^{\infty} (-1)^k \delta(t - s_k) \quad \text{and} \quad y = \left\langle \sum_{k=0}^{s(t)} (-1)^k \right\rangle. \tag{19}$$

Representing $\pm\delta(t - \alpha) = \pm D\mathsf{T}_\alpha$ by arrows of appropriate direction, the series of impulses (19) can be depicted as follows:

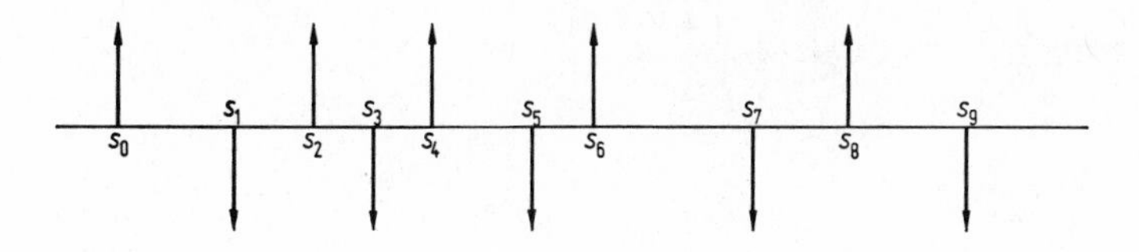

The graph of $y(\,)$ must look like this:

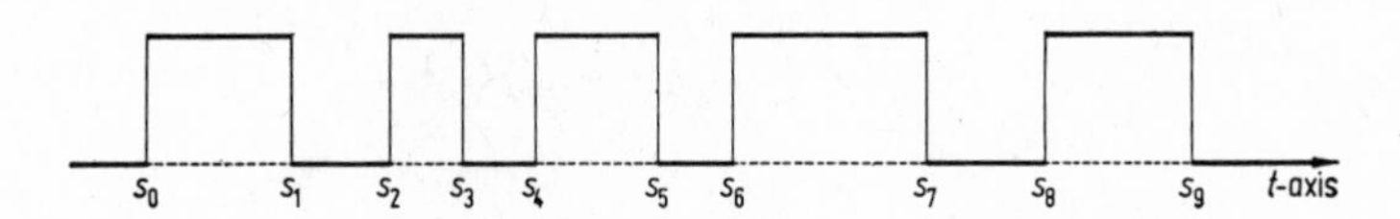

11.16 **Theorem.** *If* $s_k = k\alpha$ *for some* $\alpha > 0$. *Then the number* $s(x)$ *is the integer* m *such that* $m \leq x/\alpha < m - 1$.

Proof. It suffices to recall that $m = s(x)$ is the integer determined by the inequalities

$$m\alpha = s_m \leq x < s_{m+1} = (m+1)\,\alpha \qquad \text{(see 11.10)}.$$

11.17 **Consequences. Thus,** $s(t)$ **is the greatest integer** $< t/\alpha$: *it is often denoted* $[t/\alpha]$.

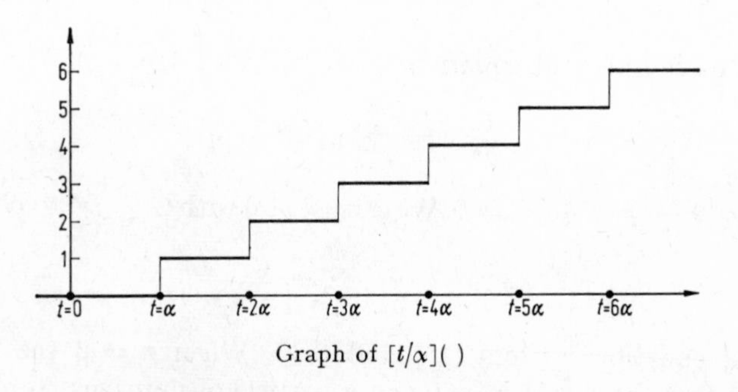

Graph of $[t/\alpha](\,)$

Since $\mathsf{T}_{k\alpha} = \langle \mathsf{T}_{k\alpha}(t)\rangle = \langle \mathsf{T}_0(t - k\alpha)\rangle$, we have

$$\text{(11.18)} \qquad \mathsf{T}_\alpha^k = \langle \mathsf{T}_0(t - k\alpha)\rangle \qquad \text{(by 5.3)}.$$

Setting $s_k = k\alpha$ in 11.12, we see that

$$\text{(11.19)} \qquad \boxed{\sum_{k=0}^{\infty} \mathsf{T}_\alpha^k f_k = \left\langle \sum_{k=0}^{[t/\alpha]} f_k(t - k\alpha) \right\rangle}.$$

Setting $f_k = w_k$ in 11.19:

$$\text{(11.20)} \qquad \sum_{k=0}^{\infty} w_k \mathsf{T}_\alpha^k = \left\langle \sum_{k=0}^{[t/\alpha]} w_k \right\rangle,$$

so that 11.14 gives

$$\boxed{\sum_{k=0}^{\infty} w_k \delta(t - k\alpha) = D\left(\sum_{k=0}^{\infty} w_k \mathsf{T}_\alpha^k\right)} . \tag{11.21}$$

It is easily verified that

$$\sum_{k=0}^{[t/\alpha]} 1 = 1 + [t/\alpha];$$

consequently, 11.20—21 imply that

$$\sum_{k=0}^{\infty} \delta(t - k\alpha) = D\langle 1 + [t/\alpha]\rangle . \tag{11.22}$$

From 11.22 it follows easily that $y = [t/\alpha]$ is the solution of the initial-value problem

$$y(0+) = 0, \quad \frac{\mathrm{d}}{\mathrm{dt}} y = \sum_{k=1}^{\infty} \delta(t - k\alpha).$$

Thus, the function $[t/\alpha]()$ describes the velocitiy of a particle of unit mass initially at rest subject to impulses (e.g., hammer blows) of magnitude $= 1$ applied at the times $t = k\alpha$ ($k = 0, 1, 2, \ldots$).

Analytic Functions of h/D

11.23 Let $\mathscr{A}_0$ be the family of all the functions that are analytic in some neighborhood of the origin. It is easily seen that $w \in \mathscr{A}_0$ if (and only if) there exists a number $r > 0$ and a sequence w_k ($k = 0, 1, 2, \ldots$) of numbers such that

$$w(z) = \sum_{k=0}^{\infty} w_k z^k \qquad (|z| < r). \tag{11.24}$$

If X is an operator, we set

$$w(X) = \sum_{k=0}^{\infty} w_k X^k; \tag{11.25}$$

of course, we shall only deal with operators X such that the right-hand side has a meaning. Since the coefficients w_k are uniquely determined by the function w, the operator $w(X)$ is also uniquely determined by the function w.

11.25.1 **Example.** If b is a number and if $\alpha > 0$, then the equations

$$w(b\mathsf{T}_\alpha) = \sum_{k=0}^{\infty} w_k b^k \mathsf{T}_\alpha^k = \left\langle \sum_{k=0}^{[t/\alpha]} w_k b^k \right\rangle$$

are from [11.25] and 11.19 (with $f_k() = w_k b^k$).

11.26 **Theorem.** *Suppose that h is an impulse of magnitude $= b$ applied at some point $t = \alpha > 0$. If $g()$ is a regulated $\mathcal{K}$-function and $w \in \mathcal{A}_0$, then the sequence*

$$(1) \qquad w_k h^k D^{-k} g \quad (k = 0, 1, 2, \ldots)$$

is $\mathcal{K}_0$-summable; further,

$$(11.27) \qquad \boxed{w\left(\frac{h}{D}\right) g = \left\langle \sum_{k=0}^{[t/\alpha]} w_k b^k g(t - k\alpha) \right\rangle}.$$

Proof. Note that $h = bD\mathsf{T}_\alpha$ (by 6.41), whence

$$(2) \qquad w_k h^k D^{-k} g = w_k b^k \mathsf{T}_\alpha^k g = \langle \mathsf{T}_0(t - k\alpha)\rangle\,(w_k b^k g):$$

the last equation is from 11.18. The $\mathcal{K}_0$-summability of the sequence (1) now comes from (2) and 11.11. Finally, observe that the equations

$$(3) \qquad w\left(\frac{h}{D}\right) g = \sum_{k=0}^{\infty} w_k \left(\frac{h}{D}\right)^k g = \sum_{k=0}^{\infty} w_k h^k D^{-k} g = \sum_{k=0}^{\infty} \mathsf{T}_\alpha^k (w_k b^k g)$$

come from Definition [11.25] and (2): Conclusion 11.27 is an immediate consequence of (3) and 11.19.

11.28 **The road ahead.** Until further notice, we suppose that $h()$ and $g()$ are regulated $\mathcal{K}_0$-functions. The following remarks are intended to help us prove that the sequence (1) is $\mathcal{K}_0$-summable. As in 11.6 we have

$$(11.29) \qquad |h(t-u)| \leq [h, x] \text{ and } |g(u)| \leq [g, x] \quad (0 \leq t \leq u \leq x).$$

To prepare for the next theorem, let us define a sequence $g_k()(k = 0, 1, 2, \ldots)$ as follows:

(11.29.1) $$g_0() = g(),$$

(11.29.2) $$g_k() = \left\{\int_0^t h(t-u)\, g_{k-1}(u)\, du\right\}().$$

11.29.3 **Lemma.** *If* $k \geq 1$, *then*

(11.30) $$g() \text{ is continuous},$$

(11.31) $$g_k = h^k D^{-k} g,$$

(11.32) $$|g_k(t)| \leq [g, x]\,[h, x]^k \frac{t^k}{k!} \qquad (\text{if } 0 \leq t \leq x).$$

Proof. From 11.29.2, 2.9, and 2.8.3 it follows that

(11.32.1) $$g_k = h * g_{k-1} = hD^{-1} g_{k-1}.$$

Definition 11.29.2 also implies that

$$|g_k(t)| \leq \int_0^t |h(t-u)|\, |g_{k-1}(u)|\, du,$$

whence

(11.32.2) $$|g_k(t)| \leq [h, x] \int_0^t |g_{k-1}(u)|\, du \qquad (\text{if } 0 \leq t \leq x).$$

In case $k = 1$, we can use 11.32.2 and 11.29.1 to obtain

(4) $$|g_1(t)| \leq [h, x]\,[g, x] \int_0^t du \qquad (\text{if } 0 \leq t \leq x);$$

consequently, 11.32 is at hand; moreover, 11.30 comes from 11.8; finally, 11.31 comes from 11.32.1 and 11.29.1.

We proceed by induction. Assume that all three properties 11.30—32 hold for $k-1$:

$$g_{k-1}() \text{ is continuous}, \tag{5}$$

$$g_{k-1} = h^{k-1} D^{-k+1} g, \tag{6}$$

$$|g_{k-1}(u)| \leq [h, x]^{k-1} [g, x] \frac{u^{k-1}}{(k-1)!} \quad (\text{if } 0 \leq u \leq x). \tag{7}$$

Property 11.30 is immediate from 11.32.1 and 11.8; moreover, Property 11.31 comes from (6) and 11.32.1. To prove 11.32 we combine 11.32.2 with (7):

$$|g_k(t)| \leq [h, x]\, [h, x]^{k-1} [g, x] \int_0^t \frac{u^{k-1}}{(k-1)!}\, du$$

(for $0 \leq t \leq x$); therefore, 11.32 holds. We have established that 11.30—32 hold for k whenever they hold for $k-1$: consequently, 11.30—32 hold for any integer $k \geq 1$.

11.33 **Theorem.** *Suppose that $h()$ and $g()$ are regulated $\mathcal{K}_0$-functions. If $w \in \mathcal{A}_0$, then the sequence*

$$w_k h^k D^{-k} g \qquad (k = 0, 1, 2, \ldots) \tag{11.34}$$

is $\mathcal{K}_0$-summable.

Proof. Since $w \in \mathcal{A}_0$, the series $\sum w_k z^k$ converges for some value $z = p > 0$:

$$\lim_{n\to\infty} w_n p^n = \lim_{n\to\infty} \left[\sum_{k=0}^{n} - \sum_{k=0}^{n-1} \right] w_k p^k = 0;$$

consequently, there exists an integer m such that

$$|w_n p^n| \leq 1 \text{ for all } n \geq m;$$

we therefore have

$$|w_k| \leq M p^{-k} \qquad (\text{all } k \geq 0), \tag{8}$$

where $M = 1 + |w_0| + |w_1 p| + \cdots + |w_m p^m|$. Suppose that $x \geq 0$; combining (8) and 11.31, we obtain

$$|w_k g_k(t)| \leq F_k(t, x) \stackrel{\text{def}}{=} M[g, x] \left(\frac{[h, x]}{p}\right)^k \frac{t^k}{k!} \tag{9}$$

whenever $0 \leq t \leq x$, which implies that

$$\left| \sum_{k=0}^{n} w_k g_k(t) \right| \leq \sum_{k=0}^{\infty} F_k(t, x) \qquad \text{when} \begin{cases} 0 \leq t \leq x \\ n = 0, 1, 2, \ldots \end{cases} \tag{10}$$

From the definition of $F_k(t, x)$ (see (9)), it follows that

$$\sum_{k=0}^{\infty} F_k(t, x) = M[g, x] \exp\left(\frac{[h, x]\, t}{p}\right). \tag{11}$$

On the other hand, since (9) implies

$$|w_k g_k(t)| \leq F_k(x, x) \qquad \text{whenever} \begin{cases} -x \leq t \leq x \\ n = 0, 1, 2, \ldots, \end{cases}$$

we may use the Weierstrass M-Test to conclude from (11) that the series

$$\sum_{k=1}^{\infty} w_k g_k(t)$$

converges uniformly in the interval $[-x, x]$; the functions $g_k()$ being continuous (by 11.30), the uniform convergence property guarantees that the function

$$\sum_{k=1}^{\infty} w_k g_k()$$

is likewise continuous. Since

$$\sum_{k=0}^{\infty} w_k g_k() = w_0 g_0() + \sum_{k=1}^{\infty} w_k g_k() \tag{11.35}$$

and since $g_0()$ is the regulated $\mathcal{K}_0$-function $g()$ (see 11.29.1), it follows that the sequence $w_k g_k$ $(k = 0, 1, 2, \ldots)$ is summable; further, its $\mathcal{K}_0$-summability is a direct consequence of (10)—(11): see 11.1. Since

$$w_k g_k = w_k h^k D^{-k} g \qquad \text{(by 11.31)}, \tag{11.36}$$

it follows that the sequence 11.34 is $\mathcal{K}_0$-summable.

A Functional Calculus for h/D

If $u \in \mathcal{A}_0$ and $v \in \mathcal{A}_0$ then $u = v$ if (and only if)

$$u(z) = v(z) \quad \text{(every small } z\text{)};$$

that is, $u = v$ if (and only if) there exists a number $r > 0$ such that $u(z) = v(z)$ whenever $|z| < r$.

11.37 If $u \in \mathcal{A}_0$ and $v \in \mathcal{A}_0$, then the function uv is defined by

$$uv(z) = u(z)\, v(z) \quad \text{(every small } z\text{)}. \tag{11.38}$$

It is well-known that $uv \in \mathcal{A}_0$ and

$$uv(z) = \sum_{k=0}^{\infty} [uv]_k z^k \quad \text{(every small } z\text{)}, \tag{11.39}$$

where

$$[uv]_k = u_k v_0 + u_{k-1} v_1 + \cdots + u_0 v_k. \tag{11.40}$$

Z *Until further notice, h is either the canonical operator of a regulated $\mathcal{K}_0$-function, or it is an impulse applied at some point $t = \alpha \geq 0$.*

11.41 If $w \in \mathcal{A}_0$ it follows from 11.26 and 11.33 (with $g = \mathsf{T}_0$ and $h() = \{h(t)\}()$) that the sequence

$$w_k h^k D^{-k} \qquad (k = 0, 1, 2, \ldots)$$

is $\mathcal{K}_0$-summable; by Definition [11.25] and 11.5 we therefore have

$$w\left(\frac{h}{D}\right) = \sum_{k=0}^{\infty} w_k \left[\frac{h}{D}\right]^k = \left\langle \sum_{k=0}^{\infty} w_k h^k D^{-k}(t) \right\rangle: \tag{1}$$

the second equation is from 11.5. From (1) it follows that

$$w\left(\frac{h}{D}\right) = \lim_{n\to\infty} \sum_{k=0}^{2n} w_k \left(\frac{h}{D}\right)^k. \tag{2}$$

11.42 **Theorem B.** *If $u \in \mathscr{A}_0$ and $v \in \mathscr{A}_0$, then*

$$u\left(\frac{h}{D}\right) v\left(\frac{h}{D}\right) = uv\left(\frac{h}{D}\right); \tag{11.43}$$

further, $u(h/D)$ is the canonical operator of a $\mathscr{K}_0$-function.

Proof. Since $uv \in \mathscr{A}_0$, we can apply (2):

$$uv\left(\frac{h}{D}\right) = \lim_{n\to\infty} \sum_{k=0}^{2n} [uv]_k \left(\frac{h}{D}\right)^k. \tag{3}$$

On the other hand, (1) implies

$$u\left(\frac{h}{D}\right) v\left(\frac{h}{D}\right) = \left\langle \sum_{k=0}^{\infty} u_k h^k D^{-k}(t) \right\rangle \left\langle \sum_{k=0}^{\infty} v_k h^k D^{-k}(t) \right\rangle;$$

in view of 11.41, we may apply Theorem A (11.2) to obtain

$$u\left(\frac{h}{D}\right) v\left(\frac{h}{D}\right) = \lim_{n\to\infty} \left(\sum_{k=0}^{n} u_k \left(\frac{h}{D}\right)^k\right) \left(\sum_{k=0}^{n} v_k \left(\frac{h}{D}\right)^k\right). \tag{4}$$

Since perfect operators obey the distributivity law of algebra, we have

$$\left(\sum_{k=0}^{n} u_k \left(\frac{h}{D}\right)^k\right) \left(\sum_{k=0}^{n} v_k \left(\frac{h}{D}\right)^k\right) = [uv]_0 + [uv]_1 \left(\frac{h}{D}\right) + \cdots + [uv]_{2n} \left(\frac{h}{D}\right)^{2n},$$

where $[uv]_k$ is defined by [11.40]; consequently, (4) becomes

$$u\left(\frac{h}{D}\right) v\left(\frac{h}{D}\right) = \lim_{n\to\infty} \sum_{k=0}^{2n} [uv]_k \left(\frac{h}{D}\right)^k = uv\left(\frac{h}{D}\right):$$

the last equation is from (3). To see that $u(h/D)$ is the canonical operator of a $\mathcal{K}$-function, note that the $\mathcal{K}_0$-summability (11.41) enables us to apply 11.5:

$$\left\langle \sum_{k=0}^{\infty} u_k h^k D^{-k}(t) \right\rangle = \sum_{k=0}^{\infty} u_k \left(\frac{h}{D}\right)^k = u\left(\frac{h}{D}\right);$$

thus, $u(h/D)$ is the canonical operator of the function

$$\sum_{k=0}^{\infty} u_k h^k D^{-k}();$$

finally, note that this is necessarily a $\mathcal{K}$-function (by 11.5.1).

11.44 *Remarks.* Let x be a real number. If $u \in \mathscr{A}_0$ and $u(0) \neq 0$, then the equation

$$u^x(z) = u(z)^x \qquad \text{(every small } z) \tag{11.45}$$

defines a function u^x such that $u^x \in \mathscr{A}_0$. In particular,

$$u^0(z) = 1 \quad \text{and} \quad u(z)\, u^{-1}(z) = u^0(z) \qquad \text{(every small } z), \tag{5}$$

so that

$$u^0\left(\frac{h}{D}\right) = 1 \qquad \text{(by [11.25])}, \tag{6}$$

and

$$u\left(\frac{h}{D}\right) u^{-1}\left(\frac{h}{D}\right) = u u^{-1}\left(\frac{h}{D}\right) = 1 \qquad \text{(by 11.43 and (6))}. \tag{11.46}$$

Consequently:

$$u^{-1}\left(\frac{h}{D}\right) = \left[u\left(\frac{h}{D}\right)\right]^{-1}. \tag{11.47}$$

Suppose that $w \in \mathscr{A}_0$ and $w(0) = 0$; the equations

$$v^x(z) = [1 + w(z)]^x = \sum_{k=0}^{\infty} \binom{x}{k} w^k(z) \qquad \text{(every small } z) \tag{11.48}$$

define a function v^x such that $v^x \in \mathscr{A}_0$; it is natural to set

$$\left[1 + w\left(\frac{h}{D}\right)\right]^x \stackrel{\text{def}}{=} v^x\left(\frac{h}{D}\right). \tag{11.49}$$

From [11.49], 11.48, and [11.25] it follows that

$$\left[1 + w\left(\frac{h}{D}\right)\right]^x = \sum_{k=0}^{\infty} \binom{x}{k} \left[w\left(\frac{h}{D}\right)\right]^k. \tag{11.50}$$

Since $v^\alpha(z)\, v^\lambda(z) = v^{\alpha+\lambda}(z)$, we can use 11.43 to infer that

$$\left[1 + w\left(\frac{h}{D}\right)\right]^\alpha \left[1 + w\left(\frac{h}{D}\right)\right]^\lambda = \left[1 + w\left(\frac{h}{D}\right)\right]^{\alpha+\lambda}. \tag{11.51}$$

For $\lambda = -\alpha$ this gives

$$\left[1 + w\left(\frac{h}{D}\right)\right]^\alpha \left[1 + w\left(\frac{h}{D}\right)\right]^{-\alpha} = u^0\left(\frac{h}{D}\right) = 1:$$

the last equation is from (6); we can now use 1.66 to conclude that

$$\left[1 + w\left(\frac{h}{D}\right)\right]^{-\alpha} = \frac{1}{\left[1 + w\left(\frac{h}{D}\right)\right]^\alpha}. \tag{11.52}$$

If N is an integer, 1.72 allows us to write

$$\left[1 + w\left(\frac{h}{D}\right)\right]^{-N} = \frac{1}{\left[1 + w\left(\frac{h}{D}\right)\right]^N} = \left(\frac{1}{1 + w\left(\frac{h}{D}\right)}\right)^N.$$

It is easily verified that

$$\boxed{\left[\frac{1}{1 + w(h/D)}\right]^N = \sum_{k=0}^{\infty} \frac{(k + N - 1)!}{k!\,(N-1)!} \left[-w\left(\frac{h}{D}\right)\right]^k}: \tag{11.53}$$

see 11.50 and 11.61.7. In the particular case $N = 1$ and $w(z) = z^m$ we obtain

$$\frac{1}{1 + (h/D)^m} = \sum_{k=0}^{\infty} \left[-\left(\frac{h}{D}\right)^m \right]^k \tag{11.53.1}$$

in case m is an integer ≥ 1.

11.53.2 **Important remark.** In all the preceding formulas 11.43—11.53.1, the symbol h stands for **either** the canonical operator of a regulated $\mathcal{K}$-function, **or else** $h = bD\mathsf{T}_\alpha$ for some numbers b and $\alpha > 0$.

Applications

11.54 Let $h(\,)$ and $g(\,)$ be regulated $\mathcal{K}_0$-functions; if $y(\,)$ is a $\mathcal{K}$-function such that

$$y - \int_0^t h(t-u)\, y(u) du = g,$$

then $y - hD^{-1}y = g$, so that

$$y = \frac{1}{1 - \dfrac{h}{D}}\, g = \left[1 - \frac{h}{D}\right]^{-1} g = \sum_{k=0}^{\infty} \left(\frac{h}{D}\right)^k g:$$

the second equality is from 11.53.1.

11.55 Suppose that $w \in \mathscr{A}_0$ and $\alpha \geq 0$; from 11.21 and [11.25] it follows that

$$\sum_{k=0}^{\infty} w_k \delta(t - k\alpha) = D \sum_{k=0}^{\infty} w_k \mathsf{T}_\alpha^k = D[w(\mathsf{T}_\alpha)]. \tag{1}$$

If $u \in \mathscr{A}_0$ and $u(0) \neq 0$ we see from (1) and 11.47 (with $h = D\mathsf{T}_\alpha$) that

$$\boxed{\left[\sum_{k=0}^{\infty} u_k \delta(t - k\alpha)\right]^{-1} = D^{-1}[u^{-1}(\mathsf{T}_\alpha)]}\,. \tag{2}$$

Further, (1), [8.8], and 11.43 give

$$(3) \qquad \boxed{\left(\sum_{k=0}^{\infty} u_k \delta(t - k\alpha)\right) * \left(\sum_{k=0}^{\infty} w_k \delta(t - k\alpha)\right) = D[u(\mathsf{T}_\alpha)\, w(\mathsf{T}_\alpha)]}.$$

In the particular case $u_k = b^k = w_k$ we have

$$(4) \qquad u(z) = \sum_{k=0}^{\infty} b^k z^k = w(z) = (1 - bz)^{-1} \quad \text{(every small } z\text{)},$$

so that (1) becomes

$$\sum_{k=0}^{\infty} b^k \delta(t - k\alpha) = D(1 - b\mathsf{T}_\alpha)^{-1} \qquad \text{(by 11.48)};$$

similarly, we can use (4) and 11.48 to re-write Equation (3):

$$\left(\sum_{k=0}^{\infty} b^k \delta(t - k\alpha)\right) * \left(\sum_{k=0}^{\infty} b^k \delta(t - k\alpha)\right) = D(1 - b\mathsf{T}_\alpha)^{-2}.$$

Note that the equations

$$D(1 - b\mathsf{T}_\alpha)^{-2} = D \sum_{k=0}^{\infty} b^k(k+1)\mathsf{T}_\alpha^k = \sum_{k=0}^{\infty} b^k(k+1)\,\delta(t - k\alpha)$$

are from 11.53 (with $h = D\mathsf{T}_\alpha$) and (1). Consequently,

$$\boxed{\left(\sum_{k=0}^{\infty} b^k \delta(t - k\alpha)\right) * \left(\sum_{k=0}^{\infty} b^k \delta(t - k\alpha)\right) = \sum_{k=0}^{\infty} b^k(k+1)\,\delta(t - k\alpha)}.$$

11.56 As a final application of Equation (1), consider the function

$$w(z) = \frac{1 - z}{1 + z} = 1 + 2\sum_{k=1}^{\infty} (-1)^k z^k \quad \text{(every small } z\text{)};$$

the second equality is easily verified. From 11.53 and 11.43 we have

$$Dw(\mathsf{T}_\alpha) = D\,\frac{1 - \mathsf{T}_\alpha}{1 + \mathsf{T}_\alpha} = D\left(1 + 2\sum_{k=1}^{\infty} (-1)^k \mathsf{T}_\alpha^k\right),$$

so that Equation (1) gives

$$\delta(t) + 2 \sum_{k=1}^{\infty} (-1)^k \delta(t - k\alpha) = D \frac{1 - \mathsf{T}_\alpha}{1 + \mathsf{T}_\alpha}. \tag{5}$$

Consequently, the left-hand side of (5) equals DG, where $G(\ )$ is the square-wave function (5.37).

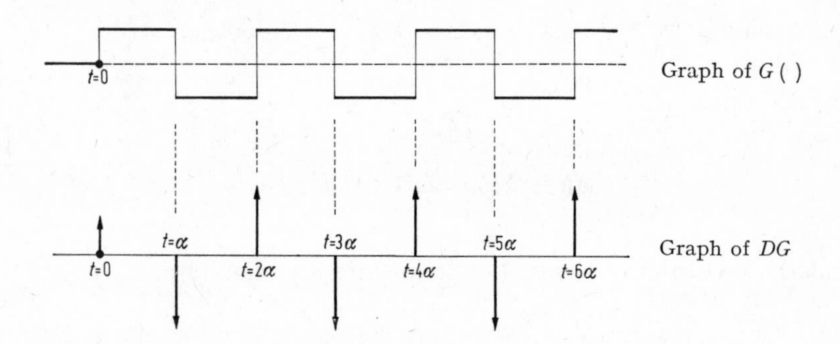

11.57 **A difference equation.** Until further notice, b is a given number and $\alpha > 0$. Let $g()$ be a $\mathcal{K}_0$-function: to find a $\mathcal{K}_0$-function $y()$ such that

$$y(t) - 2by(t-\alpha) + b^2 y(t - 2\alpha) = g(t) \qquad (-\infty < t < \infty).$$

If $y(\ )$ is such a function, we can proceed as in 5.40 to obtain

$$g = y - 2b\mathsf{T}_\alpha y + b^2 \mathsf{T}_\alpha^2 y = (1 - b\mathsf{T}_\alpha)^2 y;$$

left-multiplying by $(1 - b\mathsf{T}_\alpha)^{-2}$, we obtain

$$y = (1 - b\mathsf{T}_\alpha)^{-2} g = \sum_{k=0}^{\infty} (k+1) b^k \mathsf{T}_\alpha^k g:$$

the last equation is from 11.53 (with $h = D\mathsf{T}_\alpha$). Setting $f_k = (k+1)b^k g$ in 11.19, we get

$$y = \sum_{k=0}^{[t/\alpha]} (k+1) b^k g(t - k\alpha).$$

11.58 **Another example.** We can also solve difference equations of the form

$$y(t-2) - 2y(t-1) + y = \delta(t-1) - 1, \tag{6}$$

with the understanding that $y(t-x)$ means $\mathsf{T}_x y$ (see 5.25). From (6) and 6.54 it follows that

$$\mathsf{T}_2 y - 2\mathsf{T}_1 y + y = D\mathsf{T}_1 - 1;$$

since $\mathsf{T}_2 = (\mathsf{T}_1)^2$ (by 5.3), this means that

$$(1-\mathsf{T}_1)^2 y = -D(1-\mathsf{T}_1),$$

so that

$$y = -D(1-\mathsf{T}_1)^{-1} = \sum_{k=0}^{\infty} -D\mathsf{T}_1^k = \sum_{k=0}^{\infty} -\delta(t-k):$$

the last two equations are from 11.53 (with $h = D\mathsf{T}_1$ and $w(z) = z$) in combination with 11.21.

11.58.1 **Theorem.** *Let c be a number, and let $f()$ be a $\mathcal{K}$-function. If $x \geq 0$ and $\alpha > 0$, then the equation*

$$\boxed{\frac{\mathsf{T}_x f}{(1-c\mathsf{T}_\alpha)^{m+1}} = \left\langle \sum_{k=0}^{(t-x)/\alpha} \frac{(k+m)!}{k!\,m!} c^k f(t-x-k\alpha) \right\rangle}$$

holds for any integer $m \geq 1$.

11.58.2 *Remark.* This theorem is much more general than 5.29.

Proof of 11.58.1. From 11.53 (with $w(z) = -z$ and $h = cD\mathsf{T}_\alpha$) it follows that

$$\frac{1}{(1-c\mathsf{T}_\alpha)^{m+1}} \mathsf{T}_x f = \left(\sum_{k=0}^{\infty} \frac{(k+m)!}{k!\,m!} c^k \mathsf{T}_\alpha^k\right) \mathsf{T}_x f \tag{7}$$

$$= \sum_{k=0}^{\infty} \langle \mathsf{T}_0(t-k\alpha-x)\rangle \frac{(k+m)!}{k!\,m!} c^k f: \tag{8}$$

the last equation is from 5.3, 5.2, and 11.18. Setting

$$s_k = k\alpha + x \qquad (k = 0, 1, 2, \ldots),$$

we can combine (7)—(8) with 11.12 to obtain

$$\frac{\mathsf{T}_x f}{(1 - c\mathsf{T}_\alpha)^{m+1}} = \sum_{k=0}^{s(t)} \frac{(k+m)!}{k!\,m!} c^k f(t - k\alpha - x).$$

It now suffices to observe that $s(t)$ is the integer n determined by the inequalities

$$n\alpha + x \leq s_n < (n+1)\,\alpha + x;$$

thus, $s(t)$ is the integer n determined by

$$n \leq \frac{t-x}{\alpha} < n + 1 \qquad \text{(see 11.16—17);}$$

in consequence, $s(t)$ is the largest integer $\leq (t - x)/\alpha$.

Difference-differential Equations

11.59 Suppose that $\alpha \geq 0$, and let $y(\,)$ be a $\mathcal{K}_0$-function which is continuous in $(0, \infty)$ and such that

$$y(0+) = 0 \quad \text{with} \quad y'(t) - b\,y(t - \alpha) = g(t) \qquad (-\infty < t < \infty)$$

for a given $\mathcal{K}_0$-function $g(\,)$. From 8.45, 11.18, and 5.25 it follows that

$$\frac{\mathrm{d}}{\mathrm{d}t} y - b\mathsf{T}_\alpha y = g = Dy - b\mathsf{T}_\alpha y:$$

the second equation is from 8.33. Solving for y:

$$y = \frac{1}{D - b\mathsf{T}_\alpha} g = \frac{D^{-1}}{1 - b\mathsf{T}_\alpha/D} g;$$

since $b\mathsf{T}_\alpha(\,)$ is a regulated $\mathcal{K}_0$-function, we may use 11.53.1 (with $h = b\mathsf{T}_\alpha$) to write

$$y = \sum_{k=0}^{\infty} b^k \mathsf{T}_\alpha^k D^{-k} D^{-1} g = \sum_{k=0}^{\infty} b^k \mathsf{T}_\alpha^k \left(\left\langle \frac{t^k}{k!} \right\rangle D^{-1} g\right):$$

the second equation is from 3.15 and 3.19. Consequently, 11.19 and 3.15 gives

$$y = \left\langle \sum_{k=0}^{[t/\alpha]} b^k \int_0^{t-k\alpha} \frac{(t-k\alpha-u)^k}{k!} g(u)\, du \right\rangle.$$

11.60 Given a number c, let us find a $\mathcal{K}_0$-function $y(\,)$ such that $y'(\,)$ is continuous in $(0, \infty)$ and such that

$$y(0+) = y'(0+) = 0 \quad \text{with} \quad y''(t) - by(t-\alpha) = c \qquad (-\infty < t < \infty).$$

If $y(\,)$ is such a function, it follows from 8.47 and 5.25 that

$$\frac{d^2}{dt^2} y - b\,\mathsf{T}_\alpha y = c = D^2 y - b\,\mathsf{T}_\alpha y:$$

the second equation is from 8.31 (in view of the zero starting values). Solving for y:

$$y = \frac{c}{D^2 - b\mathsf{T}_\alpha} = \frac{c}{1 - b\mathsf{T}_\alpha D^{-1}/D} D^{-2}.$$

Setting $h = \mathsf{T}_\alpha D^{-1}$ and $m = 1$ in 11.53.1:

$$\text{(1)} \qquad y = \left[\sum_{k=0}^{\infty} b^k \mathsf{T}_\alpha^k D^{-k}/D^k\right] c D^{-2}$$

– the use of 11.53 is permissible, since $h = \mathsf{T}_\alpha D^{-1}$ is the canonical operator of the $\mathcal{K}_0$-function $\{t - \alpha\}(\,)$ (see 3.18 and 3.12). From (1) and 6.9 we obtain

$$y = \sum_{k=0}^{\infty} c b^k \mathsf{T}_\alpha^k \left\langle \frac{t^{2k+2}}{(2k+2)!} \right\rangle;$$

consequently, 11.19 gives the conclusion

$$y = \sum_{k=0}^{[t/\alpha]} c b^k \frac{(t - k\alpha)^{2k+2}}{(2k+2)!}.$$

Exercises

11.61.0 Solve the equation

$$D^4 y - a^4 y = \sum_{k=0}^{\infty} k^k \delta(t - s_k),$$

where s_k ($k = 0, 1, 2, \ldots$) is a sequence satisfying 11.9—11.9.1.

Answer:

$$\frac{1}{2a^3} \sum_{k=0}^{s(t)} k^k [\sinh(at - as_k) - \sin(at - as_k)].$$

11.61.1 Solve the equation

$$y + e^{at} \int_0^t e^{-au} y(u) \frac{(t-u)^m}{m!} du = \frac{b t^m e^{at}}{m!}.$$

Hint: use 11.54.1 with $h = -(D - a)^{-m-1}$ and 3.22.

Answer:

$$b e^{at} \sum_{k=0}^{\infty} (-1)^k \frac{t^{mk+k+m}}{(mk + k + m)!}.$$

11.61.2 Solve the problem

$$y(0+) = 0, \qquad y'(t) - by(t - \alpha) = c.$$

Answer:

$$c \sum_{k=0}^{[t/\alpha]} b^k \frac{(t - k\alpha)^{k+1}}{(k+1)!}.$$

11.61.3 Solve the equation

$$y(t) + 2y(t-1) + y(t-2) = \delta(t) + 3\delta(t-1).$$

Hint: proceed as in 11.58 and note that

$$\frac{1 + 3T_1}{(1 + T_1)^2} = \frac{3}{1 + T_1} - \frac{2}{(1 + T_1)^2}.$$

Answer:

$$y = \sum_{k=0}^{\infty} (-1)^k (1 - 2k)\, \delta(t - k).$$

11.61.4 Consider the electric circuit problem $L\,\mathrm{d}i/\mathrm{d}t + Ri = E$ in case E is a series of impulses of magnitude $= b$ applied at the times $t = k\alpha$ $(k = 1, 2, 3, \ldots)$. Solve for i in case $i(0+) = 0$.

Answer: $i = \left(\frac{b}{L}\right) \sum_{k=1}^{[t/\alpha]} \exp\left(\frac{-R(t - k\alpha)}{L}\right).$

11.61.5 The operator of the sawtooth function $F(\,)$ of period $= \alpha$ is

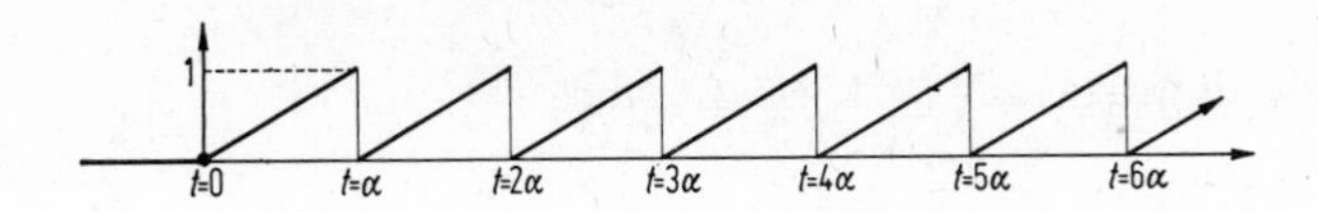

given by $F = \alpha^{-1} D^{-1} - \mathsf{T}_\alpha (1 - \mathsf{T}_\alpha)^{-1}$ (see 5.43.0): find $\mathrm{d}F/\mathrm{d}t$.

Hint: follow the procedure in 11.56.

Answer: $\frac{\mathrm{d}}{\mathrm{d}t} F = DF = \alpha^{-1} - \sum_{n=1}^{\infty} \delta(t - n\alpha).$

11.61.6 Use [6.23] to find

$$\int_{-\infty}^{\tau} \left[-\frac{1}{\alpha} + \sum_{k=0}^{\infty} \delta(t - k\alpha) \right].$$

Hint:

$$\text{if} \quad A = -\frac{1}{\alpha} + \sum_{k=0}^{\infty} \delta(t - k\alpha), \quad \text{then} \quad D^{-1}A = \frac{\mathsf{T}_\alpha - 1 + \alpha D}{\alpha D (1 - \mathsf{T}_\alpha)};$$

in view of 5.43.8, the operator $D^{-1}A$ is the canonical operator of the function whose graph is

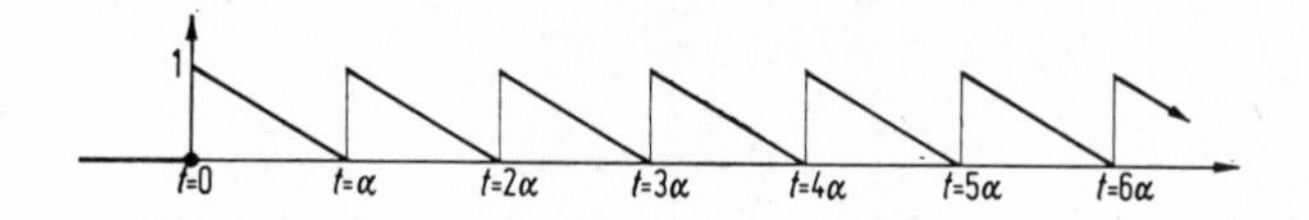

Answer: If $k = 0, 1, 2, \ldots$ and $k\alpha < \tau < k\alpha + \alpha$, then

$$\int_{-\infty}^{\tau}\left[-\frac{1}{\alpha} + \sum_{k=0}^{\infty}\delta(t - k\alpha)\right] = 1 + k - \alpha^{-1}\tau.$$

11.61.7 Prove that the equation

$$\binom{-N}{k} = \frac{(k + N - 1)!}{k!\,(N - 1)!}(-1)^k \tag{2}$$

holds for all integers $N \geq 1$. *Hints*: since

$$(1 + z)^{-1} = \sum_{k=0}^{\infty}(-1)^k z^k = \sum_{k=0}^{\infty}\binom{-1}{k} z^k,$$

it holds for $N = 1$. Let us assume that (2) holds for $N = m - 1$:

$$(1 + z)^{-m+1} = \sum_{n=0}^{\infty}\frac{(n + m - 2)!}{n!\,(m - 2)!}(-1)^n z^n.$$

Differentiating with respect to z:

$$-(m - 1)(1 + z)^{-m} = \sum_{n=1}^{\infty}\frac{(n + m - 2)!}{(n - 1)!\,(m - 2)!}(-1)^n z^{n-1};$$

setting $k = n - 1$, we obtain

$$(1 + z)^{-m} = \sum_{k=0}^{\infty}\frac{(k + m - 1)!}{k!\,(m - 1)!}(-1)^k z^k.$$

Therefore, (2) holds for $N = m$ whenever it holds for $N = m - 1$: the the induction proof is completed.

§ 12. A Functional Calculus for D

First, a few recalls. The family $\mathscr{A}_0$ has been defined in 11.23. Next, observe that Theorem 11.42 is valid not only in case h is an impulse, but also when it is the canonical operator of a regulated $\mathscr{K}_0$-function; this theorem 11.42 will now be applied to the case $h() = \mathsf{T}_0()$ (which implies $h = 1$: see 2.17). If $w \in \mathscr{A}_0$, we can therefore conclude from 11.42 (with $h = 1$) that $w(1/D)$ is the canonical operator of a $\mathscr{K}$-function; in consequence, $w(1/D)$ is a perfect operator:

$$D^n\left[w\left(\frac{1}{D}\right)\right] = \left[w\left(\frac{1}{D}\right)\right] D^n \qquad \text{(by 1.9)}. \tag{1}$$

From [11.25], 11.33, and 11.5 (with $g_k = w_k D^{-k}$) we have

$$w\left(\frac{1}{D}\right) = \sum_{k=0}^{\infty} w_k\, D^{-k} = \left\langle \sum_{k=0}^{\infty} w_k \frac{t^k}{k!} \right\rangle \qquad \text{(see 3.20)}. \tag{12.0}$$

If n is an integer, we may combine (1) and 12.0 to obtain

$$D^n\left[w\left(\frac{1}{D}\right)\right] = \left[w\left(\frac{1}{D}\right)\right] D^n = \sum_{k=0}^{\infty} w_k\, D^{-k+n}. \tag{12.1}$$

12.2 **Definition.** $Y \in \mathscr{A}_\infty$ means that there exists an integer $m(Y)$ and a sequence Y_ν ($\nu = 0, \pm 1, \pm 2, \ldots$) of numbers such that

$$Y_\nu = 0 \text{ for all } \nu > m(Y), \tag{12.3}$$

and

$$Y(p) = \sum_{\nu=-\infty}^{\infty} Y_\nu p^\nu \qquad \text{(every large } p\text{)}. \tag{12.4}$$

12.5 *Remarks.* 12.4 means that there exists a number $r > 0$ such that

$$Y(p) = \sum_{\nu=-\infty}^{\infty} Y_\nu p^\nu \qquad (|p| > r).$$

The integer $m(Y)$ may be negative. The sequence Y_ν ($\nu = 0, \pm 1, \pm 2, \ldots$) is uniquely determined by the function Y; we set

$$Y(D) \stackrel{\text{def}}{=} \sum_{\nu=-\infty}^{\infty} Y_\nu D^\nu. \tag{12.6}$$

As we shall see, the correspondence $Y \mapsto Y(D)$ (that assigns to each function Y in $\mathscr{A}_\infty$ the operator $Y(D)$) is a functional calculus for D such that

$$Y^{-1}(D) = \frac{1}{Y(D)}$$

whenever the function Y is not identically zero.

12.7 **Theorem.** *Suppose that $Y \in \mathscr{A}_\infty$. If Y is not identically zero, there exists an integer n and a function w such that $w \in \mathscr{A}_0$, $w(0) \neq 0$,*

$$(12.8)\quad Y(D) = D^n\left[w\left(\frac{1}{D}\right)\right], \quad \textit{and} \quad Y(p) = p^n\left[w\left(\frac{1}{p}\right)\right] \quad \text{(every large } p\text{)}.$$

Proof. Since Y is not identically zero, there exists an integer i such that $Y_i \neq 0$; note that $i \leq m(Y)$ (by [12.3]): let n be the largest integer such that $Y_n \neq 0$; we have

$$(2)\qquad Y_\nu = 0 \quad \text{for all} \quad \nu > n,$$

and

$$(3)\qquad Y_n \neq 0.$$

From [12.6] and (2) we see that

$$Y(D) = \sum_{\nu=-\infty}^{\infty} Y_\nu D^\nu = Y_n D^n + Y_{n-1} D^{n-1} + \cdots;$$

consequently,

$$(4)\qquad Y(D) = \sum_{k=0}^{\infty} Y_{n-k} D^{n-k}.$$

Similarly, 12.4 becomes

$$(5)\qquad Y(p) = \sum_{k=0}^{\infty} Y_{n-k} p^{n-k} \quad \text{(every large } p\text{)}.$$

If w is the function defined by

$$(6)\qquad w(z) = \sum_{k=0}^{\infty} Y_{n-k} z^k \quad \text{(every small } z\text{)},$$

then $w \in \mathscr{A}_0$, $w(0) = Y_n \neq 0$, and

$$D^n\left[w\left(\frac{1}{D}\right)\right] = \sum_{k=0}^{\infty} Y_{n-k} D^{n-k} \qquad \text{(by 12.1)}. \tag{7}$$

Conclusion 12.8 is immediate from (7), (4) and (5)—(6).

12.9 *Remark.* Let n be a (positive or negative) integer; if $w \in \mathscr{A}_0$, then the equations

$$Y(p) = p^n\left[w\left(\frac{1}{p}\right)\right] = \sum_{k=0}^{\infty} w_k p^{-k+n} \qquad \text{(see 11.25)}$$

obviously define a function Y such that $Y \in \mathscr{A}_\infty$; from [12.6] we see that

$$Y(D) = \sum_{k=0}^{\infty} w_k D^{-k+n} = D^n\left[w\left(\frac{1}{D}\right)\right]: \tag{8}$$

the last equation is from 12.1.

12.10 **Examples.** If Y is the ratio of two polynomials:

$$Y(p) = \frac{a_m p^m + \cdots + a_0}{b_s p^s + \cdots + b_0} \qquad \text{(every large } p\text{)},$$

then $Y(p) = p^{m-s} w(1/p)$, where

$$w(z) = \frac{a_m + \cdots + a_0 z^m}{b_s + \cdots + b_0 z^s} \qquad \text{(every small } z\text{)};$$

therefore, $Y \in \mathscr{A}_\infty$ and 12.9 gives

$$Y(D) = D^{m-s}\left[w\left(\frac{1}{D}\right)\right] = D^{m-s}\left[\frac{a_m + \cdots + a_0 D^{-m}}{b_s + \cdots + b_0 D^{-s}}\right]:$$

the last equation is from (8) and 11.43. Consequently,

$$Y(D) = \frac{a_m D^m + \cdots + a_0}{b_s D^s + \cdots + b_0}.$$

The Bessel function of order m is defined by

$$J_m(\tau) = \sum_{k=0}^{\infty} \frac{(-1)^k \tau^{2k+m}}{2^{2k+m}(m+k)!\,k!}. \tag{12.11}$$

Therefore, if c is a number,

$$\langle J_0(2\sqrt{ct})\rangle = \left\langle \sum_{k=0}^{\infty} \frac{(-1)^k c^k}{k!} \left(\frac{t^k}{k!}\right)\right\rangle = \sum_{k=0}^{\infty} \frac{(-1)^k c^k}{k!} D^{-k}:$$

the last equation is from 12.0. If $w(z) = \exp(z)$, we can use 12.0 once more to conclude that

$$\boxed{\langle J_0(2\sqrt{ct})\rangle = \exp(-c/D)}. \tag{12.12}$$

12.13 **Another example.** If b is a number, the analytic extension Y of the function defined by

$$Y(p) = \frac{p}{\sqrt{b^2 + p^2}} \qquad (|p| > 1) \tag{1}$$

is such that $Y \in \mathscr{A}_\infty$: note that $Y(p) = w(1/p)$, where

$$w(z) = (1 + b^2 z^2)^{-1/2} = \sum_{k=0}^{\infty} \binom{-1/2}{k} b^{2k} z^{2k} \qquad (|z| < 1). \tag{2}$$

From 12.9 (with $n = 0$) and 11.50 we see that

$$Y(D) = w\left(\frac{1}{D}\right) = \sum_{k=0}^{\infty} \binom{-1/2}{k} b^{2k} D^{-2k}. \tag{3}$$

In view of (1) and (3) it is natural to write

$$\frac{D}{\sqrt{b^2 + D^2}} = \sum_{k=0}^{\infty} \binom{-1/2}{k} b^{2k} D^{-2k}. \tag{4}$$

The equations

$$\left[\frac{D}{\sqrt{b^2 + D^2}}\right]^2 = w^2\left(\frac{1}{D}\right) = \frac{1}{1 + b^2 D^{-2}} = \frac{D^2}{b^2 + D^2} \tag{12.14}$$

are immediate from (4)—(3), 11.43, and (2). On the other hand,

$$\binom{\alpha}{k} = \frac{\alpha(\alpha-1)(\alpha-2)\cdots(\alpha-k+1)}{k!},$$

which gives

$$\binom{-1/2}{k} = \frac{(-1)^k}{k!}\left(\frac{1}{2}\right)\left(\frac{3}{2}\right)\left(\frac{5}{2}\right)\cdots\left(\frac{1}{2}+k-1\right)$$

$$= \frac{(-1)^k}{k!}\,\frac{1\cdot 3\cdot 5\cdots(2k-1)E_k}{2^k\,E_k},$$

where E_k is the product of all the even numbers up to $2k$:

$$E_k = 2\cdot 4\cdot 6\cdots(2k) = 2^k(k!).$$

Therefore,

$$\binom{-1/2}{k} = \frac{(-1)^k\,(2k)!}{k!\,(2^k)\,(2^k)\,k!} = \frac{(-1)^k\,(2k)!}{(k!)^2\,2^{2k}}, \tag{12.14.1}$$

so that (4) becomes

$$\frac{D}{\sqrt{b^2+D^2}} = \sum_{k=0}^{\infty}\frac{(-1)^k(2k)!}{(k!)^2\,2^{2k}}\,b^{2k}D^{-2k},$$

so that 12.0 gives

$$\frac{D}{\sqrt{b^2+D^2}} = \left\langle \sum_{k=0}^{\infty}\frac{(-1)^k\,(2k)!}{(k!)^2\,2^{2k}}\,b^{2k}\,\frac{t^{2k}}{(2k)!}\right\rangle$$

$$= \left\langle \sum_{k=0}^{\infty}\frac{(-1)^k}{(k!)^2}\left(\frac{bt}{2}\right)^{2k}\right\rangle.$$

From [12.11] it now follows that

$$\boxed{\frac{D}{\sqrt{b^2+D^2}} = \langle J_0(bt)\rangle}. \tag{12.15}$$

12.16 **Theorem.** *Suppose that $G\in\mathscr{A}_\infty$. If G is not identically zero, there exists a function G^{-1} in $\mathscr{A}_\infty$ such that*

$$G^{-1}(D) = [G(D)]^{-1}; \tag{12.17}$$

further, if $G(D)$ is the canonical operator of a $\mathcal{K}$-function, then $|G(\infty)| < \infty$ and there exists a function v in $\mathcal{A}_0$ such that

$$(12.18) \qquad G(p) = \sum_{k=0}^{\infty} v_k p^{-k} = v\left(\frac{1}{p}\right) \qquad \text{(every large } p\text{)}.$$

Proof. From 12.7 it follows the existence of an integer n and a function u in $\mathcal{A}_0$ such that $u(0) \neq 0$,

$$(1) \quad G(D) = D^n \left[u\left(\frac{1}{D}\right)\right] \quad \text{and} \quad G(p) = p^n \left[u\left(\frac{1}{p}\right)\right] \quad \text{(every large } p\text{)}.$$

Since $u \in \mathcal{A}_0$ and $u(0) \neq 0$, the function u^{-1} (see 11.45) belongs to $\mathcal{A}_0$ and

$$(2) \qquad D^{-n} \left[u^{-1}\left(\frac{1}{D}\right)\right] D^n \left[u\left(\frac{1}{D}\right)\right] = 1 \qquad \text{(by 11.46)}.$$

The function G^{-1} is defined by the equation

$$G^{-1}(p) = p^{-n} \left[u^{-1}\left(\frac{1}{p}\right)\right] \qquad \text{(every large } p\text{)};$$

consequently, 12.9 implies that $G^{-1} \in \mathcal{A}_\infty$ and

$$(3) \qquad G^{-1}(D) = D^{-n} \left[u^{-1}\left(\frac{1}{D}\right)\right].$$

The conclusion

$$(4) \qquad G^{-1}(D)\, G(D) = D^{-n} \left[u^{-1}\left(\frac{1}{D}\right)\right] D^n \left[u\left(\frac{1}{D}\right)\right] = 1$$

is now immediate from (3), (1), and (2). Having thus established 12.17, we consider the case where the equation $G(D) = g$ holds for some $\mathcal{K}_0$-function $g(\,)$; therefore,

$$1 = G^{-1}(D)\, g = D^{-n} \left[u^{-1}\left(\frac{1}{D}\right)\right] g \qquad \text{(by (4) and (3))},$$

so that $D^n = u^{-1}(1/D)\, g$; since $u^{-1}(1/D)$ is a function-operator (by 11.42), it follows from 6.36.1 that $n \leq 0$:

$$(5) \qquad n = -m \text{ for some } m \geq 0.$$

From (5) and (1) we obtain

$$(6) \qquad G(p) = p^{-m} \left[u\left(\frac{1}{p}\right)\right] \qquad \text{(every large } p\text{)}.$$

If v is the function defined by

(7) $$v(z) = z^m u(z) \quad \text{(every small } z\text{)},$$

then $v \in \mathcal{A}_0$ (since $m \geq 0$); moreover, the equations

(8) $$G(p) = v\left(\frac{1}{p}\right) = \sum_{k=0}^{\infty} v_k p^{-k}$$

are from (6)—(7) and 11.24. Consequently, $|G(\infty)| = |v_0| < \infty$: Conclusion 12.18 is merely a copy of (8).

12.19 **Theorem.** *Suppose that* $g(\,)$ *is a* $\mathcal{K}$*-function such that* $g = G(D)$ *for some* G *in* $\mathcal{A}_\infty$; *then*

(12.20) $$g(0+) = G(\infty) \qquad \text{(if } g\,() \text{ is regulated)},$$

and

(12.21) $$\boxed{\langle t g(t)\rangle = -D\left[\frac{\mathrm{d}}{\mathrm{d}p}\frac{G(p)}{p}\right]_{p=D}}.$$

Further, if $\{t^{-1} g(t)\}(\,)$ *is a regulated function, then*

(12.22) $$\boxed{\langle t^{-1} g(t)\rangle = D\left[\int_x^\infty \frac{G(p)}{p}\,\mathrm{d}p\right]_{x=D}}.$$

Note: we shall prove that the equations

$$F(p) = \frac{\mathrm{d}}{\mathrm{d}p}\frac{G(p)}{p} \quad \text{and} \quad Y(x) = \int_x^\infty \frac{G(p)}{p}\,\mathrm{d}p$$

(every large p and $x > 0$) define two functions F and Y in $\mathcal{A}_\infty$ such that

$$\langle t g(t)\rangle = -D[F(D)] \quad \text{and} \quad \langle t^{-1} g(t)\rangle = D[Y(D)];$$

Equations 12.21—22 *are to be interpreted as concise formulations of the above property.*

Proof. If G is identically zero, then $G(D) = 0 = g$: our conclusions are obviously true in this case. The rest of this proof concerns the case where G is not identically zero. From 12.18 and 12.9 we see that

(9) $$g = G(D) = v\left(\frac{1}{D}\right) = \sum_{k=0}^{\infty} v_k D^{-k}.$$

From (9) and 12.0 we see that

$$g = \left\langle \sum_{k=0}^{\infty} v_k \frac{t^k}{k!} \right\rangle;$$

from 2.3 it therefore results that the equation

(12.22.1) $$g(\tau) = \sum_{k=0}^{\infty} v_k \frac{\tau^k}{k!}$$

holds for all points $\tau > 0$ at which $g()$ is continuous. If m is any integer ≥ -1, it follows easily from 12.22.1 and 2.4 that

(10) $$\langle t^m g(t) \rangle = \left\langle \sum_{k=0}^{\infty} v_k \frac{t^{k+m}}{k!} \right\rangle.$$

If $g()$ is regulated, then $g()$ is continuous in some interval of the form $(0, \lambda)$ with $\lambda > 0$: letting $\tau \to 0+$ in 12.22.1, we obtain

(11) $$g(0+) = v_0 = G(\infty):$$

the second equation is from 12.18. It only remains to prove 12.21—22. Let w be the function defined by

$$w(z) = \sum_{k=0}^{\infty} v_k (k+m)! \frac{z^{k+m}}{k!} \qquad \text{(every small } z\text{)};$$

consequently, the equations

(12) $$w\left(\frac{1}{D}\right) = \left\langle \sum_{k=0}^{\infty} v_k \frac{t^{k+m}}{k!} \right\rangle = \langle t^m g(t) \rangle$$

are from 12.0 and (10) (in case $m = -1$ we need to suppose that $v_0 = 0$). From (12) and 12.0 it follows immediately that

(13) $$\langle t^m g(t) \rangle = \sum_{k=0}^{\infty} v_k \frac{(k+m)!}{k!} D^{-k-m}.$$

On the other hand, from 12.18 we see that

$$\frac{G(p)}{p} = \sum_{k=0}^{\infty} v_k p^{-k-1}; \tag{14}$$

since this series converges uniformly, we may differentiate it termwise:

$$-D\left[\frac{\mathrm{d}}{\mathrm{d}p}\frac{G(p)}{p}\right]_{p=D} = -D\sum_{k=0}^{\infty}(-k-1)\,v_k D^{-k-2}; \tag{15}$$

Conclusion 12.21 is immediate from (15) and (13) (with $m = 1$). It only remains to prove 12.22 under the assumption that $\{t^{-1}g(t)\}(\,)$ is a regulated function $f(\,)$:

$$t^{-1}g(t) = f(t) \qquad (\text{all } t > 0).$$

Accordingly, $g(t) = tf(t)$, whence $g(0+) = 0$, which gives $v_0 = 0$ (by (11)) and therefore

$$\frac{G(p)}{p} = \sum_{k=1}^{\infty} v_k p^{-k-1} \qquad \text{(by (14))}. \tag{16}$$

Once more, the uniform convergence property enables us to integrate termwise the series (16):

$$\int_x^{\infty} \frac{G(p)}{p}\,\mathrm{d}p = \sum_{k=1}^{\infty} v_k \int_x^{\infty} p^{-k-1}\mathrm{d}p = \sum_{k=1}^{\infty} k^{-1} v_k x^{-k},$$

so that

$$D\left[\int_x^{\infty}\frac{G(p)}{p}\,\mathrm{d}p\right]_{x=D} = \sum_{k=1}^{\infty}\frac{(k-1)!}{k!}\,v_k D^{-k+1}. \tag{17}$$

Conclusion 12.22 is now obtained by comparing (17) with the result of substituting $m = -1$ in (13).

Applications

The following properties are easy consequences of Definition [12.11]:

$$J_0(0-) = 1 \quad \text{and} \quad J_0'(\,) = -J_1(\,). \tag{1}$$

If $y(t) = J_0(bt)$ then $y'(t) = bJ_0'(bt) = -bJ_1(bt)$ (by (1)), whence

$$\langle -bJ_1(bt)\rangle = y' = Dy - y(0-)\,D \qquad \text{(by 3.8.1)}$$

$$= D\langle J_0(bt)\rangle - J_0(0-)\,D$$

$$= D\langle J_0(bt)\rangle - D \qquad \text{(by (1))};$$

we may therefore use 12.15 to conclude that

$$\text{(12.23)} \qquad \boxed{\langle J_1(bt)\rangle = b^{-1}\left[D - \frac{D^2}{\sqrt{b^2 + D^2}}\right]}.$$

From [12.11] it also follows that $\{t^{-1}J_1(bt)\}(\,)$ is a regulated $\mathcal{K}_0$-function; accordingly, we may use 12.22:

$$\langle t^{-1}J_1(bt)\rangle = D\left[\int_x^\infty \frac{b^{-1}}{p}\left(p - \frac{p^2}{\sqrt{b^2+p^2}}\right)\mathrm{d}p\right]_{x=D}$$

$$= b^{-1}D\left[\lambda - \left(x - \sqrt{b^2 + x^2}\right)\right]_{x=D},$$

where

$$\lambda = \lim_{p\to\infty}\left(p - \sqrt{b^2+p^2}\right) = \lim_{p\to\infty}\frac{p^2 - (b^2+p^2)}{p + \sqrt{b^2+p^2}} = 0.$$

Consequently,

$$\text{(12.24)} \qquad \boxed{\langle t^{-1}J_1(bt)\rangle = b^{-1}\left[-D^2 + D\sqrt{b^2 + D^2}\right]},$$

which can also be written

$$\text{(12.25)} \qquad \sqrt{b^2 + D^2} = D + D^{-1}\left\langle \frac{b}{t}J_1(bt)\right\rangle.$$

12.26 **An integral equation.** Let $y(\,)$ and $g(\,)$ be $\mathcal{K}$-functions such that

$$\int_0^t J_0(bt - bu)\,y(u)\,\mathrm{d}u = g;$$

consequently,

$$g = \langle J_0(bt)\rangle\,D^{-1}y = \frac{y}{\sqrt{b^2 + D^2}} \qquad \text{(by 3.15 and 12.15)};$$

solving for y:

$$(2) \qquad y = g\sqrt{b^2 + D^2} = Dg + gD^{-1}\left\langle\frac{bJ_1(bt)}{t}\right\rangle;$$

the second equation is from 12.25. If $g()$ has no jumps on $[0, \infty)$ and if $g'()$ is a $\mathcal{K}$-function, we may use 3.8.4 and 3.15 to conclude from (2) that

$$y = g' + g(0+)D + \int_0^t \frac{bJ_1(bt - bu)}{t - u} g(u)\,\mathrm{d}u:$$

note that y is a function-operator if (and only if) $g(0+) = 0$.

12.27 **Two formulas.** Since $\langle \sin bt\rangle = G(D)$ with $G(p) = bp/(b^2 + p^2)$ (see 3.27), Formula 12.22 gives

$$\left\langle\frac{\sin bt}{t}\right\rangle = D\left[\int_x^\infty \frac{b\,\mathrm{d}p}{b^2 + p^2}\right]_{x=D} = D\left[-0 + \tan^{-1}\frac{b}{x}\right]_{x=D}$$

$$= D\tan^{-1}\frac{b}{D}.$$

Similarly, since $\langle e^{at} - e^{bt}\rangle = G(D)$ with $G(p) = p/(p - a) - p/(p - b)$:

$$\left\langle\frac{e^{at} - e^{bt}}{t}\right\rangle = D\left[\int_x^\infty \left(\frac{1}{p - a} - \frac{1}{p - b}\right)\mathrm{d}p\right]_{x=D} = D\log\frac{D - a}{D - b}.$$

12.28 As an application of 12.21, observe that

$$\langle t\sin bt\rangle = -D\left[\frac{\mathrm{d}}{\mathrm{d}p}\frac{b}{b^2 + p^2}\right]_{p=D} = \frac{2bD^2}{(b^2 + D^2)^2},$$

which yields Formula 3.26 — which has already been proved by other means.

12.29 **An integral formula.** From 12.15 and 12.21 we see that

$$(12.30) \qquad \langle tJ_0(bt)\rangle = -D\left[\frac{\mathrm{d}}{\mathrm{d}p}(b^2 + p^2)^{-1/2}\right]_{p=D} = \frac{D^2}{(b^2 + D^2)^{3/2}};$$

consequently,

$$\langle t J_0(bt)\rangle = \frac{D^2}{(b^2 + D^2)} D^{-1} \frac{D}{\sqrt{b^2 + D^2}} = \langle \cos bt\rangle D^{-1} \langle J_0(bt)\rangle :$$

the last equation is from 12.15. In view of 3.15, we have obtained the integral equation

$$\boxed{t J_0(bt) = \int_0^t \cos(bt - bu)\, J_0(bu)\, du}\,.$$

12.31 We shall now follow HEAVISIDE's procedure for finding the function whose operator is $D/(D - a)^{m+1}$:

$$\frac{D}{(D-a)^{m+1}} = \frac{D^{-m}}{(1 - aD^{-1})^{m+1}} \qquad \text{(by 1.73)}$$

$$= D^{-m} \sum_{k=0}^{\infty} \frac{(k+m)!}{k!\, m!}\, a^k \left(\frac{1}{D}\right)^k \qquad \text{(by 11.53)}$$

$$= \sum_{k=0}^{\infty} \frac{(k+m)!}{k!\, m!}\, a^k D^{-m-k}$$

$$= \sum_{k=0}^{\infty} \frac{(k+m)!}{k!\, m!}\, a^k \frac{t^{m+k}}{(m+k)!} \qquad \text{(by 12.0)}$$

$$= \frac{t^m}{m!} \sum_{k=0}^{\infty} \frac{1}{k!}\, a^k \frac{t^k}{1} = \frac{t^m}{m!} \exp(at),$$

which agrees with 3.22.

Exercises

12.32.0 Find the $\mathcal{K}_0$-function $y(\,)$ such that

$$y = \int_0^t J_0(t-u)\, J_0(u)\, du\,.$$

Answer: $\{\sin t\}(\,)$.

12.32.1 Solve the equation

$$y - \int_0^t J_1(t-u)\, y(u)\, du = \cos t.$$

Answer: $\langle J_0(t)\rangle$.

12.32.2 Solve the equation

$$\int_0^t \frac{J_1(t-u)}{t-u}\, y(u)\, du = J_1(t).$$

Answer: $\langle J_0(t)\rangle$.

12.32.3 Prove that

(1) $$\left\langle \int_0^t \frac{\sin u}{u}\, du \right\rangle = \tan^{-1}\left(\frac{1}{D}\right),$$

(2) $$\left\langle \frac{1-\cos bt}{t/2} \right\rangle = D \log\left(1 + \frac{b^2}{D^2}\right),$$

(3) $$\langle 4t^{-1} \sin^2 t\rangle = D \log(1 + 4D^{-2}).$$

Hints: Use 3.14 and 12.27 to obtain (1), use 12.22 to obtain (2), and use (2) to obtain (3).

12.32.4 Given an integer $m \geq 1$, use 11.53 to prove the equation

$$\left\langle \sum_{n=1}^{\infty} \frac{t^{mn}}{(mn)!} \right\rangle = (D^m - 1)^{-1}.$$

12.32.5 Solve the equation

$$\int_0^t y(t-u)\, J_0(u)\, du = t J_0(t).$$

Answer: $\langle \cos t\rangle$.

12.32.6 Suppose that $g = G(D)$ for $G \in \mathscr{A}_\infty$ and for some $\mathcal{K}$-function $g(\,)$. Prove "Heaviside's Second Shifting Rule":

$$\boxed{\langle e^{at} g(t)\rangle = \langle e^{at}\rangle G(D-a)}.$$

12.32.7 With the same hypotheses as in 12.32.6, prove that

$$\boxed{\langle g(at)\rangle = G(a^{-1}D)} \qquad (\text{if } a > 0).$$

§ 13. Non-linear Equations

This section begins by establishing some further properties of our functional calculus for D (that is, the correspondence $Y \mapsto Y(D)$ that assigns to each function Y in $\mathscr{A}_\infty$ the perfect operator $Y(D)$); next, we shall solve equations of the form

$$cy + a\int_0^t y(t-u)\, y(u)\, \mathrm{d}u = f.$$

13.0 **Theorem.** *If $F \in \mathscr{A}_\infty$ and $G \in \mathscr{A}_\infty$, then the four functions F', $F+G$, $F-G$, and FG are also in $\mathscr{A}_\infty$; further,*

$$(13.1) \qquad (F \pm G) = F(D) \pm G(D) \quad \textit{and} \quad FG(D) = F(D)\, G(D).$$

Proof. By hypothesis, there exists a number $r > 0$ such that

$$F(p) = \sum_{\nu=-\infty}^{\infty} F_\nu p^\nu \qquad (|p| > r);$$

therefore, we may differentiate termwise:

$$F'(p) \overset{\text{def}}{=} \frac{\mathrm{d}}{\mathrm{d}p} F(p) = \sum_{\nu=-\infty}^{\infty} \nu F_\nu p^{\nu-1} \qquad (|p| > r),$$

which implies that $F' \in \mathscr{A}_\infty$. Next, note that

$$(1) \qquad (F \pm G)(p) \overset{\text{def}}{=} F(p) \pm G(p) = \sum_{\nu=-\infty}^{\infty} (F_\nu \pm G_\nu)\, p^\nu;$$

the second equality is from [12.4]. Consequently,

$$(2) \qquad (F \pm G)(p) = \sum_{\nu=-\infty}^{\infty} Y_\nu p^\nu \quad (\text{every large } p),$$

where $Y_\nu = F_\nu \pm G_\nu$; moreover, it follows from [12.3] that

$$(3) \qquad Y_\nu = 0 \quad \text{for all} \quad \nu > m(F) + m(G).$$

From (2)—(3) we see immediately that the functions $F \pm G$ are in $\mathscr{A}_\infty$. In view of (1), Definition [12.6] gives

$$(F \pm G)(D) = \sum_{\nu=-\infty}^{\infty} (F_\nu \pm G_\nu) D^\nu$$

$$= \sum_{\nu=-\infty}^{\infty} F_\nu D^\nu \pm \sum_{\nu=-\infty}^{\infty} G_\nu D^\nu = F(D) \pm G(D);$$

the last equation is from [12.6]: note that all three series converge (by 12.7). We can use 12.8 to write

$$(4) \qquad FG(p) \overset{\text{def}}{=} F(p)\, G(p) = p^m u\Big(\frac{1}{p}\Big)\, p^n v\Big(\frac{1}{p}\Big) = p^{m+n}\Big[uv\Big(\frac{1}{p}\Big)\Big]$$

and

$$(5) \qquad F(D)\, G(D) = D^m u\Big(\frac{1}{D}\Big)\, D^n v\Big(\frac{1}{D}\Big) = D^{m+n}\Big[uv\Big(\frac{1}{D}\Big)\Big]:$$

the last equation is from 11.43. On the other hand, (4) and 12.9 give

$$(6) \qquad FG(D) = D^{m+n}\Big[uv\Big(\frac{1}{D}\Big)\Big].$$

The last equation in 13.1 is now immediate from (5)—(6).

13.2 **Theorem.** *If* $F \in \mathscr{A}_\infty$ *and* $G \in \mathscr{A}_\infty$, *then*

$$F \neq G \quad \textit{implies} \quad F(D) \neq G(D);$$

moreover,

$$(13.3) \qquad \{F(D) = G(D)\} \Leftrightarrow \{F(p) = G(p) \quad (\text{every large } p > 0)\}.$$

13.3.1 **Terminology.** The expression

$$F(p) = G(p) \qquad\qquad (\text{every large } p > 0)$$

means that $F(p) = G(p)$ *for all sufficiently large positive values of* p. Conclusion 13.3 can be re-phrased as follows: if there exists a number $r > 0$ such that

$$F(p) = G(p) \qquad \text{for every } p > r,$$

then $F(D) = G(D)$; conversely, the equation $F(D) = G(D)$ implies the existence of a number $r > 0$ such that the equation $F(p) = G(p)$ holds for all $p > r$.

Proof. The function $F - G$ belongs to the family $\mathscr{A}_\infty$ (because of 13.0), and our hypothesis $F \neq G$ implies that $F - G$ is not identically zero; consequently,

$$\text{the operator } (F - G)(D) \text{ is invertible} \qquad \text{(by 12.17)},$$

which implies that

$$(F - G)(D) \neq 0 \qquad \text{(by 1.75)};$$

this in turn gives

$$F(D) - G(D) \neq 0 \qquad \text{(by 13.1)},$$

whence our conclusion $F(D) \neq G(D)$. If $F(p) = G(p)$ for every large $p > 0$, then $F_\nu = G_\nu$ (see Footnote*), which implies that $F(D) = G(D)$ (by Definition [12.6]): this proves one of the implications of our conclusion 13.3. Conversely, if $F(D) = G(D)$ then $F = G$ (since the alternative $F \neq G$ would — as we have just seen — imply the contradiction $F(D) \neq G(D)$).

13.4 **Definition.** Let $\mathscr{L}(D)$ be the family of all the operators of the form $Y(D)$, where $Y \in \mathscr{A}_\infty$.

Z 13.5 *Remarks.* Thus, $y \in \mathscr{L}(D)$ if (and only if) the equation $y = Y(D)$ holds for some Y such that $Y \in \mathscr{A}_\infty$. The family $\mathscr{L}(D)$ is precisely the image of the family $\mathscr{A}_\infty$ under the mapping $Y \mapsto Y(D)$. As we shall see, $\mathscr{L}(D)$ is what the algebraists call a *field*.

13.6 **Theorem.** *If $y \in \mathscr{L}(D)$ and $y \neq 0$, then $y^{-1} \in \mathscr{L}(D)$.*

Proof. Observe that $y = Y(D)$ for some Y in $\mathscr{A}_\infty$: from the hypothesis $y \neq 0$ it follows that $Y(D) \neq 0$, whence $Y \neq 0$ (by 13.3); therefore 12.17 gives

$$Y^{-1}(D) = [Y(D)]^{-1} = y^{-1},$$

which concludes the proof (since $Y^{-1} \in \mathscr{A}_\infty$ by 12.16).

---

* In other words, there exists a number $r > 0$ such that $F(p) = G(p)$ for all $p > r$. Since both F and G are analytic; for all $|p|$ sufficiently large, we see that $F(p) = G(p)$ for all p in an interval (R, ∞) on which both functions are analytic; by the principle of analytic continuation, the functions F and G must be identical.

13.7 **Theorem.** *Suppose that* f_k *and* g_k $(k=0,1)$ *belong to* $\mathscr{L}(D)$: *then* $g_0 g_1$ *and* $g_0 \pm g_1$ *also belong to* $\mathscr{L}(D)$,

$$\text{(13.8)} \qquad \textit{if} \;\; f_0 f_1 = 0, \;\; \textit{then either} \;\; f_1 = 0 \;\textit{or}\; f_0 = 0,$$

and

$$\text{(13.9)} \qquad \boxed{\textit{if} \;\; g_0^2 = g_1^2, \quad \textit{then either} \quad g_0 = g_1 \;\textit{or}\; g_0 = -g_1}\,.$$

Proof. Property 13.8 is proved by contradiction: if $f_k \neq 0$ for both $k=0$ and $k=1$, then $f_k^{-1} \in \mathscr{L}(D)$ (by 13.6), so that the contradiction

$$0 = (f_1^{-1} f_0^{-1})\, 0 = (f_1^{-1} f_0^{-1})\, (f_0 f_1) = 1$$

now comes from our hypothesis $0 = f_0 f_1$. If $g_k \in \mathscr{L}(D)$ then $g_k = G_k(D)$ for some G_k in $\mathscr{A}_\infty$: since 13.1 gives

$$g_0 g_1 = G_0 G_1(D) \;\text{ and }\; g_0 \pm g_1 = (G_0 \pm G_1)\,(D),$$

these operators belong to $\mathscr{L}(D)$. If $g_0^2 = g_1^2$ then

$$(g_0 + g_1)\,(g_0 - g_1) = g_0^2 - g_1^2 = 0.$$

and from 13.8 it now follows that either

$$\text{(7)} \qquad g_0 + g_1 = 0 \;\text{ or }\; g_0 - g_1 = 0.$$

Conclusion $g_1 = \pm g_0$ is immediate from (7).

Applications

From 13.6—9 it follows that, algebraically speaking, $\mathscr{L}(D)$ is a *field*: elements of $\mathscr{L}(D)$ are subject to all the usual algebraic laws. This fact will enable us to solve quadratic equations with coefficients in $\mathscr{L}(D)$: see 13.16.

Let a_k $(k = 0, 1, 2, \ldots, m)$ and c_k $(k = 1, 2, 3, \ldots, n)$ be arbitrary finite sequences of numbers; from 12.10 it follows that the operator

$$\frac{a_0 + a_1 D + \cdots + a_{m-1} D^{m-1} + a_m D^m}{(1 + c_1 D + \cdots + c_{n-1} D^{n-1} + c_n D^n)^r}$$

belongs to $\mathscr{L}(D)$ (when $r = 0, \pm 1, \pm 2, \ldots$).

13.10 *Some outstanding elements of the family* $\mathscr{L}(D)$:

(13.11) $$J_0\left(2\sqrt{ct}\right) = \exp(-c/D)$$ (see 12.12),

(13.12) $$J_0(bt) = \frac{D}{\sqrt{b^2 + D^2}}$$ (see 12.15),

(13.12.1) $$t\,J_0(bt) = D^2(b^2 + D^2)^{-3/2}$$ (see 12.29),

(13.12.2) $$-t^{-1}J_1(bt) = b^{-1}\left[D^2 - D\sqrt{b^2 + D^2}\right]:$$

see 12.24.

(13.13) $$\sqrt{b^2 + D^2} = D + bD^{-1}\langle t^{-1}J_1(bt)\rangle$$ (see 12.25),

(13.14) $$\boxed{J_1(bt) = b^{-1}\left[D - \frac{D^2}{\sqrt{b^2 + D^2}}\right]}$$ (see 12.23).

13.15 **Recall:** if $y = Y(D)$ with $Y \in \mathscr{A}_\infty$ and $|Y(\infty)| = \infty$, then y is not the canonical operator of a $\mathscr{K}$-function (see 12.16).

13.16 **First application.** Given two numbers (a, b) and a regulated $\mathscr{K}_0$-function $h(\,)$. Suppose that $h \in \mathscr{L}(D)$; further, let $y(\,)$ be a $\mathscr{K}$-function such that $y \in \mathscr{L}(D)$ and

(1) $$2by + a\int_0^t y(t-u)\,y(u)\,du = -h.$$

Consequently, $2by + ayD^{-1}y = -h$ and

$$y^2 + \frac{2bD}{a}\,y = -\frac{Dh}{a}.$$

Completing the square on the left-hand side:

(2) $$\left(y + \frac{bD}{a}\right)^2 = \frac{b^2D^2}{a^2} - \frac{Dh}{a} = \frac{b^2D^2}{a^2}\left(1 - \frac{ah}{b^2D}\right).$$

Set

$$\sqrt{1 - \frac{ah}{b^2D}} \overset{\text{def}}{=} \left(1 - \frac{ah}{b^2D}\right)^{1/2},$$

the right-hand side being defined by 11.49 (with $w(z) = az/b^2$ and $x = 1/2$); from 11.51 it follows that

$$\left(\frac{bD}{a}\sqrt{1-\frac{ah}{b^2D}}\right)^2 = \frac{b^2D^2}{a^2}\left(1-\frac{ah}{b^2D}\right). \tag{3}$$

In view of (2)—(3), we may use 13.9 to conclude that

$$y + \frac{bD}{a} = \pm\frac{bD}{a}\sqrt{1-\frac{ah}{b^2D}}:$$

that is,

$$y = \frac{bD}{a}\left(-1 \pm \sqrt{1-\frac{ah}{b^2D}}\right). \tag{13.17}$$

In the particular case

$$2y - \int_0^t y(t-u)\,y(u)\,du = \sin t,$$

we have $b = 1$, $a = -1$, and $-h = D/(D^2+1)$. Since

$$1 - \frac{ah}{b^2D} = 1 - \frac{1}{D^2+1} = \frac{D^2}{D^2+1} = (1+D^{-2})^{-1},$$

we have

$$\sqrt{1-\frac{ah}{b^2D}} = (1+D^{-2})^{-1/2} = \frac{D}{\sqrt{1+D^2}}; \tag{4}$$

the last equation is obtained as in 12.13.

Substituting (4) into 13.17, we find that

$$y = D\left[1 \pm D\sqrt{1+D^2}\right].$$

Thus, either

$$y = Y(D) \text{ with } Y(D) = D + \frac{D^2}{\sqrt{1+D^2}} \tag{5}$$

or

$$y = D - \frac{D^2}{\sqrt{1+D^2}} = \langle J_1(t)\rangle \qquad \text{(by 13.14)}. \tag{6}$$

Since $|Y(\infty)| = \infty$, we see from 13.15 that (5) is not the operator of a $\mathcal{K}$-function: the solution is given by (6).

13.18 **Second application.** Suppose that $y \in \mathscr{L}(D)$,

$$y(0+) = -1, \qquad \text{and} \qquad \frac{\mathrm{d}}{\mathrm{d}t} y = \int_0^t y(t-u)\, y(u)\, \mathrm{d}u.$$

Consequently, [8.33] gives $Dy + D = y^2 D^{-1}$, so that $y^2 - D^2 y = D^2$; completing the square:

$$\left(y - \frac{1}{2} D^2\right)^2 = \frac{D^4}{4} + D^2 = \left(\frac{D}{2}\sqrt{4 + D^2}\right)^2:$$

the second equation is from 11.51. From 13.9 we see that we have two possibilities; either

(7) $$y = Y(D) \text{ with } Y(D) = 2^{-1}\left[D^2 + D\sqrt{4 + D^2}\right]$$

or

(8) $$y = 2^{-1}\left[D^2 - D\sqrt{4 + D^2}\right] = \langle -t^{-1} J_1(2t)\rangle \quad \text{(by 13.13)}.$$

Since $|Y(\infty)| = \infty$, we see from 13.15 that (7) is not the operator of a $\mathscr{K}$-function: if there is a $\mathscr{K}$-function satisfying our problem, it is given by (8).

13.19 **Partial-fraction decomposition.** As in 4.23, let $z_0 = 0, z_1, z_2, \ldots, z_n$ be distinct numbers, and let G be a real polynomial such that

(1) $$G(D) = D^m V_1 V_2 \cdots V_n,$$

where m is an integer ≥ 0 and

(2) $$V_k = (D - z_k)^{m(k)} \qquad \text{(with } m(k) \geq 1\text{)}.$$

In view of 13.1, the equation

(3) $$G(p) = p^m (p - z_1)^{m(1)} (p - z_2)^{m(2)} \cdots (p - z_n)^{m(n)}$$

(every large p) defines a function G in $\mathscr{A}_\infty$ such that Equation (1) is satisfied. If H is a polynomial of degree not exceeding the degree of G, then the equation

(4) $$Y(p) = \frac{H(p)}{p\, G(p)} \qquad \text{(every large } p\text{)}$$

defines a function Y in $\mathscr{A}_\infty$: note that the degree of the denominator *exceeds* the degree of the numerator. The standard decomposition theorem asserts the existence of polynomials F_s ($s = 0, 1, 2, \ldots, n$) such that

$$\text{degree}(F_0) \leq m, \tag{5}$$

$$\text{degree}(F_s) \leq m(s) - 1, \tag{13.19.1}$$

and

$$\frac{H(p)}{pG(p)} = \frac{F_0(p)}{pp^m} + \sum_{s=1}^{n} \frac{F_s(p - z_s)}{(p - z_s)^{m(s)}}. \tag{6}$$

From (6) and 13.1 it follows immediately that

$$\frac{H(D)}{DG(D)} = \frac{F_0(D)}{DD^m} + \sum_{s=1}^{n} \frac{F_s(D - z_s)}{V_s} \qquad \text{(see also (2))}. \tag{7}$$

If we can prove that

$$F_0(D) = \{\varphi_0 \mid D^{1+m}\} \tag{8}$$

and

$$F_k(D - z_k) = \{\varphi_k \mid V_k\} \qquad (\text{for } k = 1, 2, 3, \ldots, n), \tag{9}$$

then we can multiply by D both sides of (7) to obtain the decomposition formula 4.25.

13.20 *Proof of* (8). From (6) it follows immediately that

$$F_0(p) = \frac{p^m H(p)}{G(p)} - p^{m+1} \sum_{s=0}^{n} \frac{F_s(p - z_s)}{(p - z_s)^{m(s)}}. \tag{10}$$

From 4.28 we see that

$$\frac{p^m H(p)}{G(p)} = \varphi_0(p) = \sum_{i=0}^{\infty} a_i (p - 0)^i: \tag{11}$$

the last equation comes from the fact that $\varphi_0(\,)$ is analytic at the point $p = 0$ (in view of (3), the factor p^m can be cancelled from the denominator): it can therefore be expanded in powers of p; note that

$$a_i = \frac{1}{i!} \varphi_0^{(i)}(0) \qquad (i = 0, 1, 2, \ldots). \tag{12}$$

Combining (10)—(11):

$$F_0(p) = \sum_{i=0}^{\infty} a_i p^i - p^{m+1} f_0(p), \tag{13}$$

where

$$f_0(p) = \sum_{s=1}^{n} \frac{F_s(p - z_s)}{(p - z_s)^{m(s)}} = \sum_{r=0}^{\infty} c_r p^r: \tag{14}$$

the last equation comes from the fact that $z_s \neq 0$ for $s \neq 0$, which implies that the function $f_0(\,)$ is analytic at the point $p = 0$. Combining (13) and (14):

$$F_0(p) = \sum_{i=0}^{\infty} a_i p^i - \sum_{r=0}^{\infty} c_r p^{r+m+1}$$

$$= \sum_{i=0}^{m} a_i p^i + \sum_{i=m+1}^{\infty} [a_i - c_{i-m-1}]\, p^i.$$

Since the degree of F_0 is $\leq m$ (see (5)), we infer that

$$F_0(p) = \sum_{i=0}^{m} a_i p^i = \sum_{i=0}^{m} \frac{1}{i!} \varphi_0^{(i)}(0)\, p^i = \{\varphi_0 \mid p^{1+m}\}:$$

the last two equations are from (12) and Definition [4.22]: Conclusion (8) is now at hand.

13.21 *The proof of* (9) is entirely similar. In view of (9) and (2), we must prove that

$$F_k(D - z_k) = \{\varphi_k \mid (D - z_k)^{m(k)}\}. \tag{15}$$

From (6) it follows that

$$F_k(p - z_k) = \frac{(p - z_k)^{m(k)} H(p)}{p G(p)} - (p - z_k)^{m(k)} \sum_{s \neq k} \frac{F_s(p - z_s)}{(p - z_s)^{m(s)}}, \tag{16}$$

where $m(s) = m + 1$ in case $s = 0$. Again we can expand in Taylor series (about $p = z_k$) both the first fraction and the function defined by the summation on the right-hand side:

$$F_k(p - z_k) = \sum_{i=0}^{\infty} a_i (p - z_k)^i - (p - z_k)^{m(k)} \sum_{r=0}^{\infty} c_r (p - z_k)^r.$$

Thus, setting $x = p - z_k$:

(17)
$$F_k(z) = \sum_{i=0}^{m(k)-1} a_i x^i - \sum_{m(k)}^{\infty} b^i x_i.$$

As in 13.20, it can easily be verified that

(18)
$$a_i = \frac{1}{i!} \varphi_k^{(i)}(z_k).$$

Since the degree of the polynomial F_k is $\leq m(k) - 1$ (see 13.19.1), we can infer exactly as in 13.20 that

(19)
$$F_k(x) = \sum_{i=0}^{m(k)-1} a_i x^i.$$

Since $x = p - z_k$, Formulas (19)—(18) imply that

$$F_k(p - z_k) = \sum_{i=0}^{m(k)-1} \frac{1}{i!} \varphi_k^{(i)}(z_k)(p - z_k)^i;$$

replacing p by D, Conclusion (15) is immediate from 4.30 (with $m = m(k)$ and $a = z_k$).

Exercises

13.22.0 Solve the equation

$$y + \int_0^t y(t-u)\, y(u)\, du = e^{-t} + te^{-t}.$$

Answer: $\langle \exp(-t) \rangle$.

13.22.1 Find the $\mathcal{K}_0$-function $y(\,)$ such that

$$\int_0^t J_0(t-u)\, y(u)\, du = \sin t.$$

Answer: $\{J_0(t)\}()$.

13.22.2 Find the $\mathcal{K}_0$-function $y(\,)$ such that

$$\int_0^t y(t-u)\,y(u)\,\mathrm{d}u = y - \frac{1}{2}\sin 2t.$$

Answer: $\{J_1(2t)\}(\,)$.

13.22.3 Find both operators y in $\mathcal{L}(D)$ such that

$$y * y = -\,2^{-1}\sin 2t + y \qquad \text{(see [8.8])}.$$

Answers: $y = \langle -J_1(2t)\rangle + \delta(t)$, and $y = \langle J_1(2t)\rangle$.

§ 14. Differential Equations with Polynomial Coefficients

In this section we shall solve problems such as

$$\left[t\,\frac{\mathrm{d}^2}{\mathrm{d}t^2} + (2t+3)\,\frac{\mathrm{d}}{\mathrm{d}t} + t + 3\right] y = f;$$

in case $f = 0$ it turns out that this problem (subject to the starting condition $y(0+) = 1$) has the following family of solutions

$$y = \mathrm{e}^{-t} + c\,\delta(t) + c\,\frac{\mathrm{d}}{\mathrm{d}t}\,\delta(t) \qquad \text{(see 14.26.1)}.$$

Let T be the mapping that assigns to each test-function $\varphi(\,)$ the function $T\varphi(\,)$ defined by

$$T\varphi(\tau) = \tau\varphi(\tau) \qquad (-\infty < \tau < \infty).$$

If V is an operator, we set

$$\frac{\mathrm{d}V}{\mathrm{d}D} \overset{\text{def}}{=\!=} -\,TV + VT. \tag{14.0}$$

Concerning our functional calculus for D (that is, the correspondence $Y \mapsto Y(D)$ that assigns to each function Y in $\mathcal{A}_\infty$ the perfect operator $Y(D)$), this section establishes one final property:

$$\boxed{\frac{\mathrm{d}}{\mathrm{d}D}\,Y(D) = \left[\frac{\mathrm{d}}{\mathrm{d}p}\,Y(p)\right]_{p=D}} \qquad \text{(see 14.8)}.$$

It is easily verified that

$$\frac{\mathrm{d}}{\mathrm{d}D}[fg] = \left[\frac{\mathrm{d}f}{\mathrm{d}D}\right] g + f\left[\frac{\mathrm{d}g}{\mathrm{d}D}\right], \tag{14.1}$$

$$\frac{\mathrm{d}}{\mathrm{d}D}[f+g] = \frac{\mathrm{d}f}{\mathrm{d}D} + \frac{\mathrm{d}g}{\mathrm{d}D}, \tag{14.2}$$

$$\frac{\mathrm{d}c}{\mathrm{d}D} = 0, \qquad \text{and} \qquad \frac{\mathrm{d}}{\mathrm{d}D}[cf] = c\,\frac{\mathrm{d}f}{\mathrm{d}D} \qquad \text{(any number } c\text{)}. \tag{14.3}$$

It is also easy to prove that

$$DT = 1 + TD. \tag{1}$$

The equations

$$\frac{\mathrm{d}D}{\mathrm{d}D} = -TD + DT = 1 \tag{2}$$

are immediate from [14.0] and (1). Combining (1) with 14.1, we see that $\mathrm{d}D^2/\mathrm{d}D = 2D$; proceeding by induction, we obtain

$$\frac{\mathrm{d}}{\mathrm{d}D} D^m = mD^{m-1} \qquad (m = 1, 2, 3, \ldots). \tag{14.4}$$

Let R be a perfect operator; to prove the equation

$$-D\frac{\mathrm{d}}{\mathrm{d}D}\left(\frac{R}{D}\right) = \boldsymbol{t}R, \tag{14.5}$$

we note that

$$-D\frac{\mathrm{d}}{\mathrm{d}D}\left(\frac{R}{D}\right) = -D\left[-T\left(\frac{R}{D}\right) + \frac{R}{D}T\right] \qquad \text{by [14.0]}$$

$$= DT\left(\frac{R}{D}\right) - \left(\frac{R}{D}\right)DT \qquad \text{by 1.10—1.9}$$

$$= (1+TD)\left(\frac{R}{D}\right) - \left(\frac{R}{D}\right)(1+TD) \qquad \text{(by (1))}$$

$$= \frac{R}{D} + TR - \frac{R}{D} - RD^{-1}TD = TR - RD^{-1}TD;$$

Conclusion 14.5 is now immediate from [6.58].

14.6 Suppose that $f(\,)$ is a $\mathcal{K}$-function: we saw in 6.62 that

$$tf = \langle tf(t)\rangle .$$

14.7 **Lemma.** *If* $w \in \mathcal{A}_0$, *then*

$$\frac{\mathrm{d}}{\mathrm{d}D}\left(\frac{w\left(\frac{1}{D}\right)}{D}\right) = \left[\frac{\mathrm{d}}{\mathrm{d}p}\left(\frac{w\left(\frac{1}{p}\right)}{p}\right)\right]_{p=D}. \tag{1}$$

Proof. Setting $h = 1$ in 11.42, we see that the equation

$$w\left(\frac{1}{D}\right) = f \tag{2}$$

holds for some $\mathcal{K}$-function $f(\,)$. Note that

$$\frac{\mathrm{d}}{\mathrm{d}D}\left(\frac{w\left(\frac{1}{D}\right)}{D}\right) = \frac{\mathrm{d}}{\mathrm{d}D}\left(\frac{f}{D}\right) \qquad \text{(by (2))}$$

$$= -D^{-1}[tf] \qquad \text{(by 14.5)}$$

$$= -D^{-1}\langle tf(t)\rangle \qquad \text{(by 14.6)}:$$

Conclusion (1) is now immediate from (2) and 12.21 (with $g = f$ and $G(D) = w(1/D)$).

14.8 **Theorem.** *If* $Y \in \mathcal{A}_\infty$, *then* $\frac{\mathrm{d}}{\mathrm{d}D}\, Y(D) = \left[\frac{\mathrm{d}}{\mathrm{d}p}\, Y(p)\right]_{p=D}$.

Proof. Since $Y_\nu = 0$ for all $\nu > m(Y)$ (see [12.3]), there exists an integer $m \geq 1$ such that $Y_\nu = 0$ for all $\nu \geq m - 1$; from [12.4] we have

$$Y(p) = p^{m-1} \sum_{\nu=-\infty}^{m-1} Y_\nu p^{\nu-m+1} \qquad \text{(every large } p). \tag{3}$$

The equality

$$w(z) = Y_{m-1} + Y_{m-2}z + Y_{m-3}z^2 + \cdots$$

defines a function w such that $w \in \mathscr{A}_0$ and

$$w\left(\frac{1}{p}\right) = \sum_{\nu=-\infty}^{m-1} Y_\nu p^{\nu-m+1}. \tag{4}$$

Combining (3) and (4):

$$Y(p) = p^m \left(\frac{u\left(\frac{1}{p}\right)}{p}\right) \qquad \text{(every large } p\text{)}. \tag{5}$$

From (5) and 12.9 we see that

$$\begin{aligned}\frac{\mathrm{d}}{\mathrm{d}D} Y(D) &= \frac{\mathrm{d}}{\mathrm{d}D}\left[D^m \left(\frac{w\left(\frac{1}{D}\right)}{D}\right)\right] \\ &= m D^{m-1} \left(\frac{w\left(\frac{1}{D}\right)}{D}\right) + D^m \frac{\mathrm{d}}{\mathrm{d}D}\left(\frac{w\left(\frac{1}{D}\right)}{D}\right):\end{aligned}$$

the last equation is from 14.1 and 14.4; we may now use 14.7 to write

$$\frac{\mathrm{d}}{\mathrm{d}D} Y(D) = m D^{m-1} \left(\frac{w\left(\frac{1}{D}\right)}{D}\right) + D^m \left[\frac{\mathrm{d}}{\mathrm{d}p} \frac{w\left(\frac{1}{p}\right)}{p}\right]_{p=D}. \tag{6}$$

From (6) and 13.1 it now follows that

$$\begin{aligned}\frac{\mathrm{d}}{\mathrm{d}D} Y(D) &= \left[m p^{m-1} \left(\frac{w\left(\frac{1}{p}\right)}{p}\right) + p^m \frac{\mathrm{d}}{\mathrm{d}p}\left(\frac{w\left(\frac{1}{p}\right)}{p}\right)\right]_{p=D} \\ &= \left[\frac{\mathrm{d}}{\mathrm{d}p} p^m \left(\frac{w\left(\frac{1}{p}\right)}{p}\right)\right]_{p=D} = \left[\frac{\mathrm{d}}{\mathrm{d}p} Y(p)\right]_{p=D}:\end{aligned}$$

the last equation is from (5). The conclusion

$$\boxed{\frac{\mathrm{d}}{\mathrm{d}D} Y(D) = Y'(D)} \qquad \text{(when } Y \in \mathscr{A}_\infty\text{)} \tag{14.8.1}$$

is now at hand.

14.9 The road ahead. Henceforth, we shall write t V instead of $\boldsymbol{t}\, V$; if V is an operator, we have

$$\mathrm{t}V = -D\,\frac{\mathrm{d}}{\mathrm{d}D}\left(\frac{V}{D}\right) \qquad (14.10) \qquad \text{(see 14.5)}.$$

In particular, it is easily verified that

$$\mathrm{t}D^{m+1} = -mD^m \qquad (14.11.0) \qquad (m = 0, 1, 2, \ldots);$$

moreover, it follows easily from 14.10, 14.1, and 14.8 that

$$\boxed{\mathrm{t}V = \frac{V}{D} - \frac{\mathrm{d}V}{\mathrm{d}D}} \qquad (14.11.1)$$

Recall that

$$\frac{\mathrm{d}^m}{\mathrm{d}\mathrm{t}^m}\,y = D^m y - \sum_{k=0}^{m-1} y^{(k)}(0+)\,D^{m-k} \qquad (14.11.2) \qquad \text{(by [8.31])};$$

in particular,

$$\frac{\mathrm{d}}{\mathrm{d}\mathrm{t}}\,y = Dy - y(0+)D, \qquad (14.11.3)$$

and

$$\frac{\mathrm{d}^2}{\mathrm{d}\mathrm{t}^2}\,y = D^2 y - y(0+)\,D^2 - y'(0+)\,D. \qquad (14.11.4)$$

The formulas

$$\mathrm{t}\,\frac{\mathrm{d}y}{\mathrm{d}\mathrm{t}} = -D\,\frac{\mathrm{d}y}{\mathrm{d}D}, \qquad (14.12)$$

$$\mathrm{t}\,\frac{\mathrm{d}^2y}{\mathrm{d}\mathrm{t}^2} = -D^2\,\frac{\mathrm{d}y}{\mathrm{d}D} - \frac{\mathrm{d}y}{\mathrm{d}\mathrm{t}}, \qquad (14.13)$$

and

$$\mathrm{t}^2\,\frac{\mathrm{d}y}{\mathrm{d}\mathrm{t}} = D\,\frac{\mathrm{d}^2y}{\mathrm{d}D^2} \qquad (14.14)$$

are readily proved. For example, to prove 14.12; note that the equations

$$\mathrm{t}\,\frac{\mathrm{d}y}{\mathrm{d}\mathrm{t}} = -D\,\frac{\mathrm{d}}{\mathrm{d}D}\left[\frac{Dy - y(0+)\,D}{D}\right] = -D\,\frac{\mathrm{d}}{\mathrm{d}D}\,[y - y(0+)]$$

are immediate from 14.10 and 14.11.3: Equation 14.12 is an immediate consequence of 14.2 and 14.3. As a final example, let us prove 14.14:

$$t^2 \frac{dy}{dt} \overset{\text{def}}{=} t\left(t \frac{d}{dt} y\right)$$

$$= t\left(-D \frac{d}{dD} y\right) \qquad \text{by 14.12}$$

$$= -D \frac{d}{dD} \left[\frac{-D \frac{d}{dD} y}{D}\right] \qquad \text{by 14.10}$$

$$= -D \frac{d}{dD} \left[-\frac{d}{dD} y\right] = D \frac{d^2}{dD^2} y.$$

Formulas 14.12—14 could be written symbolically as follows:

$$\boxed{\begin{aligned} t \frac{d}{dt} &= -D \frac{d}{dD}, \\ t \frac{d^2}{dt^2} &= (-D^2) \frac{d}{dD} - \frac{d}{dt}, \\ t^2 \frac{d}{dt} &= D \frac{d^2}{dD^2}. \end{aligned}}$$

Many other such equations can be easily obtained from 14.10 and 14.8. Since $t\,D\mathsf{T}_\lambda = \lambda D\mathsf{T}_\lambda$ (by 6.64), Formula 14.10 gives

$$\lambda D\mathsf{T}_\lambda = t[D\mathsf{T}_\lambda] = -D \frac{d}{dD} \mathsf{T}_\lambda,$$

that is,

$$\frac{d}{dD} \mathsf{T}_\lambda = -\lambda \mathsf{T}_\lambda. \tag{1}$$

Formula (1) combines with the fact that $\mathsf{T}_0 = 1$ to suggest the notation

$$\mathsf{T}_\lambda = e^{-\lambda D}:$$

we shall not use this notation. As we shall now see, 14.10—14 can be used to solve equations with polynomial coefficients.

14.15 First application. Let us find an operator y such that

$$y(0+) = a \quad \text{with} \quad \left[\mathrm{t}\,\frac{\mathrm{d}^2}{\mathrm{dt}^2} + 2\,\frac{\mathrm{d}}{\mathrm{dt}} + \mathrm{t}\right] y = 0. \tag{2}$$

If y is such an operator, 14.13 and 14.11.1 give

$$-D^2\,\frac{\mathrm{d}y}{\mathrm{d}D} - \frac{\mathrm{d}y}{\mathrm{dt}} + 2\,\frac{\mathrm{d}y}{\mathrm{dt}} + \frac{y}{D} - \frac{\mathrm{d}y}{\mathrm{d}D} = 0;$$

since $\mathrm{d}y/\mathrm{dt} = Dy - aD$, we obtain

$$Dy - aD + \frac{y}{D} = (D^2 + 1)\,\frac{\mathrm{d}y}{\mathrm{d}D},$$

whence

$$(D^2 + 1)\,\frac{y}{D} - (D^2 + 1)\,\frac{\mathrm{d}y}{\mathrm{d}D} = aD;$$

that is,

$$\frac{aD}{1 + D^2} = \frac{y}{D} - \frac{\mathrm{d}y}{\mathrm{d}D} = \mathrm{t}y: \tag{3}$$

the last equation is from 14.11.1.

Suppose that $a \neq 0$. If y is the canonical operator of a $\mathcal{K}$-function $y()$, then (3), 3.27, and 14.6 give

$$\langle a \sin t\rangle = \langle t y(t)\rangle;$$

from 2.3 it therefore follows that the equation

$$a \sin \tau = \tau y(\tau)$$

holds at each point $\tau > 0$ where $y()$ is continuous; dividing by τ:

$$y() = \left\{\frac{a}{t}\sin t\right\}(), \tag{4}$$

and

$$y = \left\langle\frac{a}{t}\sin t\right\rangle \qquad (\text{in case } a \neq 0).$$

Suppose that $y \in \mathscr{L}(D)$ and $a = 0$: since the equation $y = Y(D)$ holds for some function Y in $\mathscr{A}_\infty$, Equations (3) and 14.18.1 give

$$p^{-1}Y(p) - Y'(p) = 0 \quad (\text{every large } p),$$

whence

$$\frac{dp}{p} = \frac{dY(p)}{Y(p)}$$

which integrates to give $\log p = \log Y(p) + b$, and therefore yields $Y(p) = cp$. Consequently,

$$y = Y(D) = cD = c\delta(t) \qquad \text{(in case } a = 0\text{)}.$$

Note that c is an arbitrary number. It is easily verified that this impulse $y = cD$ is indeed a solution of the starting-value problem

$$y(0+) = 0 \qquad \text{with} \qquad \left[\mathrm{t}\frac{\mathrm{d}^2}{\mathrm{dt}^2} + 2\frac{\mathrm{d}}{\mathrm{dt}} + \mathrm{t}\right] y = 0:$$

consequently, the one-parameter family $y = cD$ is the only family of solutions of this problem that belongs to $\mathscr{L}(D)$.

14.16 Another application. If $y(\,)$ is a $\mathscr{K}$-function such that

$$y(0+) = a \text{ with } \left[\mathrm{t}\frac{\mathrm{d}^2}{\mathrm{dt}^2} + (\mathrm{t} + 2)\frac{\mathrm{d}}{\mathrm{dt}} + 1\right] y = 0, \tag{5}$$

the equation

$$-D^2\frac{dy}{dD} - D\frac{dy}{dD} + \frac{dy}{dt} + y = 0$$

is from 14.13 and 14.12; since $dy/dt = Dy - aD$ (by [8.33]), it implies

$$-D^2\frac{dy}{dD} - D\frac{dy}{dD} + Dy + y = aD,$$

whence

$$\frac{a}{D+1} = -\frac{dy}{dD} + \frac{y}{D} = \mathrm{t}y:$$

the last equation is from 14.11.1. From 3.24 it now follows that

$$\left(\frac{e^{-t} - 1}{-1}\right) a = \mathrm{t}y = \langle ty(t)\rangle:$$

the last equation is from 14.6. The conclusion

$$y(t) = \frac{a}{t}(1 - e^{-t}) \qquad \text{(all } t > 0\text{)} \tag{6}$$

is now at hand.

14.17 *Remarks.* Thus, if $y(\,)$ is any $\mathcal{K}$-function satisfying (5), then $y(\,)$ is the function determined by (6). Similarly, if $a \neq 0$ and if $y(\,)$ is any $\mathcal{K}$-function satisfying (2), then $y(\,)$ is given by (4); for the case $a = 0$ we found it expedient to suppose that $y \in \mathcal{L}(D)$. The condition $y \in \mathcal{L}(D)$ is slightly more restrictive than the Laplace transformability of $y(\,)$, which is the consistently required restriction for Laplace transformation techniques. In what follows, we shall consistently restrict ourselves to operators in $\mathcal{L}(D)$ (this is to make things easier: our present attitude is in contrast with our previous philosophy of rejecting un-necessary hypotheses). In comparison with other methods (such as the FROBENIUS method of undetermined coefficients), the method that we are using can be much simpler, but it can also be much more laborious. A fairly general problem will be worked out in 14.27.

Table of Formulas

$$\left\langle \frac{e^{at} - e^{bt}}{t} \right\rangle = D \log \frac{D-a}{D-b} \qquad \text{(see 12.27)}, \tag{14.18}$$

$$\left\langle \frac{1 - \cos bt}{t} \right\rangle = \frac{D}{2} \log \left(1 + \frac{b^2}{D^2}\right) \qquad \text{(see 12.32.3)}, \tag{14.19}$$

$$\left\langle \frac{J_1(bt)}{t} \right\rangle = \frac{1}{b} \left(-D^2 + D\sqrt{b^2 + D^2}\right) \qquad \text{(see 12.24)}, \tag{14.20}$$

$$\langle L_m(t) \rangle = \left(1 - \frac{1}{D}\right)^m \qquad \text{(see 14.29)}. \tag{14.21}$$

14.22 **The anti-derivative symbol.** We shall define the symbol $\int P(p)\,dp$ for three choices of $P(\,)$. If $m \geq 0$ and $P(p) = (p-b)^m$ we set

$$\int (p-b)^m \, dp \stackrel{\text{def}}{=} \frac{(p-b)^{m+1}}{m+1}.$$

If $P(p) = 1/(p-a)$ we set

$$\int \frac{dp}{p-a} \stackrel{\text{def}}{=} \log (p-a).$$

If $n \geq 2$ and $P(p) = (p - a)^{-n}$ we set

$$\int \frac{\mathrm{d}p}{(p - a)^n} \stackrel{\text{def}}{=} \frac{1}{(-n + 1)\,(p - a)^{n-1}}.$$

If $R(\,)$ is a linear combination

$$R(\,) = \sum_{k=0}^{n} c_k P_k(\,)$$

of functions $P_k(\,)$ of the above type, we set

$$\int R(p)\,\mathrm{d}p \stackrel{\text{def}}{=} \sum_{k=0}^{n} c_k \int P_k(p)\,\mathrm{d}p.$$

Note that

$$\frac{\mathrm{d}}{\mathrm{d}p} \int R(p)\,\mathrm{d}p = R(p). \tag{14.23}$$

14.24 **Linear differential equations of the first order.** Given two functions $P(\,)$ and $Q(\,)$, let us find an operator y in $\mathscr{L}(D)$ such that

$$\frac{\mathrm{d}y}{\mathrm{d}D} + P(D)\,y = Q(D): \tag{1}$$

we suppose that $P \in \mathscr{A}_\infty$ and $Q \in \mathscr{A}_\infty$. Since the equation $y = Y(D)$ holds for some Y in $\mathscr{A}_\infty$ (by 13.5), Equations (1) and 14.8 imply

$$Y'(D) + P(D)\,Y(D) = Q(D). \tag{2}$$

Since $Y' \in \mathscr{A}_\infty$ (by 13.0), we can use 13.1 to infer that

$$Y'(D) + P(D)\,Y(D) = [Y' + PY](D). \tag{3}$$

In view of 13.3, Equations (2)—(3) imply that

$$Y'(p) + P(p)\,Y(p) = Q(p) \qquad \text{(every large } p\text{)}. \tag{4}$$

To solve (4), set

$$A(p) = \int P(p)\,\mathrm{d}p, \tag{14.25}$$

and note that $A'(p) = P(p)$; further,

$$\frac{\mathrm{d}}{\mathrm{d}p}\,\mathrm{e}^{A(p)} = P(p)\,\mathrm{e}^{A(p)}.$$

Multiplying by $\exp(A(p))$ both sides of (4), we obtain

$$e^{A(p)} \frac{d}{dp} Y(p) + Y(p) \frac{d}{dp} e^{A(p)} = Q(p) e^{A(p)},$$

so that

$$\frac{d}{dp} \{e^{A(p)} Y(p)\} = Q(p) e^{A(p)},$$

which implies the existence of a number c such that

$$Y(p) = e^{-A(p)} \{c + \int e^{A(p)} Q(p) \, dp\}.$$

Since $y = Y(D)$, this gives

$$(14.26) \qquad y = \left[c e^{-A(p)} + e^{-A(p)} \int e^{A(p)} Q(p) \, dp\right]_{p=D}.$$

Illustrative Examples

14.26.1 **First problem.** Let us find a family of operators y such that $y \in \mathscr{L}(D)$,

$$y(0+) = a, \quad \text{and} \quad \left[t \frac{d^2}{dt^2} + (2t + 3) \frac{d}{dt} + t + 3\right] y = 0.$$

If y is such an operator, then 14.13 and 14.11.1 give

$$-D^2 \frac{dy}{dD} - \frac{dy}{dt} + 3 \frac{dy}{dt} + 2t \frac{dy}{dt} + \left(\frac{y}{D} - \frac{dy}{dD}\right) + 3y = 0;$$

re-arranging, we can apply 14.12 to obtain

$$(-D^2 - 2D - 1) \frac{dy}{dD} + 2 \frac{dy}{dt} + yD^{-1} + 3y = 0;$$

therefore, 14.11.3 gives

$$-(D + 1)^2 \frac{dy}{dD} + 2Dy + yD^{-1} + 3y = 2a,$$

so that

$$\frac{dy}{dD} - \frac{2D^2 + 3D + 1}{D(D + 1)^2} y = \frac{-2aD}{(D + 1)^2}.$$

To solve this equation we apply 14.24 with

$$A(p) = -\int \frac{2p^2 + 3p + 1}{p(p+1)^2}\,\mathrm{d}p = -\int \left(\frac{1}{p} + \frac{1}{p+1}\right)\mathrm{d}p,$$

whence

$$A(p) = -\log[p(p+1)] \quad \text{and} \quad \mathrm{e}^{A(p)} = p^{-1}(p+1)^{-1}.$$

Substituting into 14.26:

$$y = \left[c p(p+1) + p(p+1)\int \frac{-2ap}{p(p+1)^3}\,\mathrm{d}p\right]_{p=D},$$

which gives

$$y = cD(D+1) + \frac{aD}{D+1} = cD^2 + c\,\delta(t) + \langle a\mathrm{e}^{-t}\rangle.$$

The only function-operator of this family is $\langle a\exp(-t)\rangle$.

14.27 **Second problem.** Let β and λ be integers; we are given a number $\alpha \neq 0$ and an operator f in $\mathscr{L}(D)$. Let us find an operator y in $\mathscr{L}(D)$ such that

$$\text{(5)} \qquad \mathrm{t}\frac{\mathrm{d}^2y}{\mathrm{dt}^2} + (\alpha\mathrm{t} + \beta)\frac{\mathrm{d}y}{\mathrm{dt}} + \lambda\alpha y = f \quad \text{and} \quad y(0+) = a,$$

where a is also a given number. If y is such an operator, the equation

$$\frac{\mathrm{d}y}{\mathrm{d}D} + \left[\frac{(1-\beta)D - \lambda\alpha}{D(D+\alpha)}\right]y = \frac{(1-\beta)aD - f}{D(D+\alpha)}$$

is readily obtained from 14.11—13. Since y and f are in $\mathscr{L}(D)$, the equations $y = Y(D)$ and $f = F(D)$ hold for some Y and F in $\mathscr{A}_\infty$: we may apply 14.24 with

$$\text{(6)} \qquad Q(D) = \frac{(1-\beta)aD - F(D)}{D(D+\alpha)}$$

and

$$P(D) = \frac{(1-\beta)D - \lambda\alpha}{D(D+\alpha)} = \frac{-\lambda}{D} + \frac{1-\beta+\lambda}{D+\alpha}.$$

Consequently, if p is a large positive number, then

$$A(p) = \int P(p)\,\mathrm{d}p = -\lambda \log p + (\lambda + 1 - \beta) \log (p + \alpha),$$

so that

$$\mathrm{e}^{A(p)} = \frac{(p+\alpha)^{\lambda+1-\beta}}{p^{\lambda}}. \tag{7}$$

Combining (6)—(7) with 14.26:

$$y = \frac{D^{\lambda}}{(D+\alpha)^{\lambda+1-\beta}} \left[c + \int \frac{(p+\alpha)^{\lambda+1-\beta}}{p^{\lambda}} \left\{ \frac{(1-\beta)\,a p - F(p)}{p(p+\alpha)} \right\} \mathrm{d}p \right]_{p=D}.$$

That is:

$$y = \frac{D^{\lambda}}{(D+\alpha)^{\lambda+1-\beta}} \left[c + \int \frac{(p+\alpha)^{\lambda-\beta} \{(1-\beta)\,a p - F(p)\}}{p^{\lambda+1}} \mathrm{d}p \right]_{p=D}: \tag{8}$$

note that c is an arbitrary number.

14.28 **Particular case:**

$$\mathrm{t}\,\frac{\mathrm{d}^2 y}{\mathrm{d}\mathrm{t}^2} + (\alpha\,\mathrm{t} + 5)\,\frac{\mathrm{d}y}{\mathrm{d}\mathrm{t}} + 3\alpha y = 0 \quad \text{and} \quad y(0+) = 0. \tag{9}$$

In view of (9) and (5), we can apply (8) with $\lambda = 3, \beta = 5$, and $f = a = 0$: this gives

$$y = c D^3 (D + \alpha) = c D^4 + c\alpha D^3.$$

14.29 **Laguerre differential equation.** Given an integer m, let y be an operator in $\mathscr{L}(D)$ such that $y(0+)$ exists and

$$\mathrm{t}\,\frac{\mathrm{d}^2 y}{\mathrm{d}\mathrm{t}^2} + (1 - \mathrm{t})\,\frac{\mathrm{d}y}{\mathrm{d}\mathrm{t}} + m y = 0. \tag{10}$$

In view of (10) and (5), we can apply (8) with $\alpha = -1, \beta = 1, \lambda = -m$, and $f = 0$:

$$y = c(D-1)^m D^{-m} = c(1 - D^{-1})^m$$

$$= c \sum_{k=0}^{m} \binom{m}{k} (-1)^k D^{-k} = c \sum_{k=0}^{m} \binom{m}{k} (-1)^k \frac{t^k}{k!}:$$

the last equation is from 12.0. Setting $c = 1$, we obtain the Laguerre polynomial of degree m:

$$(11) \qquad L_m(t) \overset{\text{def}}{=} \sum_{k=0}^{m} \binom{m}{k} \frac{(-t)^k}{k!} = \left(1 - \frac{1}{D}\right)^m = (D-1)^m \frac{t^m}{m!};$$

the last equation is from 3.20. In a few books (e.g., [V 1]), this Laguerre polynomial is obtained by setting $c = m!$.

14.29.1 **A Rodriguez-type formula.** Formula (11) states that $L_m = G(D)$ with $G(p) = (1 - p^{-1})^m$; from Heaviside's Second Shifting Rule (12.32.6) we have

$$\langle e^{at} L_m(t) \rangle = \langle e^{at} \rangle \left(1 - \frac{1}{D-a}\right)^m = \frac{D(D-a-1)^m}{(D-a)^{m+1}}:$$

the second equation is from 3.21. Setting $a = -1$:

$$\langle e^{-t} L^m(t) \rangle = D^m \frac{D}{(D+1)^{m+1}} = D^m \left\langle e^{-t} \frac{t^m}{m!} \right\rangle \qquad \text{(by 3.22)}.$$

Let us call V the above operator; from 6.17 and 6.13 we find that

$$e^{-\tau} L_m(\tau) = \{V\}(\tau) = \frac{d^m}{d\tau^m} \left(e^{-\tau} \frac{\tau^m}{m!}\right) \qquad (\text{all } \tau > 0),$$

whence

$$L_m = e^t \left(\frac{d}{dt}\right)^m \left[e^{-t} \frac{t^m}{m!}\right] \qquad \text{(by 2.4)}.$$

Exercises

14.30 Each of the following problems can be solved by utilizing the formulas contained in this § 14. The answers involve arbitrary numbers c, c_0, c_1.

14.30.0 Solve the following three equations

$$t \frac{d}{dt} y = 0, \qquad t \frac{dy}{dt} + y = 0, \qquad \text{and} \qquad t^2 \frac{d^2y}{dt^2} = 0.$$

Answers: $y = c$, $\quad y = c\delta(t)$, $\quad y = c_0 + c_1 \delta(t)$.

14.30.1 Solve the problem

$$\left[t\,\frac{d^2}{dt^2} + t\,\frac{d}{dt} - 1\right] y = 0 \quad \text{with} \quad y(0+) = 0.$$

Answer: $y = ct$.

14.30.2 Solve the problem

$$t\,\frac{d^2y}{dt^2} + m\,\frac{dy}{dt} = t^2 \quad \text{with} \quad y(0+) = a,$$

where m is an integer ≤ 0 such that $m \neq -2$.

Answer: $y = a + c\,\dfrac{t^{1-m}}{(1-m)!} + \dfrac{t^3}{3m+6}$.

14.30.3 Solve the problem

$$t\,\frac{d^2y}{dt^2} + (t-1)\,\frac{dy}{dt} - y = 0 \quad \text{with} \quad y(0+) = a.$$

Answer: $y = c(t-1) + (c+a)\,e^{-t}$.

14.30.4 Solve the problems

$$t\,\frac{d^2y}{dt^2} + (1-m-t)\,\frac{dy}{dt} + my = t - 1 \quad \text{with} \quad y(0+) = 0$$

and

$$t\,\frac{d^2y}{dt^2} + (t-2)\,\frac{dy}{dt} - 3y = 0 \quad \text{with} \quad y(0+) = 0.$$

Answers: $y = c\,\dfrac{t^m}{m!} + \dfrac{t}{m-1}$ and $y = ct^3$.

14.30.5 Solve the problem

$$t\,\frac{d^2y}{dt^2} - (t+1)\,\frac{dy}{dt} + y = 0 \quad \text{with} \quad y(0+) = a.$$

Answer: $y = -c(t+1) + (c+a)\,e^t$.

14.30.6 Solve the problems

$$t \frac{d^2y}{dt^2} + (1 - 2t) \frac{dy}{dt} - 2y = 0 \quad \text{with} \quad y(0+) = 0$$

and

$$t \frac{d^2y}{dt^2} + (t + 3) \frac{dy}{dt} + y = 0 \quad \text{with} \quad y(0+) = 0.$$

Answers: $y = c \langle \exp (2t) \rangle$ and $y = c D\delta(t) + c\delta(t)$.

14.30.7 Prove Formula 14.10.0.

14.30.8 Solve the problem

$$\left[t \frac{d^2}{dt^2} + (t + 1) \frac{d}{dt} + 1\right] y = t + \frac{1}{2}.$$

Answer: $y = c e^{-t} + \frac{t}{2}$.

Appendix

§ 15. Theorems

15.0 **Theorem.** *The function* $\varphi(\ )$ *defined by*

$$\varphi(t) = t^{1/2}\,\mathsf{T}_0(t)\,\frac{t^{-2}}{\sqrt{\pi}}\exp\left(\frac{-1}{t}\right) \qquad (-\infty < t < \infty) \tag{15.1}$$

is a test-function.

Proof. First, we shall establish the existence of a sequence P_k ($k = 0, 1, 2, \ldots$) of polynomials such that the equation

$$\varphi^{(k)}(t) = t^{1/2}\mathsf{T}_0(t)\,P_k(t)\,t^{-2k-2}\exp\left(\frac{-1}{t}\right) \qquad (-\infty < t < \infty) \tag{1}$$

holds for every integer $k \geq 0$. For $k = 0$ we clearly have $P_0(\ ) = 1/\sqrt{\pi}$. We proceed by induction: assuming that (1) holds for $k = n$, it is easily verified by differentiation that the polynomial defined by the equation

$$P_{n+1}(t) = \left(\frac{1}{2} - 2n - 2\right)P_n(t)\,t + P_n'(t)\,t^2 + P_n(t)$$

is such that (1) is satisfied for $k = n + 1$.

Next, observe that

$$0 = \lim_{t \to 0} \mathsf{T}_0(t)\,t^{-2n-3}\exp\left(\frac{-1}{t}\right) \qquad (\text{any integer } n \geq 1); \tag{2}$$

indeed, the Taylor series expansion of the function exp clearly implies that

$$e^{1/t} \geq 1 + \frac{t^{-2n-4}}{(2n+4)!} \qquad (\text{for all } t > 0),$$

whence

$$t^{-2n-3}e^{-1/t} \leq \frac{t\,(2n+4)!}{1 + t^{2n+4}(2n+4)!} \qquad (\text{all } t > 0),$$

which obviously approaches zero as $t \to 0+$.

Finally, the proof of 15.0 is achieved by showing that the equation

$$\varphi^{(k)}(0) = 0 \tag{3}$$

holds for any integer $k \geq 0$ (we indicated in 0.3 that it would have sufficed to establish the existence of all the derivatives at $t = 0$). Equation (3) clearly holds for $k = 0$ (by Definition 15.1). Again proceeding by induction, we assume that (3) holds for $k = n$:

$$\varphi^{(n)}(0) = 0; \tag{4}$$

the equations

$$\varphi^{(n+1)}(0) = \lim_{t\to 0} \frac{\varphi^{(n)}(t) - \varphi^{(n)}(0)}{t - 0} = \lim_{t\to 0} t^{-1}\varphi^{(n)}(t)$$

$$= \left\{\lim_{t\to 0} t^{1/2} P_n(t)\right\} \left[\lim_{t\to 0} \mathsf{T}_0(t)\, t^{-2n-3} \exp\left(\frac{-1}{t}\right)\right]$$

are from (4) and (1) (with $k = n$); the conclusion $\varphi^{(n+1)}(0) = 0$ is now immediate from (2).

15.2 **Warning:** the remainder of this section requires some acquaintance with Lebesgue integration theory.

15.3 **Theorem.** *Let* $\delta_1(\,)$ *be a non-negative function such that* $\delta_1(t) = 0$ *for* $t \leq 0$ *and such that*

$$1 = \int_{-\infty}^{\infty} \delta_1(x)\, dx. \tag{15.4}$$

For any integer $n \geq 1$, *the equation*

$$\delta_n(t) = n\,\delta_1(nt) \qquad (-\infty < t < \infty) \tag{15.5}$$

defines a sequence $\delta_n(\,)$ $(n = 1, 2, 3, \ldots)$ *of entering functions satisfying the equation*

$$\mathsf{T}_0 * g(\,) = \lim_{n\to\infty} [(\mathsf{T}_0 * \delta_n) * g(\,)] \tag{15.6}$$

for every entering function $g(\,)$.

Proof. Set $h_n = \mathsf{T}_0 * \delta_n$; the equations

$$h_n(t) = \int_{-\infty}^{t} \delta_n(u)\,du = \int_{-\infty}^{t} n\,\delta_1(nu)\,du = \int_{-\infty}^{nt} \delta_1(x)\,dx \tag{5}$$

are from 0.26 and 15.5. If $t > 0$, then (5) implies that

$$\lim_{n\to\infty} h_n(t) = \lim_{n\to\infty} \int_{-\infty}^{nt} \delta_1(x)\,dx = \int_{-\infty}^{\infty} \delta(x)\,dx = 1 \qquad (\text{all } t > 0); \tag{6}$$

the last equation is from 15.4. Let τ be any real number: we shall now verify that

$$\lim_{n\to\infty} \int_{-\infty}^{\tau} h_n(\tau - x)\,g(x)\,dx = \int_{-\infty}^{\tau} \lim_{n\to\infty} h_n(\tau - x)\,g(x)\,dx. \tag{7}$$

To that effect, observe that the inequalities

$$0 \le h_n(t) = \int_{-\infty}^{nt} \delta_1(x)\,dx \le \int_{-\infty}^{\infty} \delta_1(x)\,dx = 1$$

are immediate from (5), 15.4, and the fact that $\delta_1(\,) \ge 0$; therefore

$$|h_n(t)| \le 1 \qquad (\text{all } t > 0),$$

whence

$$|h_n(\tau - x)\,g(x)| \le |g(x)| \qquad (\text{for } x < \tau). \tag{8}$$

Clearly, the entering function $|g()|$ is integrable in the interval $(-\infty, \tau)$, in view of (8) and (6) we may apply the Lebesgue Dominated Convergence Theorem to obtain (7).

Since $h_n = \mathsf{T}_0 * \delta_n$ and $\delta_n(t) = 0$ for $t \le 0$, it follows from 0.22 (with $\alpha = 0$ and $\beta = 0$) that

$$h_n(t) = \mathsf{T}_0 * \delta_n(t) = 0 \qquad (\text{for } t \le 0);$$

we can therefore set $\alpha = 0$ in 0.21 to write

$$h_n * g(\tau) = \int_{-\infty}^{\tau} h_n(\tau - x)\,g(x)\,dx. \tag{9}$$

Equations (9), (7) and (6) now give

$$\lim_{n\to\infty} h_n * g(\tau) = \int_{-\infty}^{\tau} g(x)\,dx = \mathsf{T}_0 * g(\tau): \tag{10}$$

the last equation is from 0.26. Since $h_n = \mathsf{T}_0 * \delta_n$ and since (10) holds for all real τ, we have obtained our conclusion 15.6.

15.7 The Lebesgue Dominated Convergence Theorem applies to the infinite interval $(-\infty, \tau)$: see p. 151 in [D 2]. However, we obviously need only use it for the finite interval (β, τ), where β is one of the numbers such that $g(t) = 0$ for $t \leq \beta$.

15.8 *Remarks.* It follows from 15.0 that the equation

$$\delta_1(t) = \frac{1}{\sqrt{\pi}} \mathsf{T}_0(t)\, t^{-3/2} e^{-1/t} \qquad (-\infty < t < \infty) \tag{15.9}$$

defines a test-function $\delta_1(\;)$. The equations

$$\int_0^\infty \delta_1(t)\,dt = \frac{1}{\sqrt{\pi}} \int_0^\infty e^{-t}[-2d(t^{-1/2})] = \frac{2}{\sqrt{\pi}} \int_0^\infty e^{-x^2}\,dx = 1$$

are obtained by the change of variable $x = t^{-1/2}$; the last equality is proved in elementary calculus.

Consequently, $\delta_1(\;)$ satisfies the hypotheses of 15.3: the equality

$$\mathsf{T}_0 * g(\;) = \lim_{n\to\infty} [(\mathsf{T}_0 * \delta_n) * g(\;)] \tag{15.10}$$

therefore obtains for any entering function $g(\;)$.

Three Basic Theorems

15.11 **Lemma.** *If $f(\;)$ is integrable on an interval (α, ω), then*

$$0 = \lim_{\varepsilon\to 0} \int_\alpha^\omega |f(x+\varepsilon) - f(x)|\,dx: \tag{15.12}$$

see Theorem 248 of [K 1] (or Theorem 4.3c of [W 5]).

15.13 Lemma. *Let $g(\,)$ be a regulated entering function: to every finite interval (α, ω) there corresponds a number $g[\alpha, \omega]$ such that*

$$|g(t)| \leq g[\alpha, \omega] \qquad \text{(whenever } \alpha \leq t \leq \omega). \tag{1}$$

Proof. Since $g(\,)$ has at most a finite number of discontinuities inside the interval (α, ω) (see Definition 0.23), this interval can be subdivided into a finite number of closed intervals J_k ($k = 1, 2, 3, \ldots, n$) such that $g(\,)$ is continuous inside each interval. Let (x_{k-1}, x_k) be the end-points of the interval J_k: these points form a partition

$$\alpha = x_0 < x_1 < \cdots < x_{k-1} < x_k < \cdots < x_{n-1} < x_n = \omega.$$

Let $g_k(\,)$ ($k = 1, 2, 3, \ldots, n$) be the functions defined by

$$g_k(\tau) = \begin{cases} g(x_{k-1}+) & \text{if } \tau = x_{k-1} \\ g(\tau) & \text{if } x_{k-1} < \tau < x_k \\ g(x_k-) & \text{if } \tau = x_k: \end{cases}$$

Definition [2.59] guarantees the existence of the above limits. Clearly, the function $g_k(\,)$ is continuous on the closed interval J_k: it therefore attains a maximum M_k on J_k: consequently, the relation

$$|g(t)| \leq g[\alpha, \omega] \overset{\text{def}}{=} \sum_{k=1}^{n} M_k + \sum_{k=0}^{n} |g(x_k)|$$

holds for any t in the closed interval $[\alpha, \omega]$. This concludes the proof.

15.14 Theorem A. *Suppose that $f(\,)$ and $g(\,)$ are entering functions. If $g(\,)$ is regulated (or continuous), then $f * g(\,)$ is a continuous entering function.*

Proof. Since continuous functions are regulated, it suffices to consider the case where $g(\,)$ is regulated. Let α be one of the negative numbers such that

$$f(t) = g(t) = 0 \qquad \text{(for } t \leq \alpha), \tag{2}$$

and let ω be any positive number; we plan to demonstrate that $f * g(\,)$ is continuous inside the interval

$$T = (\alpha - \omega,\ \alpha + \omega). \tag{3}$$

Until further notice, let τ be a point inside the interval T. Clearly, $\tau - \alpha < \omega$ (since $\tau < \alpha + \omega$); from (2) and 0.20 we see that

$$f * g(\tau) = \int_{\alpha}^{\omega} f(\tau - u)\, g(u)\, du. \tag{4}$$

First, we shall verify that this integral exists; to that effect, let $h_\tau(\,)$ be the function defined by

$$h_\tau(u) = g[\alpha, \omega]\, |f(\tau - u)| \qquad (\text{when } \alpha < u < \omega):$$

the inequalities $\alpha < u < \omega$ should be interpreted in the sense of holding for almost-all values of u. Since $f(\,)$ is locally integrable, the function $h_\tau(\,)$ is integrable on the interval (α, ω), and the inequality

$$|f(\tau - u)\, g(u)| \leq h_\tau(u) \qquad (\text{when } \alpha < u < \omega) \tag{5}$$

is immediate from (1): it implies that the function $F_\tau(\,)$ defined by

$$F_\tau(u) = f(\tau - u)\, g(u) \qquad (\text{when } \alpha < u < \omega) \tag{6}$$

is dominated by the integrable function $h_\tau(\,)$. Observe that $F_\tau(\,)$ is the integrand of our convolution (4); since $F_\tau(\,)$ is continuous almost-everywhere (because $f(\,)$ and $g(\,)$ are continuous almost-everywhere), the existence of the integral (4) is guaranteed by a well-known theorem (Theorem 7 p. 84 in [B 2], or Theorem 201 in [K 1]).

It remains to show that the function $f * g(\,)$ is continuous. Let t and $t + \varepsilon$ be inside the interval T: a double application of (4) gives

$$\begin{aligned} |f * g(t + \varepsilon) - f * g(t)| &\leq \int_{\alpha}^{\omega} |f(t + \varepsilon - u) - f(t - u)|\, |g(u)|\, du \\ &\leq g[\alpha, \omega] \int_{\alpha}^{\omega} |f(t - u + \varepsilon) - f(t - u)|\, du \\ &= g[\alpha, \omega] \int_{t-\omega}^{t-\alpha} |f(x + \varepsilon) - f(x)|\, dx: \end{aligned} \tag{7}$$

the second inequality is from (1); the last equation is obtained by the change of variable $x = t - u$. Since $t < \alpha + \omega$ we have

$$t - \alpha < \omega \quad \text{and} \quad t - \omega < \alpha,$$

whence

$$(8) \qquad \int_{t-\omega}^{t-\alpha} |f(x+\varepsilon) - f(x)|\, dx \leq \int_{\alpha}^{\omega} |f(x+\varepsilon) - f(x)|\, dx.$$

Combining (7) and (8):

$$(9) \qquad |f * g(t+\varepsilon) - f * g(t)| \leq g[\alpha, \omega] \int_{\alpha}^{\omega} |f(x+\varepsilon) - f(x)|\, dx:$$

this holds for any positive number ω and any two points $(t, t+\varepsilon)$ inside the interval T. Letting ε approach zero, it now follows from 15.12 and (9) that

$$\lim_{\varepsilon \to 0} |f * g(t+\varepsilon) - f * g(t)| = 0;$$

consequently,

$$(10) \qquad f * g(t) = \lim_{\varepsilon \to 0} f * g(t+\varepsilon) \qquad (\text{when } \alpha - \omega < t < \alpha + \omega).$$

Since ω is any positive number, Equation (10) shows that $f * g(\,)$ is continuous everywhere.

15.15 **Theorem.** *Let $F_1(\,)$, $F_2(\,)$, and $H(\,)$ be entering functions. If either $H(\,)$ or $F_2(\,)$ is continuous, then*

$$(15.16) \qquad \int_{-\infty}^{\infty} dx \int_{-\infty}^{\infty} v_\tau(x, y)\, dy = \int_{-\infty}^{\infty} dy \int_{-\infty}^{\infty} v_\tau(x, y)\, dx \quad (-\infty < \tau < \infty),$$

where

$$(15.17) \qquad v_\tau(x, y) = F_1(\tau - x)\, F_2(x - y)\, H(y).$$

Proof. If $H(\,)$ is continuous, we set $g(\,) = |H(\,)|$ in 15.14: this shows that the function $|F_2| * |H|(\,)$ is a continuous entering function; the same conclusion is obtained from 15.14 and 0.35 in case $F_2(\,)$ is continuous. Note that $|H|(\,)$ is the function defined by $|H|(\,) = |H(\,)|$. Similarly, we use the notations

$$(1) \qquad |F_1|(\,) = |F_1(\,)| \text{ and } |F_2|(\,) = |F_2(\,)|.$$

Another application of 15.14 (with $g = |F_2| * |H|$) now shows that the function $|F_1| * (|F_2| * |H|)(\,)$ is continuous, whence

$$|F_1| * (|F_2| * |H|)(\tau) < \infty \qquad (-\infty < \tau < \infty);$$

this enables us to obtain the inequality

$$(2) \qquad \int_{-\infty}^{\infty} \mathrm{d}x \int_{-\infty}^{\infty} |F_1|(\tau - x)\ \{|F_2|(x - y)\}\ |H|(y)\ \mathrm{d}y < \infty$$

by means of a double application of Definition [0.16]. From (2), (1), and [15.17] we see that

$$(3) \qquad \int_{-\infty}^{\infty} \mathrm{d}x \int_{-\infty}^{\infty} |v_\tau(x, y)|\ \mathrm{d}y < \infty.$$

Conclusion 15.16 now follows from (3) and Tonelli's theorem (see p. 631 in [H 1], or p. 194 in [D 2]); observe that the function $v_\tau(\,,\,)$ is measurable in the plane (this fact is proved on p. 634 of [D 2] and on p. 31 of [W 5]).

15.18.0 **First Fundamental Theorem of Calculus.** If $f(\,)$ is locally integrable and $-\infty < a < t$, then the equation

$$\frac{\mathrm{d}}{\mathrm{d}t} \int_a^t f(u)\ \mathrm{d}u = f(t)$$

holds whenever t is a point at which the function $f(\,)$ is continuous: see Theorem 252 in [K 1] (or p. 141 in [K 2]).

15.18.1 *Remark.* If $\varphi(\,)$ is a test-function, then $\mathsf{T}_0 * \varphi(\,)$ is also a test-function: this is an immediate consequence of 0.43 (with $g(\,) = \mathsf{T}_0(\,)$). This can also be seen by referring to 0.42 and using the fact that $\mathsf{T}_0 * \varphi(\,)$ vanishes to the left of some point (by 0.22).

15.18.2 **Theorem.** *If $f_1(\,)$ and $f_2(\,)$ are entering functions such that*

$$(1) \qquad f_1^* = f_2^*,$$

then $f_1(\,) = f_2(\,)$.

Proof. Let $\delta_n()$ $(n = 1, 2, 3, \ldots)$ be the sequence of test-functions defined by 15.5 and 15.9. If $g()$ is an entering function, then

$$(2) \qquad \mathsf{T}_0 * g() = \lim_{n\to\infty} [g * (\mathsf{T}_0 * \delta_n)()] \qquad \text{by 15.10 and 0.35}$$

$$(3) \qquad = \lim_{n\to\infty} [g^* \cdot (\mathsf{T}_0 * \delta_n)()] \qquad \text{by [0.39] and 15.18.1.}$$

Consequently,

$$(4) \qquad \mathsf{T}_0 * f_1() = \lim_{n\to\infty} [f_1^* \cdot (\mathsf{T}_0 * \delta_n)()] \qquad \text{by (2)—(3)}$$

$$(5) \qquad = \lim_{n\to\infty} [f_2^* \cdot (\mathsf{T}_0 * \delta_n)()] = \mathsf{T}_0 * f_2():$$

the last two equations are from (1) and (2)—(3) (the second equation is obtained by observing that (1) implies $f_1^* \cdot \varphi() = f_2^* \cdot \varphi()$ for every test-function $\varphi()$, and then we use 15.18.1 with $\varphi() = \mathsf{T}_0 * \delta_n()$). From (4)—(5) and 0.26 (with $c = 1$) we see that

$$(6) \qquad \int_{-\infty}^{\tau} f_1(u)\,\mathrm{d}u = \int_{-\infty}^{\tau} f_2(u)\,\mathrm{d}u \qquad (-\infty < \tau < \infty).$$

Suppose that t is a point where both $f_1()$ and $f_2()$ are continuous: the equations

$$f_1(t) = \frac{\mathrm{d}}{\mathrm{d}t} \int_{-\infty}^{t} f_1(u)\,\mathrm{d}u = \frac{\mathrm{d}}{\mathrm{d}t} \int_{-\infty}^{t} f_2(u)\,\mathrm{d}u = f_2(t)$$

are from 15.18.0, (6), and 15.18.0. This establishes the equality $f_1(t) = f_2(t)$ in case t is a continuity point of both $f_1()$ and $f_2()$: in view of [0.33], this concludes the proof of $f_1() = f_2()$.

15.18.3 **Theorem.** *Suppose that $F()$ is a function such that $F'()$ is a $\mathcal{K}$-function. If $F()$ has no jumps on $[0, \infty)$, then*

$$(1) \qquad DF - F(0-)D = F'.$$

Proof. Let $f()$ be the function defined by

$$(2) \qquad f(\alpha) = \begin{cases} F(\alpha) & (\alpha < 0) \\ \frac{1}{2}[F(\alpha-) + F(\alpha+)] & (\alpha \geq 0). \end{cases}$$

By hypothesis, the function $F()$ has no jumps: this means that $F(\alpha-) = F(\alpha+)$ for all $\alpha \geq 0$; consequently, Definition (2) gives

$$(3) \qquad f(\alpha) = F(\alpha-) = F(\alpha+) \qquad (\text{all } \alpha \geq 0).$$

By hypothesis, $F'(\ )$ is a $\mathcal{K}$-function [1.36]: it is therefore continuous on the interval $(0, \infty)$ outside of a sequence of points which we denote by E; recall that there are only finitely-many points of E in each interval $(0, \lambda)$.

Let (a, b) be an interval containing no point of E: therefore, $F'(\ )$ is continuous in this interval, whence $F'(\ )$ is defined at each point in this interval, which implies that $F(\ )$ is continuous in the interval (a, b):

$$F(\tau-) = F(\tau) = F(\tau+) \qquad (a < \tau < b),$$

so that (3) implies

$$f(\tau) = F(\tau) \qquad (a < \tau < b). \tag{4}$$

If x is a point outside of the sequence E, then x lies in some interval (a, b) containing no point of E (the end-points of this interval can be taken to be the two points of E that are closest to the point x); from (4) it now follows that

$$f'(x) = F'(x) \qquad (\text{if } x \text{ is outside of } E) \tag{5}$$

and

$$f(x) = F(x) \qquad (\text{if } x \text{ is outside of } E). \tag{6}$$

Another way of looking at this: in any finite interval the equalities (5)—(6) fail at most at finitely-many values of x; it therefore follows from 2.2 that

$$f = F \text{ and } f' = F'. \tag{7}$$

We shall conclude by proving that the function $f(\ )$ is continuous in $(0, \infty)$: this will allow us to combine the Derivation Property (1.42) with 1.64 to obtain

$$f' = Df - f(0-)\,D,$$

from which (7) gives

$$F' = DF - f(0-)\,D: \tag{8}$$

Conclusion (1) is now immediate from (8) by setting $\alpha = -0$ in (2).

To show that the function $f(\ )$ is continuous in $(0, \infty)$, suppose that $0 < x < \infty$: if x is outside of E, we already know from (4) that $f(\ )$ is continuous at $\tau = x$. If x is a point of E, there are only finitely-many points of E in the interval $(0, x + 1)$: consequently, there are two intervals (a, x) and (x, b) containing no points of E: from (4) we see that

$$f(\tau) = F(\tau) \quad (\text{for } \tau \neq x \text{ and } a < \tau < b);$$

taking limits:

$$F(x-) = f(x-) \text{ and } F(x+) = f(x+) \tag{9}$$

we have

$$f(x) = F(x-) = f(x-) \qquad \text{by (3) and (9)} \tag{10}$$

and

$$f(x) = F(x+) = f(x+) \qquad \text{by (3) and (9).} \tag{11}$$

Combining (10)—(11), we conclude that $f(\,)$ is continuous in the interval $(0, \infty)$.

A Theorem for § 6

15.19 Theorem. *Let $\delta_1(\,)$ be a non-negative function of magnitude* $= 1$ *such that $\delta_1(t) = 0$ for $t \leq 0$. For $n = 1, 2, 3, \ldots$ the equation*

$$\delta_n(t) = n\,\delta_1(nt) \qquad (-\infty < t < \infty) \tag{15.20}$$

defines a sequence $\delta_n(\,)$ $(n = 1, 2, 3, \ldots)$ of $\mathcal{K}$-functions with magnitude $= 1$ *and such that*

$$D = \lim_{n\to\infty} \delta_n = \lim_{n\to\infty} \langle n\,\delta_1(nt)\rangle. \tag{15.21}$$

Proof. Let $\varphi(\,)$ be any test-function: from 15.6 we have

$$\mathsf{T}_0 * \varphi(\,) = \lim_{n\to\infty} [(\mathsf{T}_0 * \delta_n) * \varphi(\,)], \tag{1}$$

whence

$$\mathsf{T}_0^* \cdot \varphi(\,) = \lim_{n\to\infty} [(\mathsf{T}_0 * \delta_n)^* \cdot \varphi(\,)] \qquad \text{by [0.39]} \tag{2}$$

$$= \left[\lim_{n\to\infty} (\mathsf{T}_0 * \delta_n)^*\right] \cdot \varphi(\,) \qquad \text{by [6.6].} \tag{3}$$

Consequently,

$$\mathsf{T}_0^* = \lim_{n\to\infty} (\mathsf{T}_0 * \delta_n)^*. \tag{4}$$

Right-multiplying by D^2 both sides of (4), we may use 6.9 to obtain

$$DD\mathsf{T}_0^* = \lim_{n\to\infty} D[D(\mathsf{T}_0 * \delta_n)^*],$$

whence

(5) $$D\,[\![\mathsf{T}_0]\!] = \lim_{n\to\infty} D\,[\![\mathsf{T}_0 * \delta_n]\!] \qquad \text{by 1.16.}$$

From (5), 1.21, and 1.29 it follows that

(6) $$D = \lim_{n\to\infty} [\![\delta_n]\!].$$

Finally, note that $\delta_n(\,) = \{n\,\delta_1(nt)\}(\,)$ (see [1.38]); consequently, [2.1] gives

(7) $$\delta_n = [\![\delta_n]\!] = \langle n\,\delta_1(nt)\rangle:$$

see also 2.8.3; the second equation is from 15.20 and 2.1. Conclusion 15.21 is immediate from (6)—(7).

A Theorem for § 9

15.22 **Lemma.** *Consider a rule assigning to each positive number λ a pair $(h_\lambda(\,), H_\lambda(\,))$ of $\mathcal{K}$-functions. If*

$$\left|\frac{\partial}{\partial x} h_x(u)\right| \le H_\lambda(u) \qquad \text{when } \begin{cases} \lambda/2 < x < 3\lambda/2 \\ 0 < u \le m, \end{cases}$$

then

(15.23) $$\frac{\partial}{\partial x}\int_0^m h_x(u)du = \int_0^m \frac{\partial}{\partial x} h_x(u)\,du \qquad (\text{all } x > 0).$$

Proof: see Theorem 250 in [K 1] (or [H 2], bottom of p. 355).

15.24 **Theorem.** *Consider a rule assigning to each positive number λ a pair $(y_\lambda(\,), Y_\lambda(\,))$ of $\mathcal{K}$-functions. If*

(15.25) $$\left|\frac{\partial}{\partial x} y_x(u)\right| \le Y_\lambda(u) \qquad \text{when } \begin{cases} \lambda/2 < x < 3\lambda/2 \\ 0 < u < \infty, \end{cases}$$

and for all positive numbers u, then

$$\frac{d}{d\lambda} y_\lambda = \left\langle \frac{\partial}{\partial \lambda} y_\lambda(t) \right\rangle \qquad (\text{all } \lambda > 0). \tag{15.26}$$

Proof. If $f(\,)$ is a $\mathcal{K}$-function, Definitions 2.1—2 give

$$\{\langle f(t)\rangle \cdot \varphi\}(\tau) = f \cdot \varphi(\tau) = \int_0^\infty \varphi'(\tau - u)\, f(u)\, du \tag{1}$$

for any test-function $\varphi(\,)$ and for all real values of τ. Let α be one of the numbers such that $\varphi(t) = 0$ for $t \leq \alpha$; since

$$u \geq \tau - \alpha \Rightarrow \tau - u \leq \alpha \Rightarrow \varphi'(\tau - u) = 0,$$

Equation (1) becomes

$$\boxed{\{\langle f(t)\rangle \cdot \varphi\}(\tau) = f \cdot \varphi(\tau) = \int_0^{\tau-\alpha} \varphi'(\tau - u)\, f(u)\, du}\,. \tag{2}$$

If $\lambda > 0$, the equations

$$y_\lambda^{[1]} \cdot \varphi(\tau) = \frac{\partial}{\partial \lambda}\{y_\lambda \cdot \varphi(\tau)\} = \frac{\partial}{\partial \lambda} \int_0^{\tau-\alpha} \varphi'(\tau - u)\, y_\lambda(u)\, du \tag{3}$$

are from [9.4] and (2): let τ be **a fixed real number.** We set

$$h_x(u) = \varphi'(\tau - u)\, y_x(u) \qquad (\text{all } x > 0). \tag{4}$$

Observe that $h_x(\,)$ is a $\mathcal{K}$-function such that

$$\left|\frac{\partial}{\partial x} h_x(u)\right| \leq N_\tau \left|\frac{\partial}{\partial x} y_x(u)\right| \qquad (\text{when } 0 \leq u \leq \tau - \alpha), \tag{5}$$

where

$$N_\tau \stackrel{\text{def}}{=} \max_{0 \leq u \leq \tau - \alpha} |\varphi'(\tau - u)|.$$

Combining (5) with our hypothesis 15.25, we obtain

$$\left|\frac{\partial}{\partial x} h_x(u)\right| \leq N_\tau Y_\lambda(u) \qquad \text{when } \begin{cases} \lambda/2 < x < 3\lambda/2 \\ 0 < u \leq \tau - \alpha. \end{cases} \tag{6}$$

Consequently, we may apply 15.22 with $m = \tau - \alpha$ and $H_\lambda(\,) = N_\tau Y_\lambda(\,)$: Formula 15.23 gives

$$\frac{\partial}{\partial\lambda}\int_0^{\tau-\alpha} h_\lambda(u)\,du = \int_0^{\tau-\alpha}\frac{\partial}{\partial\lambda} h_\lambda(u)\,du \qquad (\text{all } \lambda > 0);$$

that is, by (4):

$$\frac{\partial}{\partial\lambda}\int_0^{\tau-\alpha}\varphi'(\tau-u)\,y_\lambda(u)du = \int_0^{\tau-\alpha}\varphi'(\tau-u)\,\frac{\partial}{\partial\lambda}y_\lambda(u)\,du,$$

so that (3) yields

$$(7) \qquad y_\lambda^{[1]}\cdot\varphi(\tau) = \int_0^{\tau-\alpha}\varphi'(\tau-u)\,\frac{\partial}{\partial\lambda}y_\lambda(u)\,du$$

$$(8) \qquad = \left\langle\frac{\partial}{\partial\lambda}y_\lambda(t)\right\rangle\cdot\varphi(\tau) \qquad \text{by (2).}$$

Equations (7)—(8) hold for any real τ; consequently,

$$(9) \qquad y_\lambda^{[1]}\cdot\varphi(\,) = \left\langle\frac{\partial}{\partial\lambda}y_\lambda(t)\right\rangle\cdot\varphi(\,).$$

Since (9) holds for any test-function $\varphi(\,)$, it follows immediately from 0.10 that

$$y_\lambda^{[1]} = \left\langle\frac{\partial}{\partial\lambda}y_\lambda(t)\right\rangle \qquad (\text{all } \lambda > 0);$$

consequently, our conclusion 15.26 comes directly from 9.10.

15.27 *Remark.* In the preceding proof we have derived the following two useful properties.

15.28 Suppose that $\varphi(\,)$ is a test-function, and let α be one of the numbers such that $\varphi(t) = 0$ for $t \leq \alpha$. If $-\infty < \tau < \infty$ there exists a number $N_\tau > 0$ such that

$$(15.29) \qquad |\varphi'(\tau-u)| \leq N_\tau \qquad (\text{when } 0 \leq u \leq \tau-\alpha);$$

in fact, N_τ is the maximum of the continuous function $F(\,)$ defined by $F(u) = |\varphi'(\tau-u)|$ on the closed and finite interval $[0, \tau-\alpha]$.

15.30 If $f(\,)$ is a $\mathcal{K}$-function, then

$$f\cdot\varphi(\tau) = \int_0^{\tau-\alpha}\varphi'(\tau-u)\,f(u)\,du:$$

see 2.2.

A Theorem for § 11

15.31 **Lemma.** *Let* $H_n(\,)$ $(n = 0, 1, 2, \ldots)$ *be a sequence of* $\mathcal{K}$*-functions such that the equation*

$$\lim_{n\to\infty} H_n(t) = H_\infty(t) \qquad (0 \le t < \infty) \tag{15.32}$$

holds for some function $H_\infty(\,)$. *If there exists a rule assigning to every positive number* x *a regulated* $\mathcal{K}$*-function* $H_x(\,)$ *such that*

$$|H_n(t)| \le H_x(t) \qquad \text{when } \begin{cases} 0 \le t \le x \\ n = 1, 2, 3, \ldots, \end{cases} \tag{15.33}$$

then

$$\lim_{n\to\infty} H_n = H_\infty. \tag{15.34}$$

Proof. Let $\varphi(\,)$ be any test-function and let τ be any real number. From 15.28 and 15.33 it results that

$$|\varphi'(\tau - u)\, H_n(u)| \le N_\tau H_x(u) \quad (\text{when} \quad 0 \le u \le \tau - \alpha) \tag{1}$$

and for every integer $n \ge 0$. On the other hand,

$$\left[\lim_{n\to\infty} H_n\right] \cdot \varphi(\tau) = \lim_{n\to\infty} H_n \cdot \varphi(\tau) \qquad (\text{by } 6.6) \tag{2}$$

$$= \lim_{n\to\infty} \int_0^{\tau-\alpha} \varphi'(\tau - u)\, H_n(u)\, du \qquad (\text{by } 15.30) \tag{3}$$

$$= \int_0^{\tau-\alpha} \lim_{n\to\infty} \varphi'(\tau - u)\, H_n(u)\, du \qquad (\text{by } (1)) \tag{4}$$

$$= \int_0^{\tau-\alpha} \varphi'(\tau - u)\, H_\infty(u)\, du \qquad (\text{by } 15.34) \tag{5}$$

$$= H_\infty \cdot \varphi(\tau) \qquad (\text{by } 15.30).$$

Since τ is any real number, we have proved that the equation

$$\left[\lim_{n\to\infty} H_n\right] \cdot \varphi(\,) = H_\infty \cdot \varphi(\,)$$

holds for any test-function $\varphi(\,)$: Conclusion 15.34 is now immediate from 1.10.

Note: we went from (3) to (4) by using the Lebesgue Dominated Convergence Theorem (this is permissible because of (1)).

15.35 **Theorem.** *If* f_k $(k = 0, 1, 2, \ldots)$ *and* g_k $(k = 0, 1, 2, \ldots)$ *are* $\mathcal{K}_0$-*summable sequences, then*

$$\lim_{n\to\infty} \left(\sum_{k=0}^{n} f_k\right)\left(\sum_{k=0}^{n} g_k\right) = \left\langle \sum_{k=0}^{\infty} f_k(t) \right\rangle \left\langle \sum_{k=0}^{\infty} g_k(t) \right\rangle. \tag{15.36}$$

15.37 *Remarks.* If $h(\,)$ is a $\mathcal{K}_0$-function, then $h(t) = 0$ for $t \leq 0$ (see [11.0]); consequently,

$$h(\,) = \{h(t)\}(\,) \qquad \text{(see [2.0])}.$$

If $F(\,)$ and $G(\,)$ are $\mathcal{K}_0$-functions, we therefore have

$$F * G(\tau) = 0 \quad \text{for } \tau \leq 0 \qquad \text{(see 2.6)} \tag{15.38}$$

and 2.7 implies that

$$F * G(\tau) = \int_0^\tau F(\tau - u)\, G(u)\, du \qquad (\text{for } \tau \geq 0). \tag{15.39}$$

Proof of 15.36. Set

$$F_n(\,) = \sum_{k=0}^{n} f_k(\,), \quad G_n(\,) = \sum_{k=0}^{n} g_k(\,), \tag{6}$$

$$F_\infty(\,) = \sum_{k=0}^{\infty} f_k(\,), \quad \text{and} \quad G_\infty(\,) = \sum_{k=0}^{\infty} g_k(\,). \tag{7}$$

In view of [11.0.4] we see that the equations

$$F_\infty(\tau) = \lim_{n\to\infty} F_n(\tau) \quad \text{and} \quad G_\infty(\tau) = \lim_{n\to\infty} G_n(\tau) \tag{8}$$

hold for $0 \leq \tau < \infty$.

From the summability of our sequences it follows that the functions $F_\infty()$ and $G_\infty()$ are regulated $\mathcal{K}_0$-functions: see [11.0.2]. We want to establish 15.36; that is, we want to prove that

$$\boxed{\lim_{n\to\infty} F_n G_n = F_\infty G_\infty} \,. \tag{9}$$

In view of the $\mathcal{K}_0$-summability of our two sequences f_k ($k = 0, 1, 2, \ldots$) and g_k ($k = 0, 1, 2, \ldots$), we know that to every $x > 0$ there corresponds a regulated $\mathcal{K}_0$-function $F_x()$ such that

$$|F_n(t-u)| \le F_x(t-u) \qquad \text{when } \begin{cases} 0 \le u \le t \le x \\ n = 1, 2, 3, \ldots : \end{cases} \tag{10}$$

to see this, it suffices to observe that

$$|F_n(\tau)| = \left|\sum_{k=0}^{n} f_k(\tau)\right| \qquad \big(\text{by } (6)\big):$$

Conclusion (10) is immediately obtained by replacing t by $t - x$ in 11.1.

Similarly, we see that to every $x > 0$ there corresponds a $\mathcal{K}_0$-function $G_x()$ such that

$$|G_n(u)| \le G_x(u) \le [G, x] \qquad \text{when } \begin{cases} 0 \le u \le x \\ n = 1, 2, 3, \ldots : \end{cases} \tag{11}$$

the last inequality comes from 11.7 (since $G_x()$ is regulated). Combining (10) and (11), we obtain

$$|F_n(t-u)\, G_n(u)| \le J_x(u) \qquad \text{when } \begin{cases} 0 \le u \le t \le x \\ n = 1, 2, 3, \ldots, \end{cases} \tag{12}$$

where

$$J_x(u) = F_x(t-u)\,[G, x].$$

The local integrability of $F_x()$ implies that the function $J_x()$ is locally integrable: the inequality (12) therefore enables us to apply the

Lebesgue Dominated Convergence Theorem:

$$\lim_{n\to\infty} \int_0^t F_n(t-u)\, G_n(u)\, du = \int_0^t \lim_{n\to\infty} F_n(t-u)\, G_n(u)\, du$$

$$= \int_0^t F_\infty(t-u)\, G_\infty(u)\, du \qquad \text{(by (8))}.$$

Consequently, we can use 15.39 to write

$$\lim_{n\to\infty} F_n * G_n(t) = F_\infty * G_\infty(t) \qquad (0 \le t \le x). \tag{13}$$

The equation

$$\lim_{n\to\infty} F_n * G_n(\tau) = F_\infty * G_\infty(\tau) \qquad (0 \le \tau < \infty) \tag{14}$$

is a consequence of (13) (with $t = \tau$ and $x = \tau + 1$). From 11.8 and (14) it follows that the equation

$$H_n(\,) = F_n * G_n(\,) \qquad (n = 0, 1, 2, \ldots) \tag{15}$$

defines a sequence of regulated $\mathcal{K}_0$-functions such that

$$\lim_{n\to\infty} H_n(t) = H_\infty(t) \qquad (0 \le t < \infty). \tag{16}$$

Second part of the proof. Observe that

$$|F_n * G_n(t)| \le F_x * G_x(t) \qquad (\text{when } 0 \le t \le x): \tag{17}$$

indeed,

$$|F * G(t)| \le \int_0^t |F_n(t-u)|\, |G_n(u)|\, du \qquad \text{(by 15.39)}$$

$$\le \int_0^t F_x(t-u)\, G_x(u)\, du \qquad \big(\text{by (10)–(11)}\big)$$

$$= F_x * G_x(t) \qquad \text{(by 15.39)}.$$

Final part of the proof. Since $F_x(\,)$ and $G_x(\,)$ are regulated $\mathcal{K}_0$-functions, it follows from 11.8 that the equation

$$H_x(\,) = F_x * G_x(\,)$$

defines a regulated $\mathcal{K}_0$-function; further, Equation (17) gives

$$|F_n * G_n(t)| \leq H_x(t) \qquad \text{when } \begin{cases} 0 \leq t \leq x \\ n = 1, 2, 3, \ldots \end{cases} \tag{18}$$

In view of (15)—(16) and (18), we may apply 15.31 to conclude that

$$\lim_{n \to \infty} H_n = H_\infty;$$

that is,

$$\lim_{n \to \infty} F_n * G_n = F_\infty * G_\infty \qquad \text{(by (15))}. \tag{19}$$

Right-multiplying by D both sides of (19), it follows from 3.15 that

$$\left[\lim_{n \to \infty} F_n D^{-1} G_n\right] D = F_\infty D^{-1}\, G_\infty D,$$

which (by 6.9) implies (9). This concludes the proof.

Glossary of Terminology and Notations

Following is a list of the most frequently occuring terminology and notational conventions; each item is given a description.

Terminology

Test-function: it is infinitely differentiable on $(-\infty, \infty)$ and vanishes to the left of some point; see [0.0].

Entering function: it is locally integrable on $(-\infty, \infty)$ and vanishes to the left of some point; see [0.23].

$\mathcal{K}$-function: it has at most a finite number of discontinuities in every finite interval and is absolutely integrable on every finite interval; see [1.36].

$\mathcal{K}_0$-function: it is a $\mathcal{K}$-function that vanishes on the negative axis; see [11.0].

Regulated function: it is defined on $(-\infty, \infty)$ and has finite limits on both sides of every point; see [2.59].

Canonical operator of a $\mathcal{K}$-function $h(\,)$: it is the operator h that assigns to every test-function $\varphi(\,)$ the function $h \cdot \varphi(\,)$ defined by

$$h \cdot \varphi(\tau) = \int_0^\infty \varphi'(\tau - u)\, h(u)\, du \qquad (-\infty < \tau < \infty);$$

this operator h is often written $\langle h(t) \rangle$; see [2.1]—2.2.

$[0, \infty)$-derivative of a $\mathcal{K}$-function $y(\,)$: it is the operator $Dy - y(0-)\,D$; it is denoted $\partial y/\partial t$; see 2.47.

Derivative on the open interval $(0, \infty)$ of a $\mathcal{K}$-function $y(\,)$: it is the operator $Dy - y(0+)\,D$; it is denoted $\mathrm{d}y/\mathrm{d}t$; see 8.33.

Notations

In order to distinguish between a function and its canonical operator, the symbol denoting the function is usually followed by a matched pair of parentheses; for example, T_0 is the canonical operator of the function $\mathsf{T}_0(\,)$.

$\mathsf{T}_0(\,)$	Heaviside's step function; see [0.2].
D	Differentiator; see [0.6].
$V \cdot \varphi(\,)$	the image of $\varphi(\,)$ under the mapping V; see [0.5].
$[\![h]\!]$	the operator of the function $h(\,)$; see [0.8].
$\mathsf{T}_\alpha(\,)$	$\mathsf{T}_\alpha(t) = \mathsf{T}_0(t - \alpha)$; see [1.30].
$\{h(t)\}(\,)$	the entering function of the function $h(\,)$; see [1.37].

$h = \langle h(t) \rangle = [\![\{h(t)\}]\!]$ see [1.40] and 2.2.

$\frac{\partial}{\partial t} y = Dy - y(0-)\,D$ see [2.47].

$\frac{\partial^2}{\partial t^2} y = D^2 y - y(0-)\,D^2 - y'(0-)\,D$ see 2.54.

$\frac{\partial^m}{\partial t^m} y = D^m y - \sum_{k=0}^{m-1} y^{(k)}(0-)\,D^{m-k}$ see [2.53].

$\left(\lim_{x \to a} V_x\right) \cdot \varphi() = \lim_{x \to a} [V_x \cdot \varphi()]$ see [6.6].

$\{V\}(t)$ and $\{V\}(\,)$ see [8.21].

$\int_{-\infty}^{\tau} A = \{D^{-1} A\}(\tau)$ see 6.23.

$\frac{\partial}{\partial t} A = DA - (\{A\}(0-))\,D$ see 6.25.

$\delta(t - \alpha) = D\mathsf{T}_\alpha$ and $\delta(t) = D\mathsf{T}_0 = D$ see 6.54.

$\boldsymbol{t} V = TV - VD^{-1}TD$ see [6.58].

$\frac{\mathrm{d}}{\mathrm{d}t} \boldsymbol{y} = Dy - \boldsymbol{y}(0)\, D$ see 7.8.

$\boldsymbol{B}^{(-1)}(\tau) = \int_{-\infty}^{\tau} \boldsymbol{B}(u)\, \mathrm{d}u$ see [7.10].

$\int \boldsymbol{y} = \boldsymbol{y}^{(-1)}(0) + D^{-1} y$ see 7.11.

$\left(\sum a_k D^k \,|||\, \sum c_k / D^k\right) = \sum_{k=1}^{n} a_k \sum_{s=0}^{k-1} c_s D^{k-s}$ see 7.23.

$\mathrm{t}^{-n-1} = \frac{-1}{n!} (-D)^{n+1} \langle \log t \rangle$ see [8.0].

$\mathrm{t}^{-3/2} = D \langle -2t^{-1/2} \rangle$ see [8.1].

$\mathrm{t}^{-5/2} = D^2 \left\langle \frac{4}{3}\, t^{-1/2} \right\rangle$ see [8.2].

$\mathrm{t}^{\alpha} = D^n \left\langle \frac{t^{\alpha+n}}{(\alpha+1)(\alpha+2)\cdots(\alpha+n)} \right\rangle$ see 8.14.3.

$\tan = D \langle -\log|\cos t| \rangle$ see 8.14.2.

$y * V = y D^{-1} V$ see [8.8].

$\mathrm{PF} \int_0^t \frac{f(u)}{(t-u)^p}\, \mathrm{d}u = f * \mathrm{t}^{-p}$ see [8.11].

$y^{(k)}(0+) = \lim_{\tau \to 0+} \{D^k y\}(\tau)$ see [8.30].

$y(0+) = \lim_{\tau \to 0+} \{y\}(\tau).$

$y'(0+) = \lim_{\tau \to 0+} \{Dy\}(\tau).$

$\frac{\mathrm{d}^m}{\mathrm{d}t^m} y = D^m y - \sum_{k=0}^{m-1} y^{(k)}(0+) D^{m-k}$ see [8.31].

$$\frac{\mathrm{d}}{\mathrm{dt}}\, y = Dy - y(0+)\, D$$ see 8.33.

$$\frac{\mathrm{d}^2}{\mathrm{dt}^2}\, y = D^2 y - y(0+)\, D^2 - y'(0+)\, D.$$

$$y_\lambda^{[1]} = \lim_{x\to\lambda} \frac{y_x - y_\lambda}{x - \lambda}$$ see [9.3].

$$\frac{\mathrm{d}}{\mathrm{d}x}\, y_{ax+b} = a\, y_{ax+b}^{[1]}$$ see [9.6].

$$p_s^u = \mathrm{e}^{-spu}\, \mathsf{T}_{su}$$ see [9.18].

$$\operatorname{cerf} \lambda = \frac{2}{\sqrt{\pi}} \int_\lambda^\infty \exp(-u^2)\, \mathrm{d}u$$ see [10.0].

$$\hat{p}_x(\,) = \left\{\mathrm{e}^{-pt} \operatorname{cerf} \frac{x}{2\sqrt{t}}\right\}(\,)$$ see [10.1].

$$\sqrt{D} = \left\langle \frac{1}{\sqrt{\pi t}} \right\rangle$$ see [10.25].

$$\mathrm{q}^\lambda = \left\langle \operatorname{cerf} \frac{\lambda}{2\sqrt{t}} \right\rangle$$ see [10.26].

$$\frac{\mathrm{d}V}{\mathrm{d}D} = -TV + VT$$ see [14.0].

Summary of Results and Table of Formulas

(0.26) $$c\mathsf{T}_0 * g(\tau) = c \int_{-\infty}^{\tau} g(u)\, du \qquad (-\infty < \tau < \infty).$$

When $h(\,)$ is an entering function, then $[\![h]\!]$ is the operator that assigns to every test-function $\varphi(\,)$ the function $[\![h]\!] \cdot \varphi(\,)$ defined by the equation

(1.17) $$[\![h]\!] \cdot \varphi(\tau) = \int_{-\infty}^{\infty} \varphi'(\tau - u)\, h(u) du \qquad (-\infty < \tau < \infty);$$

this operator $[\![h]\!]$ is perfect. Further,

(0.27) $$[\![c\mathsf{T}_0]\!] \cdot \varphi(\,) = c\varphi(\,) \qquad \big(\text{for every test-function } \varphi(\,)\big).$$

If V, V_1, V_2 are perfect operators, then

(1.10) $$V(V_1 + V_2) = VV_1 + VV_2 = V_1V + V_2V = (V_1 + V_2)\, V$$

and

(0.28) $$V[\![\mathsf{T}_0]\!] = V = [\![\mathsf{T}_0]\!]\, V;$$

further,

(1.19) $$[\![V \cdot \varphi]\!] = V[\![\varphi]\!] \qquad \big(\text{for every test-function } \varphi(\,)\big).$$

(1.23—24) $$f(\,) = g(\,) \textit{ if } \quad (\textit{and only if}) \quad [\![f]\!] = [\![g]\!];$$

(1.27) $$[\![f * g]\!] = [\![f]\!]\, D^{-1}[\![g]\!] = [\![f]\!] * [\![g]\!] \qquad (\text{see } [8.8])$$

and

(1.29) $$[\![g]\!] = D[\![\mathsf{T}_0 * g]\!].$$

If α and λ are real numbers, then

(1.34) $$[\![\mathsf{T}_\alpha]\!]\,[\![\mathsf{T}_\lambda]\!] = [\![\mathsf{T}_{\alpha+\lambda}]\!],$$

and

(1.35) $$[\![\mathsf{T}_\alpha]\!] \cdot \varphi(\tau) = \varphi(\tau - \alpha) \qquad (-\infty < \tau < \infty)$$

for every test-function $\varphi(\,)$. **If c is a number, we agree to write**

(1.51) $$c = \langle c \rangle \qquad \text{(see also 2.16).}$$

In consequence:

(1.56) $$cV = \langle c \rangle\, V = V \langle c \rangle \qquad \text{(for any perfect operator } V\text{);}$$

see also 2.22. In particular,

(1.57) $$V1 = V = 1V \qquad \text{(for any operator } V\text{),}$$

and

(1.55) $$[\![\mathsf{T}_0]\!] = \mathsf{T}_0 = 1 = \langle 1 \rangle \qquad \text{(see also 2.11).}$$

(2.1) $$h = \langle h(t) \rangle = [\![\{h(t)\}]\!];$$

(2.3) $$F = G \quad \textit{if (and only if)} \quad \{F(t)\}(\,) = \{G(t)\}(\,);$$

(2.5) $$\langle aF(t) + bG(t) + c \rangle = aF + bG + c;$$

(2.23) $$VD^nF = D^nFV = D^nVF = VFD^n;$$

(2.8) $$\langle F(t) \rangle\, D^{-1} \langle G(t) \rangle = FD^{-1}G = \{F(t)\} * \{G(t)\};$$

(2.9) $$FD^{-1}G = F * G = \left\langle \int_0^t F(t-u)G(u)\,du \right\rangle.$$

We shall omit the angular brackets on the right-hand side of each of the following four equations

(2.28) $$D^{-1}h + \left\langle \int_{-\infty}^{0} h(u)\, \mathrm{d}u \right\rangle = \int_{-\infty}^{t} h(u)\, \mathrm{d}u;$$

(3.14) $$D^{-1}G = \int_{0}^{t} G(u)\, \mathrm{d}u;$$

$$D^{-n}H = \int_{0}^{t} \frac{(t-u)^{n-1}}{(n-1)!} H(u)\, \mathrm{d}u;$$

(11.58.1) $$\frac{H}{1 - b\mathsf{T}_\lambda} = \sum_{k=0}^{[t/\lambda]} b^k H(t - k\lambda) \qquad \text{(for } \lambda > 0\text{)}.$$

If $-\infty < \alpha < \infty$ and $h_\alpha(t) = h(t-\alpha)$ for $-\infty < t < \infty$, then

(1.32) $$[\![\mathsf{T}_\alpha]\!]\,[\![h]\!] = [\![h_\alpha]\!].$$

Moreover, if $\alpha \geq 0$ then

(8.28) $$\{\mathsf{T}_\alpha W\}(\tau) = \{W\}(\tau - \alpha) \qquad (-\infty < \tau < \infty)$$

and

(2.11) $$\mathsf{T}_\alpha = \langle \mathsf{T}_\alpha(t) \rangle = [\![\mathsf{T}_\alpha]\!],$$

so that

(2.12) $$\mathsf{T}_\alpha F = \langle \mathsf{T}_\alpha(t)\, F(t-\alpha) \rangle;$$

that is, if we omit the angular brackets from the right-hand side:

$$\mathsf{T}_\alpha F = \mathsf{T}_\alpha(t)\, F(t-\alpha)$$

and

(5.6) $$\mathsf{T}_\alpha F = \mathsf{T}_\alpha \langle F(t) \rangle = \mathsf{T}_0(t-\alpha)\, F(t-\alpha).$$

Next,

(2.14) $$\mathsf{T}_\alpha \mathsf{T}_\lambda = \mathsf{T}_{\alpha+\lambda} \qquad \text{(if } \alpha \geq 0 \text{ and } \lambda \geq 0\text{)},$$

(2.15) $$\mathsf{T}_\alpha [\![\mathsf{T}_{-\alpha}]\!] = 1 = \mathsf{T}_0 \qquad \text{(if } \alpha \geq 0\text{)},$$

and

(2.19) $$\mathsf{T}_\alpha^{-1} = \frac{1}{\mathsf{T}_\alpha} = [\![\mathsf{T}_{-\alpha}]\!] \qquad (\text{if } \alpha \geq 0).$$

If $k = 1, 2, 3, \ldots$ then

(6.14.2) $$\{D^k F\}(\tau) = F^{(k)}(\tau)$$

and

(6.20) $$\{D^k \mathsf{T}_\alpha\}(\tau) = 0 \qquad (\text{if } \alpha \geq 0 \text{ and } \tau \neq \alpha);$$

further,

(8.44) $$\left\{\frac{\mathrm{d}^k}{\mathrm{d}t^k}\, y\right\}(\tau) = \left\{\frac{\partial^k}{\partial t^k}\, y\right\}(\tau) = y^{(k)}(\tau) \qquad (\text{every } \tau > 0).$$

Next,

(6.17) $$\{\langle F(t)\rangle\}(\tau) = F(\tau) \qquad (\text{every } \tau > 0),$$

and

(6.63) $$\{\boldsymbol{t} f\}(\tau) = \tau f(\tau),$$

(6.62) $$\boldsymbol{t} f = \langle t f(t)\rangle;$$

further,

(6.66) $$\boldsymbol{t}\,\delta(t-\alpha) = \alpha\,\delta(t-\alpha) \qquad (\text{for } \alpha \geq 0),$$

and

(6.67) $$\boldsymbol{t}\,[D\delta(t-\alpha)] = \alpha\, D\delta(t-\alpha) - \delta(t-\alpha) \qquad (\text{for } \alpha \geq 0).$$

Some more formulas:

(6.61) $$D[\boldsymbol{t}A] = A + \boldsymbol{t}[DA];$$

(6.54) $$D\mathsf{T}_\alpha = \frac{\mathrm{d}}{\mathrm{d}t}\,\mathsf{T}_\alpha = \delta(t-\alpha) \qquad (\text{for } \alpha \geq 0);$$

(6.81) $$\int_{-\infty}^{\infty} \boldsymbol{t}\,(mD^2\mathsf{T}_\alpha) = -m \qquad (\text{for } \alpha \geq 0);$$

(6.27) $$\int_{-\infty}^{\infty} c\,\delta(t-\alpha) = c \qquad (\text{for } \alpha \geq 0).$$

If $g()$ is a $\mathcal{K}$-function such that $g = G(D)$ with $G \in \mathcal{A}_\infty$, then

(12.20) $$g(0+) = G(\infty);$$

(12.32.6) $$\langle e^{at} g(t)\rangle = \langle e^{at}\rangle [G(D-a)];$$

(12.32.7) $$\langle g(\alpha t)\rangle = G(\alpha^{-1} D) \qquad (\text{if } \alpha > 0);$$

$$\left\langle e^{at} g\left(\frac{t}{\lambda}\right)\right\rangle = G(\lambda D - \lambda a)\,\frac{D}{D-a} \qquad (\text{if } \lambda > 0);$$

(12.21) $$\langle t g(t)\rangle = -D\left[\frac{d}{dp}\,\frac{G(p)}{p}\right]_{p=D}.$$

If $g()$ is a $\mathcal{K}$-function such that $\{g(t)/t\}()$ is a regulated function and such that $g = G(D)$ with $G \in \mathcal{A}_\infty$, then

(12.22) $$\left\langle\frac{g(t)}{t}\right\rangle = D\left[\int_x^\infty \frac{G(p)}{p}\,dp\right]_{x=D}.$$

The following equations hold when $y()$ and $y'()$ are $\mathcal{K}$-functions such that $y()$ has no jumps on $[0, \infty)$:

(3.8.0) $$\frac{\partial}{\partial t} y = \frac{d}{dt} y = y' = \left\langle\frac{d}{dt} y(t)\right\rangle;$$

(3.8.1) $$y' = Dy - y(0-)\,D \quad \text{and} \quad y(0-) = y(0+);$$

(3.8.2) $$y = D^{-1} y' + y(0-);$$

(3.8.3) $$D^{-1} y' = y - y(0-);$$

(3.8.4) $$\boxed{Dy = y' + y(0-)\,D}.$$

The following three equations hold when $y()$ is continuous in the open interval $(0, \infty)$ and if $y'()$ is a $\mathcal{K}$-function:

(8.46) $$\boxed{\frac{d}{dt} y = \langle y'(t)\rangle = y' = \left\langle\frac{d}{dt} y(t)\right\rangle}.$$

Suppose that m is an integer ≥ 1, and let $y(\,)$ be a function whose m^{th} derivative $y^{(m)}(\,)$ is a $\mathcal{K}$-function; if

$$y^{(m-1)}(\lambda) = \lim_{t\to\lambda-} y^{(m-1)}(t) = \lim_{t\to\lambda+} y^{(m-1)}(t)$$

for every $\lambda > 0$, then

$$\frac{\mathrm{d}^k}{\mathrm{d}t^k}\, y = \langle y^{(k)}(t)\rangle = y^{(k)} = \left\langle \frac{\mathrm{d}^k}{\mathrm{d}t^k}\, y(t)\right\rangle \qquad (0 \leq k \leq m). \tag{8.48}$$

Corollary of 9.12. Given a number $y(x, t)$ depending on the parameters $x > 0$ and $t > 0$: if

$$\lim_{\substack{x\to\alpha\\ x<\alpha}} y(x, t) = y(\alpha, t) = \lim_{\substack{x\to\alpha\\ x>\alpha}} y(x, t)$$

for all $\alpha > 0$ and $t > 0$, then

$$\frac{\mathrm{d}}{\mathrm{d}x}\, y_x = \left\langle \frac{\partial}{\partial x}\, y(x, t)\right\rangle,$$

where $y_x(t) = y(x, t)$. Moreover, if the function $\partial y_x(\,)/\partial t$ is continuous in the open interval $(0, \infty)$, then

$$\boxed{\frac{\mathrm{d}}{\mathrm{d}t}\, y_x = \left\langle \frac{\partial}{\partial t}\, y(x, t)\right\rangle}\,.$$

The following equations hold when p is a number and $s > 0$:

$$\frac{\mathrm{d}}{\mathrm{d}x}\, p_s^{\pm x+b} = \pm\, s\,(D + p)\, p_s^{\pm x+b}; \tag{9.19}$$

$$p_s^0 = 1; \tag{9.43}$$

$$p_s^x\, p_s^\lambda = p_s^{x+\lambda} \qquad (\text{for } x \geq 0 \text{ and } \lambda \geq 0); \tag{9.44}$$

$$\frac{1}{p_s^x} = \mathrm{e}^{spx}\, [\![\mathsf{T}_{-sx}]\!] \qquad (\text{for } x \geq 0); \tag{9.45}$$

$$p_s^x \langle g(t)\rangle = \langle \mathrm{e}^{-spx} g(t - sx)\rangle. \tag{9.47}$$

If $p = 0$ then

(9.48) $$p_s^x = \mathsf{T}_{xs}$$

and

(9.52) $$\frac{p_s^\lambda G}{1 - b p_s^m} = \left\langle \sum_{k=0}^{-\lambda/m + t/ms} b^k G\left(\tau - \frac{km + \lambda}{1/s}\right)\right\rangle :$$

see 9.49 and 11.58.1.

In the following equations, $\hat{p}_x = \langle \hat{p}_x(t) \rangle$, where

$$\hat{p}_x(t) = \int_{x/2\sqrt{t}}^{\infty} \exp(-pt - u^2)\, \mathrm{d}u;$$

(10.2) $$\hat{p}_0 = \frac{D}{D + p} \quad \text{and} \quad \hat{p}_\infty = 0;$$

(10.3) $$\hat{p}_x(0+) = 0;$$

(10.4) $$\frac{\mathrm{d}}{\mathrm{dt}}\, \hat{p}_x = D\hat{p}_x;$$

(10.5) $$\frac{\mathrm{d}}{\mathrm{d}x}\, \hat{p}_x = \frac{\partial}{\partial x}\, \hat{p}_x(t) = \frac{-1}{\sqrt{\pi t}} \exp\left(-pt - \frac{x^2}{4t}\right).$$

Further,

(10.6) $$\frac{\mathrm{d}^2}{\mathrm{d}x^2}\, \hat{p}_x = \frac{\partial^2}{\partial x^2}\, \hat{p}_x(t) = \frac{x t^{-3/2}}{2\sqrt{\pi}} \exp\left(-pt - \frac{x^2}{4t}\right)$$

and

(10.7) $$\frac{\mathrm{d}}{\mathrm{dt}}\, \hat{p}_x + p\hat{p}_x = \frac{\mathrm{d}^2}{\mathrm{d}x^2}\, \hat{p}_x.$$

Moreover,

(10.17)

$$\frac{D + p}{D}\, \hat{p}_\lambda = \frac{\mathrm{e}^{\lambda\sqrt{p}}}{2} \operatorname{cerf}\left(\frac{\lambda}{2\sqrt{t}} + \sqrt{pt}\right) + \frac{\mathrm{e}^{-\lambda\sqrt{p}}}{2} \operatorname{cerf}\left(\frac{\lambda}{2\sqrt{t}} - \sqrt{pt}\right).$$

Henceforth,

(10.25) $$\sqrt{D} = \frac{1}{\sqrt{\pi t}},$$

and

$$\mathrm{q}^{\lambda} = \operatorname{cerf} \frac{\lambda}{2\sqrt{t}}. \tag{10.26}$$

Symbolically:

$$\mathrm{q}^{\lambda} = \mathrm{e}^{-\lambda\sqrt{D}}.$$

We have:

$$\langle t^{-1/2}\rangle\, \mathrm{q}^{\lambda} = \frac{1}{\sqrt{t}} \exp\left(\frac{-\lambda^2}{4t}\right), \tag{10.30}$$

$$\frac{\mathrm{d}}{\mathrm{d}x}\, \mathrm{q}^{sx} = -s\sqrt{D}\, \mathrm{q}^{sx}, \tag{10.31}$$

and

$$\frac{\mathrm{d}}{\mathrm{d}x}\, \mathrm{q}^{sl-sx} = s\sqrt{D}\, \mathrm{q}^{sl-sx}; \tag{10.43}$$

further,

$$D^{-1}\sqrt{D}\, \mathrm{q}^{\lambda} = 2\left(\frac{t}{\pi}\right)^{1/2} \exp\left(\frac{-\lambda^2}{4t}\right) - \lambda \operatorname{cerf} \frac{\lambda}{2\sqrt{t}} \tag{10.32}$$

and

$$\frac{2D}{D+p}\, \mathrm{q}^{\lambda} = \exp(pt + \lambda\sqrt{p}) \operatorname{cerf}\left(\frac{\lambda}{2\sqrt{t}} + \sqrt{pt}\right) + \exp(pt - \lambda\sqrt{p}) \operatorname{cerf}\left(\frac{\lambda}{2\sqrt{t}} - \sqrt{pt}\right). \tag{10.36}$$

We also have

$$\frac{a^2}{D^2 - a^2}\, \mathrm{q}^{sx} = -\mathrm{q}^{sx}(t) + \frac{\exp(a^2 t + s x a)}{2} \operatorname{cerf}\left(\frac{sx}{2\sqrt{t}} + a\sqrt{t}\right) + \frac{\exp(a^2 t - s x a)}{2} \operatorname{cerf}\left(\frac{sx}{2\sqrt{t}} - a\sqrt{t}\right), \tag{10.37}$$

and

$$\frac{\sqrt{D} - a}{D - a^2}\, \mathrm{q}^{sx} = \frac{\mathrm{q}^{sx}(t)}{a} - \frac{\exp(a^2 t + s x a)}{a} \operatorname{cerf}\left(\frac{sx}{2\sqrt{t}} + a\sqrt{t}\right). \tag{10.38}$$

Elementary Formulas

(14.10) $$(-\mathrm{t})^k = D \frac{\mathrm{d}^k}{\mathrm{d}D^k} D^{-1} \qquad (k = 1, 2, 3, \ldots).$$

(14.12) $$\left(-\mathrm{t} \frac{\mathrm{d}}{\mathrm{dt}}\right)^k = \left(D \frac{\mathrm{d}}{\mathrm{d}D}\right)^k \qquad (k = 1, 2, 3, \ldots).$$

(14.13) $$\mathrm{t} \frac{\mathrm{d}^2}{\mathrm{dt}^2} = (-D^2) \frac{\mathrm{d}}{\mathrm{d}D} - \frac{\mathrm{d}}{\mathrm{dt}}.$$

(14.14) $$\mathrm{t}^2 \frac{\mathrm{d}}{\mathrm{dt}} = D \frac{\mathrm{d}^2}{\mathrm{d}D^2}.$$

Next,

(3.16) $$D = \left\langle \frac{1}{\sqrt{\pi t}} \right\rangle^2,$$

$$D^0 = 1 = \mathsf{T}_0 = \langle \mathsf{T}_0(t) \rangle,$$

and

(3.18) $$D^{-1} = \frac{1}{D} = \langle t \rangle.$$

Henceforth, as has been so often the case, **the angular brackets are omitted from the right-hand side of the equations:**

(3.20) $$D^{-n} = \frac{1}{D^n} = \frac{t^n}{n!} \qquad (n = 0, 1, 2, \ldots).$$

(3.21) $$\frac{D}{D-a} = \mathrm{e}^{at}.$$

(3.22) $$\frac{(n!)\, D}{(D-a)^{n+1}} = t^n \mathrm{e}^{at} \qquad (n = 0, 1, 2, \ldots).$$

(3.23) $$\frac{D}{(D-a)^m} = \mathrm{e}^{at} \frac{t^{m-1}}{(m-1)!} \qquad (m = 1, 2, 3, \ldots).$$

(3.24) $$\frac{1}{D-a} = D^{-1} \mathrm{e}^{at} = \frac{-1}{a} [1 - \mathrm{e}^{at}].$$

$$\frac{1}{(D-a)^2} = \frac{1}{a^2} [1 - \mathrm{e}^{at}(1 - at)].$$

$$\frac{1}{(D-a)^m} = \left(\frac{-1}{a}\right)^m \left[1 - \mathrm{e}^{at}\left(1 - at + \frac{(at)^2}{2!} + \cdots + \frac{(-at)^{m-1}}{(m-1)!}\right)\right].$$

In all that follows, $\alpha > 0, \beta > 0$, and a, b, c are numbers (possibly *complex* numbers):

(3.25) $$\frac{D}{(D-c)^2 \pm \beta^2} = \left(\frac{e^{ct}}{\beta}\right) {\sin \atop \sinh} \beta t.$$

(3.26) $$\left[\frac{D}{D^2+\beta^2}\right]^2 = \left(\frac{t}{2\beta}\right) \sin \beta t.$$

(3.27) $$\frac{D}{D^2+\beta^2} = \left(\frac{1}{\beta}\right) \sin \beta t.$$

(3.28) $$\frac{D^2}{D^2 \pm \beta^2} = {\cos \atop \cosh} \beta t.$$

(3.29) $$D\left[\frac{D}{D^2+\beta^2}\right]^2 = \left(\frac{1}{2\beta}\right) \sin \beta t + \left(\frac{t}{2}\right) \cos \beta t.$$

(3.30) $$\frac{1}{D^2+\beta^2} = \frac{1-\cos \beta t}{\beta^2}.$$

(3.31) $$\frac{(D-c)\, D}{(D-c)^2+\beta^2} = e^{ct} \cos \beta t.$$

(3.32) $$D\left[\frac{1}{(D-c)^2+\beta^2}\right]^2 = \frac{e^{ct}}{2\beta^2}\left[(-t) \cos \beta t + \frac{\sin \beta t}{\beta}\right].$$

$$\left[\frac{D}{(D-c)^2+\beta^2}\right]^2 = \frac{e^{ct}}{2\beta^2}\left[(-ct) \cos \beta t + \left(\frac{c}{\beta} + \beta t\right) \sin \beta t\right].$$

(3.33) $$\frac{D}{(D^2+\alpha^2)(D^2+\beta^2)} = \frac{\alpha \sin \beta t - \beta \sin \alpha t}{\alpha\beta(\alpha^2-\beta^2)}.$$

(12.12) $$\exp\left(\frac{-\alpha}{D}\right) = J_0(2\alpha t)$$ (also: 13.11).

(12.15) $$\frac{D}{\sqrt{D^2+\alpha^2}} = J_0(\alpha t)$$ (also: 13.12).

(12.23) $$D - \frac{D}{\sqrt{D^2+\alpha^2}} = \alpha J_1(\alpha t)$$ (also: 13.14).

(12.24) $$-D^2 + D\sqrt{D^2+\alpha^2} = \frac{\alpha J_1(\alpha t)}{t}.$$

(12.27) $$D \tan^{-1} \frac{\alpha}{D} = \frac{\sin \alpha t}{t}.$$

(12.27) $$D \log \frac{D-a}{D-b} = \frac{e^{at} - e^{bt}}{t} \qquad \text{(also: 14.18)}.$$

(12.30) $$\frac{D^2}{(D^2+\alpha^2)^{3/2}} = t J_0(\alpha t).$$

(12.32.3) $$\text{arc tan} \frac{1}{D} = \int_0^t \frac{\sin u}{u} \, du.$$

(12.32.3) $$D \log \left(1 + \frac{\alpha^2}{D^2}\right) = \frac{2 - 2\cos \alpha t}{t} \qquad \text{(also: 14.19)}.$$

(14.29.1) $$(1 - D^{-1})^m = L_m(t) \qquad \text{(Laguerre polynomial)}.$$

(5.9) $$\frac{(1 + \mathsf{T}_\pi) D}{1 + D^2} = \begin{cases} \sin t & (0 < t < \pi) \\ 0 & \text{(otherwise)}. \end{cases}$$

(5.24.5) $$\alpha + \frac{2\mathsf{T}_\alpha - 1}{D} = |t - \alpha|.$$

(5.43.2) $$\frac{D(1 + \mathsf{T}_\pi)}{(D^2 + 1)(1 - \mathsf{T}_\pi)} = |\sin t|.$$

(5.13) $$c(\mathsf{T}_\alpha - \mathsf{T}_\beta) = \begin{cases} c & (\alpha < t \le \beta) \\ 0 & \text{(otherwise)}. \end{cases}$$

$$\mathsf{T}_0 = 1 \text{ and } \quad \text{(symbolically)} \quad \mathsf{T}_\lambda = e^{-\lambda D} \quad \text{when } \lambda \ge 0.$$

Note also that

$$1 - \mathsf{T}_\lambda = \begin{cases} 1 & (0 < t \le \lambda) \\ 0 & \text{(otherwise)}. \end{cases}$$

If $0 < \alpha \leq \lambda$ then

(5.20) $$\frac{\beta}{\alpha D}(1 - \mathsf{T}_\alpha) - \beta \mathsf{T}_\lambda = \begin{cases} \beta t/\alpha & (0 < t \leq \alpha) \\ \beta & (\alpha < t \leq \lambda) \\ 0 & (t > \lambda). \end{cases}$$

(5.19) $$\frac{\beta}{\alpha D}(1 - \mathsf{T}_\alpha) - \beta \mathsf{T}_\alpha = \begin{cases} \beta t/\alpha & (0 < t \leq \alpha) \\ 0 & (t > \alpha). \end{cases}$$

(5.21) $$\frac{\beta}{\alpha D}(1 - \mathsf{T}_\alpha) = \begin{cases} \beta t/\alpha & (0 < t \leq \alpha) \\ \beta & (t > \alpha). \end{cases}$$

(5.43.7) $$\frac{\mathsf{T}_\alpha - 1}{\alpha D} - 1 = \begin{cases} (-t + \alpha)/\alpha & (0 < t \leq \alpha) \\ 0 & (t < \alpha). \end{cases}$$

(5.22) $$\frac{m\mathsf{T}_\alpha}{D} + b\mathsf{T}_\alpha = \begin{cases} 0 & (0 < t \leq \alpha) \\ mt - m\alpha + b & (t > \alpha). \end{cases}$$

(5.17) $$\frac{(1 - \mathsf{T}_{2\alpha})^2}{D} = \begin{cases} t & (0 < t \leq 2\alpha) \\ (-t + 4\alpha)/2\alpha & (2\alpha < t \leq 4\alpha) \\ 0 & (t > 4\alpha). \end{cases}$$

(11.22) $$D\left\langle 1 + \left[\frac{t}{\alpha}\right]\right\rangle = \frac{D}{1 - \mathsf{T}_\alpha} = \sum_{k=0}^{\infty} \delta(t - k\alpha).$$

(11.55) $$\frac{D}{1 - b\mathsf{T}_\alpha} = \sum_{k=0}^{\infty} b^k \, \delta(t - k\alpha).$$

(11.55) $$\frac{D}{(1 - b\mathsf{T}_\alpha)^2} = \sum_{k=0}^{\infty} (k + 1)\, b^k \, \delta(t - k\alpha).$$

(11.53) $$\frac{D}{(1 - b\mathsf{T}_\alpha)^n} = \sum_{k=0}^{\infty} \frac{(k + n - 1)!}{k!\,(n - 1)!} b^k \, \delta(t - k\alpha).$$

(11.56) $$D\frac{1 - \mathsf{T}_\alpha}{1 + \mathsf{T}_\alpha} = \delta(t) + 2 \sum_{k=1}^{\infty} (-1)^k \, \delta(t - k\alpha).$$

$$(5.36)\qquad \frac{1-\mathsf{T}_\alpha}{1+\mathsf{T}_\alpha} = 1 + 2\sum_{k=1}^{\infty}(-1)^k \mathsf{T}_0(t-k\alpha).$$

$$(5.32.3)\qquad \frac{D}{(1+D^2)(1-\mathsf{T}_\pi)} = \sum_{k=0}^{\infty} \mathsf{T}_0(t-k\pi)\sin(t-k\pi).$$

$$(5.43.1)\qquad \frac{1-\mathsf{T}_1}{D(1+\mathsf{T}_1)} = t + 2\sum_{k=1}^{\infty}(-1)^k(t-k)\,\mathsf{T}_0(t-k).$$

$$(11.58.1)\qquad \frac{\mathsf{T}_x f}{(1-c\mathsf{T}_\alpha)^{m+1}} = \sum_{k=0}^{(t-x)/\alpha}\frac{(k+m)!}{k!\,m!}c^k f(t-x-k\alpha).$$

Periodic Functions

If V is an operator and $\tau > 0$, then $\{V\}(\tau)$ is the value at τ of the operator V (see 6.16): we conclude this table by giving the function $\{V\}()$ for various choices of V. For example, if $f()$ is continuous on the interval $(0,\infty)$, then

$$\left\{\frac{\mathsf{T}_x f}{1-c\mathsf{T}_\alpha}\right\}(\tau) = \sum_{k=0}^{n} c^k f(\tau-x-k\alpha) \qquad \left(\text{when } n < \frac{\tau-x}{\alpha} < n+1\right)$$

and $n = 0, 1, 2, \ldots$: this is an immediate consequence of the preceding formula 11.58.1.

If $h()$ is an integrable function that vanishes outside the interval $(0,\lambda)$, then

$$\left\{\frac{h}{1-\mathsf{T}_\lambda}\right\}(\tau) = h(\tau-k\lambda) \qquad \left(\text{when } k < \frac{\tau}{\lambda} < k+1\right),$$

— and $k = 0, 1, 2, \ldots$: see 5.35.

$$\left\{\frac{1}{\alpha D} - \frac{\mathsf{T}_\alpha}{1-\mathsf{T}_\alpha}\right\}(\tau) = \frac{\tau}{\alpha} - k \qquad \left(\text{when } k < \frac{\tau}{\alpha} < k+1\right),$$

— and $k = 0, 1, 2, \ldots$: see 5.43.0.

$$\left\{\frac{\mathsf{T}_\alpha - 1 + \alpha D}{\alpha D (1 + \mathsf{T}_\alpha)}\right\}(\tau) = 1 + k - \frac{\tau}{\alpha} \qquad \left(\text{when} \quad k < \frac{\tau}{\alpha} < k + 1\right),$$

— and $k = 0, 1, 2, \ldots$: see 5.43.8.

$$\left\{\frac{1 - \mathsf{T}_1}{D(1 + \mathsf{T}_1)}\right\}(\tau) = \begin{cases} \tau - 2k & (\text{when } 2k < \tau < 2k + 1) \\ 2k + 2 - \tau & (\text{when } 2k + 1 < \tau < 2k + 2), \end{cases}$$

— and $k = 0, 1, 2, \ldots$: see 5.43.1.

$$\left\{\frac{1 - \mathsf{T}_\alpha}{1 + \mathsf{T}_\alpha}\right\}(\tau) = \begin{cases} 1 & \left(\text{when } 2k < \frac{\tau}{\alpha} < 2k + 1\right) \\ -1 & \left(\text{when } 2k + 1 < \frac{\tau}{\alpha} < 2k + 2\right), \end{cases}$$

— and $k = 0, 1, 2, \ldots$: see 5.36.

$$\left\{\frac{D}{(1 + D^2)(1 - \mathsf{T}_\pi)}\right\}(\tau) = \begin{cases} \sin\tau & \left(\text{when } 2k < \frac{\tau}{\pi} < 2k + 1\right) \\ 0 & \left(\text{when } 2k + 1 < \frac{\tau}{\pi} < 2k + 2\right), \end{cases}$$

— and $k = 0, 1, 2, \ldots$: see 5.32.3.

$$\left\{\frac{D(1 + \mathsf{T}_\pi)}{(D^2 + 1)(1 - \mathsf{T}_\pi)}\right\}(\tau) = \sin(\tau - k\pi) \qquad \left(\text{when } k < \frac{\tau}{\pi} < k + 1\right),$$

— and $k = 0, 1, 2, \ldots$: see 5.43.2.

Bibliography

[B 1] Berberian, S. K.: Measure and integration, New York: Macmillan 1965.

[B 2] Berg, L.: Introduction to the operational calculus, New York: Interscience, John Wiley 1967.

[D 1] Ditkin, V. A., and A. P. Prudnikov: Integral transforms and operational calculus, London-New York-Paris-Los Angeles: Pergamon Press 1965.

[D 2] Dunford, N., and J. T. Schwartz: Linear operators, Part I: General theory, New York: Interscience 1958.

[D 3] Doetsch, G.: Einführung in Theorie und Anwendung der Laplace-Transformation (Mathematische Reihe Bd. 24), Basel: Birkhäuser 1958.

[E 1] Erdélyi, A.: Operational calculus and generalized functions, New York: Holt, Rinehart & Winston 1962.

[H 1] Hobson, E. W.: The theory of functions of a real variable and the theory of Fourier series, vol. I, New York: Dover 1957.

[H 2] Hobson, E. W.: The theory of functions of a real variable and the theory of Fourier series, vol. II, New York: Dover 1957.

[J 1] Jeffery, R. L.: The theory of functions of a real variable, Toronto: University of Toronto Press 1953.

[K 1] Kestelman, H.: Modern theories of integration, New York: Dover 1960.

[K 2] Korevaar, J.: Mathematical methods, vol. I, New York: Academic Press 1960.

[M 1] Mikusiński, J.: Operational calculus, London-New York-Paris-Los Angeles: Pergamon Press 1959.

[R 1] Rjabcev, I. I.: On the structure of Mikusiński operators in a pseudo-normed space. Amer. Math. Soc. Transl. **53**, 1–11 (1966).

[R 2] Rjabcev, I. I.: Local properties of Mikusiński operators. Amer. Math. Soc. Transl. **53**, 13–22 (1966).

[S 1] Schwartz, L.: Théorie des distributions, Paris: Hermann 1950.

[S 2] Sauer, R., und I. Szabó: Mathematische Hilfsmittel des Ingenieurs, Teil I, Berlin-Heidelberg-New York: Springer 1967.

[V 1] Van der Pol, B., and H. Bremmer: Operational calculus based on the two-sided Laplace integral, New York: Cambridge University Press 1955.

[W 1] Weston, J. D.: An extension of the Laplace transform calculus. Rend. Mat. Palermo (2) **6**, 325–333 (1957).

[W 2] Weston, J. D.: Operational calculus and generalized functions. Proc. Roy. Soc., Ser. A, **250**, 460–471 (1959).

[W 3] Weston, J. D.: Characterizations of Laplace transforms and perfect operators. Arch. Rat. Mech. Anal. **3**, 348–354 (1959).

[W 4] Weston, J. D.: Positive perfect operators. Proc. London Math. Soc., Ser. 3, **10**, 545–565 (1960).

[W 5] Williamson, J. H.: Lebesgue integration, New York: Holt, Rinehart & Winston 1962.

Bibliographical Comments

This is added in correction proof, December 1969. At the present time, HARRIS SHULTZ has incorporated his "unpublished results" (pp. 109–110, 127) into a paper entitled "Linear operators and operational calculus, Part II" submitted for publication in the Studia Mathematica; this paper deals with operational calculus on an interval $[a, b)$ with $b \leq \infty$; see also "Linear operators and operational calculus, Part I" by KRABBE (submitted for publication in the Studia Mathematica in November 1969).

Next, a chronological remark concerning the proof of the fact that perfect operators can be identified with the SCHWARTZ convolution algebra $\mathscr{D}'_+$. HARRIS SHULTZ proved the surjectivity (pp. 126–127) early in 1968; the same result appeared later in the article by RAIMOND A. STRUBLE entitled "On operators and distributions" [Canad. Math. Bull. **11**, No. 1, 61–64 (1968)].

Other research papers besides RJABCEV's [R 1]–[R 2] have characterized a subalgebra of the Mikusiński field that is essentially the algebra of perfect operators; these research papers are

WLOKA, J.: Distributionen und Operatoren. Math. Annalen **140**, 227–244 (1960).

and

NORRIS, D. O.: A topology for Mikusiński operators. Studia Mathematica **24**, 245–255 (1964).

Subject and Author Index